Lise Meitner
and the Dawn
of the Nuclear Age

Patricia Rife

Lise Meitner and the Dawn of the Nuclear Age

Foreword by

John Archibald Wheeler

Birkhäuser
Boston • Basel • Berlin

Patricia Rife
University of Hawaii-Manoa, Maui Program
Matsunaga Institute for Peace
Kihei, HI 96753

Library of Congress Cataloging-in-Publication Data
Rife, Patricia.
  Lise Meitner and the dawn of the nuclear age / Patricia Rife ;
with a foreword by John Archibald Wheeler.
     p.   cm.
  Includes bibliographical references and index.
  ISBN 0-8176-3732-X (alk. paper).—ISBN 3-7643-3732-X (alk.
paper)
  1. Meitner, Lise, 1878-1968. 2. Physics—History. 3. Nuclear
physics—History. 4. Nuclear physicists—Austria—Biography.
5. Women physicists—Austria—Biography. I. Meitner, Lise,
1878-1968. II. Title.
QC774.M4R54 1998
530'.092—dc21
[B]   97-52076
                                                              CIP

Printed on acid-free paper.
© 1999 Birkhäuser Boston   *Birkhäuser* ®

© 1990 Claassen, original publisher of related biography in German (1st edition)
© 1992 Claassen (2nd edition)

ISBN 0-8176-3732-X
ISBN 3-7643-3732-X

Cover design by Joseph Sherman, Dutton & Sherman Design, New Haven, CT.
Typeset by Martin Stock, Cambridge, MA.
Printed and bound by Edwards Brothers, Inc., Ann Arbor, MI.
Printed in the United States of America.

9 8 7 6 5 4 3 2 1

To all those working for world peace through paths of wisdom

Contents

Contents

Foreword

Excelsior! Or, in everyday language, ever higher. That, in a word, says to me what I remember of Lise Meitner as unique. Identify a central and decidable issue of physics, devise a way to go after it, and pour on the attack.

I first met Lise Meitner in early 1935, not in Berlin but in Copenhagen, after a dinner given by Niels Bohr and his wife at Carlsberg, the House of Honor awarded him by the Danish Academy. I remember Meitner and Bohr seated in reddish brown, leather-covered armchairs, facing each other without ever looking out on the garden and the distant Carlsberg brewery. Oh, so serious their talk. And, from what I could hear of it, about serious matters: from conditions in Germany to the anomalous scattering of light by lead.

Already in my first weeks in Copenhagen in the fall of 1934, I had come to know Lise Meitner's nephew, Otto Robert Frisch, who gave a seminar in Danish after having had only one month to learn the language. He taught me in the laboratory how a Geiger counter could stutter — the radioactive source originally so powerful as to saturate the counter except for statistically infrequent intervals. I never learned whether he and his aunt had ever exchanged information on this tricky point, but everything about them and their way of analyzing issues made me believe in their reliability — to be tested to the hilt a few years later, in the affair of the uranium nucleus and its fission by a neutron.

Know, or Know Who Knows was the motto of the distinguished American public figure, John W. Gardner, and the title of his book. Identify the Central Puzzle and

Clear It Up—an apt motto for Meitner. Identify scientists who can do that, bring them to the Copenhagen Institute for an annual conference and there identify the central issue and help to clear it up: that, in brief, was the work of Niels Bohr and his Institute for Theoretical Physics in Copenhagen.

Then, why wasn't Meitner at the great International Conference on Physics held in London and Cambridge in October 1934, the first such conference since the end of World War I, an occasion in which Enrico Fermi impressed the world by what neutrons could do and broke open the paths to the discovery of fission, and where Gray and Tarrant reported strange results on the back-scattering of gamma ray radiation? Bohr was trying to heal the postwar rift in Germany's relations with the rest of Europe, a rift that arose out of many Germans' sense of having been unjustly treated. About reparations, inflation, and Germany's financial mess, Bohr could do nothing. About its assignment to an inferior and unrepresented place in the postwar International Union of Physics, he could, however, protest, and did, and refused to accept the presidency of the International Union himself. But, soon after I and others returned to Copenhagen from the London conference, I learned that the laboratory's senior experimentalist, H.C. Jacobsen, had already been doing experiments on the back-scattering of light by lead, inspired not least by earlier stimulus from Lise Meitner and by her published experiments with Herbert H. Hupfeld — with puzzling results widely known under the name of the Meitner-Hupfeld Effect. Meitner inspired not only experimental work but also theoretical calculations. Her student, Max Delbrück, had consulted with the leading experts on quantum electrodynamics and knew the straightforward calculations for the probability of the proton to produce a positive-negative electron pair in the field of force of an atomic nucleus. But, mathematical divergences afflicted the calculation of the processes of a higher order, such as the scattering of light by the field of force of a nucleus. Inspired by discussions with Meitner, Delbrück worked away at the calculations of the probability for this process of direct scattering of light by the nucleus, a calculation which we know today lay beyond the power of the methods of that time.

Delbrück's valiant but unsuccessful efforts to defeat the mathematical divergences that beset his calculations had become well known in Copenhagen and made him something of a hero when, out of Germany in time, he arrived in Copenhagen as a long-term guest at Bohr's institute. In the lunchroom, in those days on the top floor of the institute, the discussions between Niels Bohr and Max Delbrück that I was privileged to overhear dealt so often with questions of molecular biology that I was not surprised when Delbrück told me one day that he was going to be working in that field at the Rockefeller Institute for Medical Research (now Rockefeller University) in New York. "And I'll be doing whatever it takes to make a success of this new work, from making my own microscope observations to washing my own Petri dishes," he resolved. It was hardly unexpected when he won the Nobel Prize in later years. Lise Meitner encouraged not only herself, but her bloodhounds, to go after the most exciting open issues of science of the day.

Meitner herself escaped from Berlin in 1938. Escaped what? How can I describe the hell that I visited at Auschwitz some months after the end of World War II. Fortunately, she went not to Denmark (soon to be occupied by the Germans), but to Sweden, easy for her Copenhagen-based nephew, Otto Robert Frisch, to visit. There, on a wintry day, aunt and nephew walked and talked about the latest experiments of Otto Hahn and Fritz Strassmann, at the Kaiser Wilhelm Institut für Chemie in Berlin–Dahlem, in which they had detected barium (atomic number 56) formed in the neutron irradiation of uranium (atomic number 92). Fission! Hahn had to come to terms with fission of uranium as a new physical process. And Meitner did.

Frisch returned to Copenhagen in the holiday season of 1938–1939 to report the new finding to Niels Bohr. He arrived at the very last minute before Bohr's departure from Copenhagen for America on the Swedish-American liner S.S. Drottningholm. Bohr had been invited to the Institute for Advanced Study in Princeton for the spring semester. He was driven by a powerful motive: Come to an agreement with Einstein on what Einstein sometimes called "this incredible quantum business." Einstein was skeptical of the probability interpretation of quantum mechanics, telling me on more than one visit to his house that "The good Lord does not play dice." If Bohr and Einstein could have worked together, day after day, I am prepared to believe they would have arrived at some deep new insight which we still lack today on those greatest of questions: "How come the quantum?" and "How come existence?" But because fission posed immediate and exciting questions of principle in nuclear physics and because I had worked with nuclear physics at his institute in Copenhagen, Bohr's Princeton visit ended up with him working day after day with me on fission and only on special days — but not day after day — with Einstein on the quantum. Oh how I grieve at the loss in this way of what might well have been the greatest discovery of this century: but the energy released from fission called.

Not until after Hiroshima and Nagasaki did Lise Meitner learn what her discovery had led to — what I estimate was a shortening of the war by at least two months and a saving of at least a million lives. A postwar Hollywood filmmaker sought a representation of the part of Meitner in a film he was making — for her, uncharted and uncomfortable territory. Friends of Meitner's advised her to make me her representative in negotiating the wording for her part in the film. I worked faithfully on this task, but Meitner was not satisfied with the outcome and she withdrew. However, my work for her on the project earned me my last chance to see her, and appreciate anew the force of her character and her drive to "Identify the Central Puzzle and Clear It Up." The readers who open Patricia Rife's book have before them a story of one of the half dozen most remarkable women of the twentieth century: my friend and esteemed colleague.

John Archibald Wheeler
Princeton University
Physics Department

Introduction

Women's stories have not been told. And without stories, there is no articulation of experience. Without stories a woman is lost when she comes to make the important decisions of her life. She does not learn to value her struggles, to celebrate her strengths, to comprehend her pain. Without stories she is alienated from those deeper experiences of self and world that have been called spiritual or religious.... If women's stories are not told, the depth of women's souls will not be known.

Carol P. Christ
Diving Deep and Surfacing

THE story of Lise Meitner and her dramatic interpretation of nuclear fission is a tapestry woven together through scientific discovery, political events, social and racial prejudice, and close collegial research. It is a story illuminating historic as well as personal themes, repeating patterns and crescendos much like the classical music she so deeply loved. As a historian of science and something of a "sleuth-biographer," I have interpreted these themes in the context of Meitner's times, which changed dramatically during her 89 years. Like the Greek historian Herodotus pointed out, a life must be interpreted to be fully understood. Through my own personal and professional journey during the past ten years researching and writing this book, I have analyzed and chronicled Lise Meitner's scientific work and discoveries, which were directly impacted by the sociopolitical events of World Wars I and II.

Historians face a tremendous challenge in choosing which materials to incorporate and which to "interpret" when weaving the threads of a life story into a meaningful biography. To do justice to the broader context of Meitner's work with radioactive substances and processes, I have documented many of the amazing scientific discoveries of her era, such as the impact of the Bohr atomic model, the invention of the Geiger counter and discovery of the neutron. As you read about Meitner's lonely struggles as a shy, brilliant pioneer in the scientific community — as one of the first women to ever receive a doctorate in the sciences in the University of Vienna's 500-year history and the first woman to become a faculty member in physics at the University of Berlin[1] — you will meet such influential physicists as Boltzmann, Planck, Bohr and Einstein. Each was part of the international community who also became friends before World War I — and each had a direct impact on Lise Meitner's creative achievements in 20th century quantum physics. And what better palette from which to paint than Meitner's physics circle and their ever-changing social milieu!

Imagine yourself attending a scientific congress in the cool Swiss Alps in 1909. A quiet 31-year-old woman with a long dress, an Austrian physicist, is sitting in the audience next to Max Planck. They are waiting expectantly as 30-year-old Albert Einstein quietly approaches the podium to deliver his inaugural lecture to the august German Scientific Society. Meitner had attended all of her mentor Boltzmann's lectures during her doctoral struggles, when he gallantly struggled to defend atomism against the logical positivists. He had committed suicide a year before, and it was Planck who urged Meitner — and Einstein — to come to Berlin. Both would decide to make Berlin their new home. And Lise Meitner's choices before World War I would progress similarly to Einstein's: proximity to Planck and the University of Berlin was the central feature of their emerging careers; they moved there as single physicists, spending late nights alone with their books and ideas.

Years later, Meitner's shrewd insights into atomic structure, which led to her interpretation of fission (and, utilizing Einstein's formula $E = mc^2$, her dramatic insight into the tremendous energies which would be released from this process), turned the key in the transuranic puzzle, and thus contributed a fundamental building block to the understanding of physics. But it was also a discovery that released a tremendous force that changed our world. From Meitner's 1938 insight into the process of fission, and publications by Hahn, Strassmann, Frisch and herself, intense government funding and research efforts were focused around the world toward the development of the world's first atomic bombs. Harry Truman, in a formal dinner honoring Lise Meitner's accomplishments after the gruesome war, ironically teased, "So you're the little lady who got us into all of this!"[2] His joke could not have been further from the truth: Lise Meitner never worked on weapons research, and like Einstein, remained committed to the international peace efforts of scientists after the war.

Meitner's life story makes it clear that she was often at the center of revolutionary events, while she was forced toward the periphery through discriminatory practices and policies, excluded from decisions made about her own work, and denied just and formal recognition by her peers. She lived and worked within the contrasting mileux of scientific objectivity and social discrimination, but persevered and encouraged many others during her sixty-year career to enter radiophysics and radiochemistry research. She diligently focused upon her groundbreaking work in atomic physics, publishing over 128 articles in her lifetime, serving on scientific commissions, writing and editing, and even served on a United Nation's International Atomic Energy Agency's committee until her mid 80's.[3] She was, as Victor Weisskopf recounted to me at M.I.T. years ago, an excellent, creative experimentalist first and foremost, as well as a part-time Lecturer in the University of Berlin's lecture halls for over ten years. Hence, she influenced an entire generation of doctoral students, including Leo Szilard, Carl Friedrich von Weizsäcker, Max Delbrück and many others before her dramatic escape from Nazi Germany. Her publication and teaching records attest to her effectiveness in this dual role. She thrived and made key contributions to the "quantum revolution," debating long into the night over her coffee and cigarettes with friends Niels Bohr, Wolfgang Pauli, Werner Heisenberg, Hans Geiger, Max von Laue and Max Born during colloquia at Bohr's Copenhagen Institute for Physics or the Kaiser Wilhelm Institutes at Berlin–Dahlem.

Later, with the rise of the Nazi Party, she would break with many young German physicists, including Heisenberg. I have documented her ethical concerns through letters, sent and unsent, which she wrote to colleagues after the war. In 1945, she wrote despondently to her colleague Otto Hahn:

> When I heard a very objective report prepared by the British and Americans for the BBC about [the concentration camps] Bergen-Belsen and Buchenwald, I began to wail out loud and couldn't sleep all night. If only you had seen the people who came here from the camps. They should force a man like Heisenberg, and millions of others with him, to see these camps and the tortured people.... But you never had any sleepless nights: you didn't want to see — it was too disturbing. ... Please believe me that everything I write here is an attempt to help you. [4]

Lise Meitner's thirty years of joint research with Otto Hahn were the heart of her scientific career. This creative collaboration led to broadened experimental processes and new discoveries in the fields of radiophysics and radiochemistry: by 1910, they were known around the world as the "radioactive pioneers." Later, their competition with the Curie lab in Paris and Fermi's group in Rome would add drama and tension to unraveling the decay patterns and complex nature of radioactive processes as the "transuranics" added headaches to all of their lives. Hahn's life, letters and books are illustrated in this biography as she would have wanted it; he was her friend, colleague, confidant, and at the end, perhaps a nemesis until the wounds of World War II had healed. But always, until the end of their lives, the Meitner–Hahn team was a working partnership.

Berlin had been Lise Meitner's chosen workplace and home for nearly thirty years, until Hitler's Third Reich drove her out at the age of 60. I lived and conducted my research there twice, discovering documents, photos, and stories through interviews with retired physicists who remembered her well. I was followed by secret police in East Berlin, interrogated at the Berlin Wall, and questioned by the FBI when I returned home to San Francisco. Meitner's life path became mine for many years, and I am grateful to the many colleagues who became my friends: Eric Bohr in Denmark, Gudmund Borelius and Karl-Erik Larsson in Sweden, Dietrich Hahn in Munich, Joseph Rotblatt in London. Meitner's correspondence, now housed in Cambridge, England, where she was buried, was voluminous, and clearly illustrates an active and compassionate mind at work, reaching out to colleagues from her lonely exile during World War II in Sweden, a nation she would later call home.

Ironically, after being nominated over twelve times for the Nobel Prize, the Swedish Academy of Sciences did not recognize Lise Meitner's pivotal role in the interpretation of nuclear fission, and hence, in 1946, she sat in the audience, not on stage, at the stately Nobel Prize ceremonies while the 1944 Prize in Chemistry was awarded to Otto Hahn (who had just been released from imprisonment with other German physicists). Hence, as you read this biography, continue to ask the question: "Why was recognition often denied this pioneering woman physicist?" The answer, like gender bias worldwide, is not simple or clear, like the physics Meitner loved so well.

It is documented that both world wars entirely altered the military-industrial complex of the world, including government support for scientific research, and Meitner's letters speak out adamantly about this. She remained committed to the peaceful use of atomic research all her life, and here is where her career differed dramatically from many of her peers. Even when she was living and working in Sweden, active in the Royal Institute for Technology and International Committee of Atomic Scientists, founded by Einstein after the war, she spoke out for the education of more women in science and stressed the peaceful uses to which scientific research could be applied. Einstein had fondly dubbed Lise Meitner "our German Madame Curie" during their Berlin heyday in the Roaring Twenties. A charter member of the Emergency Committee of Atomic Scientists founded by Einstein, Hans Bethe, Harold Urey, Leo Szilard and Edward Condon after the war, she continued to speak at universities throughout her retirement (including Harvard, Princeton, Brown and others) about the social responsibilities of scientists.[5] In fact, her gravestone, proudly standing in a country cemetery in England, simply states: "A physicist who never lost her humanity."[6]

It must be remembered that Nazi Germany was an ominous place for anyone to conduct scientific research. In July 1938, persecuted but holding on to her job at the Institute for Chemistry, Meitner's escape was orchestrated by many physicists who saw darker times ahead. The amazing life and work of this pioneering scientist, who had few role models to emulate yet was herself a role model for many, sheds light upon how intellectuals, like all people, are deeply affected by their social-political milieu. Her nephew, Otto Robert Frisch, also chose physics as a career, and barely escaped from Germany in the early 1930s. When Lise Meitner's fame culminated in her interpretation of the process of nuclear fission *after* her escape from Berlin, it was Frisch, 34 years her junior, who played a key experimental role in the demonstration of the fission process at Bohr's Institute. I give him full credit throughout my narrative, as well as Hahn and Meitner's younger assistant, Fritz Strassmann, who was also (like Frisch) at the beginning of his rough entry into a scientific career as the Third Reich shattered careers and lives.

Historians face a weighty task. We must interpret the past carefully and creatively yet rest our analysis upon documented evidence of discovery priorities, scientific work, and actual "voice" through oral history interviews and written correspondence. What I have chosen to highlight, through careful selection of primary documents concerning the development of atomic bomb research in Nazi Germany, America and Russia, are summaries of nuclear research conducted in the international race to produce the first atomic weapon. New documents continue to come to light from this era, and there is much still waiting to be mined and published. Documenting the abstract tools of thought, elaborate experimental equipment, conflicting experimental data, political pressures from differing laboratories, and pressures from government funders, the "dawn" of the nuclear age is still being interpreted. This story parallels Meitner's story, since she was, in essence, also at the center of many pivotal events in the history of quantum physics and later, fission research and applications. This story, I believe, sheds important light upon the challenging dilemmas our modern world continues to face in confronting nuclear weapons and misuses of nuclear power.

Understanding the social roles of scientists as they conduct their work and impact our world continues to fascinate me. It was the German philosopher Hannah Arendt (like Meitner, also a victim of fascist policies) who wrote so eloquently about the impact of science in our so-called "nuclear age." In view of heroic scientific discoveries and achievements, Arendt writes, "performed for centuries in the unseen quiet of the laboratories, it seems only proper that their deeds should eventually have turned out to have greater news value, to be of greater political significance than the administrative and diplomatic doings of most so-called statesmen."[7] Over twenty years after Hiroshima and Nagasaki, Meitner wrote with characteristic modesty: "That life has not always been easy — the first and second world wars and their consequences saw to that — while for the fact that it has indeed been full, I have to thank the wonderful developments of physics during my lifetime and the great and lovable personalities with whom my work in physics brought me contact."[8]

Lise Meitner's quiet courage and her tenacious sense of scientific curiosity and acumen opened many doors. Her impressive publications list and commitment to world peace illustrate a balanced and deeply ethical person who made scientific research her life's focus without forgetting her role as a responsible world citizen. She and Eleanor Roosevelt made a pledge in 1945: to work for world peace. She continues to be a role model in this respect, and it is with gratitude and respect that I offer you this biography of a woman whose insights into matter changed our world.

CHAPTER I

Choosing the Path of Physics: 1878–1906

LISE MEITNER, the third child of Philipp and Hedwig (Skovran) Meitner, was born on November 7, 1878 in fin-de-siècle Vienna, capital of the Austro-Hungarian Empire. By the turn of the century the Meitners had eight children.

Growing up with her younger siblings and an elder brother and sister who adored her, Lise had a loving family, a secure childhood, and intelligent parents who challenged their children in creative ways. Music, thanks to their mother, was central to family life and all of the children learned to play the piano at an early age. Their father, a lawyer, emphasized logic and engaged tutors from whom Lise learned a great deal about the form and structure of mathematics. She also loved to read. And along with these intellectual activities, this shy, petite child performed traditional female tasks.

Tracing their family history, we know that Lise's maternal grandparents were Jewish, and that they had traveled from the Moravian steppes hundreds of miles away to the burgeoning Austrian capital. Her father's ancestors, also Jewish, had resided in the Bohemian/Moravian village of Meiethein which had been part of the Austro-Hungarian Empire.

In 1785 Kaiser Josef II declared that all families living within his realm had to register a family name. He made German the official language, lifted barriers of serfdom, and liberated the Jewish ghetto. Several hundred Jews were given permits to enter Vienna and work in the military, civil service, or other jobs related to the Empire.[1] Lise's great-great grandfather took the surname "Meietheiner," parallel to the name of their village, which was later shortened to Meitner by Philipp's father Moritz.

In the mid-1800s Moritz Meitner married a beautiful widow from Moravia. Philipp characterized his mother from two incidents in his youth that, to most, would seem tragic: when their family house burned down and when cholera broke out in their village. In both instances his mother *sang* to keep up the family spirits. This mixture of cheerfulness and determination became apparent in all of the children and grandchildren. Most certainly Lise Meitner inherited her grandmother's ability to persevere in the face of tragic events.

The Meitners easily assimilated into the cosmopolitan life of Vienna. Viennese Jews were not relegated to the social fringes of Austrian society, but were regarded by many in the bureaucracy as the "state people par excellence." Jewish emigrants did not constitute a so-called "unhistoric" nationality like the Slovaks or Ukrainians, who had been literally uprooted by Empire policies, and for many generations, the monarchy did not insist on a "declared nationality" for its Jewish citizens. Jews became, in historian Carl Schorske's words, the "supranational people of a multinational state, the one folk which, in effect, stepped into the shoes of the earlier aristocracy."[2] Some prejudices, however, continued to exist. The Austrian Constitution of 1867 proclaimed the "fundamental rights of all citizens" within the Hapsburg Empire, bestowing long-sought legal and civic equality on Austrian Jews 11 years before Lise was born.[3] But fierce debates still raged about such constitutional "rights."

Philipp Meitner was a "free thinker,"[4] respectful of the law but critical of its inconsistencies. The Meitners lived among other liberal Jewish professionals, and it was this society rather than the synagogue that dictated their behavior. Some families, like the Meitners, converted to Protestantism. These professionals used to socialize in one another's homes, in which the noisy conversations of children blended with animated adult conversation concerning the latest in Europe's politics, art, and always music. To the Viennese, the city's classical orchestra was as important as the aging Emperor's rulings. Such was the climate in which Lise was raised. As Schorske notes: "In Austria, where higher culture was greatly prized as a mark of status by the liberal, urban middle class, the Jews of that class merely shared the prevalent values" of hard work and disciplined study[5] – hallmarks of Germanic culture. The Meitner children were also urged to follow their own internal bent: "Listen to me and to your father but think for yourself!" mother Hedwig used to say in her Moravian-Austrian dialect.[6] Throughout her life Lise Meitner did just that.

From Lise's sister Frieda we learn that, at the age of eight, Lise kept a mathematics book under her pillow and used to cover the crack beneath the bedroom door so that her parents would not catch her reading late into the night. Lise had an unusual curiosity about science at an early age, although she was not encouraged like her brothers in this area: "Although I had a very marked bent for mathematics and physics during my early years," she later reflected, "I did not begin a life of study immediately. This was partly due to the ideas that were then prevalent with regard to the education of women."[7] Her parents had no objection to her studying mathematics at school, but they wanted her to pursue something "practical," for example, the teaching profession. Her sisters took up piano and flirted with boys in their spare time, but Lise, at an all-girls' school, persevered to get the highest grades.

She was often teased for "studying too much." Early photographs show her as a conservatively dressed adolescent, looking shyly away from the camera with an intense gaze. Clearly she did not fit the model of a "proper" Viennese young lady. Studies were more important to her than fashions, opera, or home life. The ideal young girl of this period would have been more like the one portrayed by Stefan Zweig in his autobiography:

> This is how the society of those days wished young girls to be: silly and untaught, well-educated and ignorant, curious and shy, uncertain and unprotected and predisposed by this education, without knowledge of the world from the beginning.... But what a tragedy if one of these young girls missed her time, if she was not yet married at 25 or 30![8]

As she neared graduation from the crowded *Mädchen-Bürgerschule* near her home, Lise was painfully aware of the boundaries of Viennese society:

> Austria was so thoroughly patriarchal that women were bound to suffer. Every part of the female anatomy had to be concealed by clothing so cumbersome that it was impossible to dress oneself without assistance. In turn, this burdensome clothing necessitated a totally artificial manner of movement and a feigned manner of decorum in excursions out of the home. The code of conduct required of women was equally artificial, on top of which society did not permit women to be educated beyond what was essential for 'good' breeding.[9]

Lise the scholar did not fit this model. From her earliest years Lise was curious about the natural world. Someone once explained to her "how a puddle with a bit of oil on it showed lovely colors," and she responded by exclaiming how exciting it was to discover "that there were such things to find out about our world," and kept asking "more and more questions of that kind."[10] Mandatory education for girls ended at age 14. Today it would be shocking for us to think that a gifted girl would not be able to graduate with a high school diploma; but while Lise Meitner learned some mathematics, her schoolmistress did not feel that algebra was a necessary subject of study for Viennese girls. Nor were young women of that period permitted to attend the *gymnasium* (college-preparatory high school) together with university-bound young men. When, in 1899, a university education for women was at last legally permitted in the Empire,[11] new opportunities for bright women seemed to be the theme of the times, although in reality few such opportunities were available .

Lise's courage and determination, however, cannot be underestimated. During adolescence she had tutored her siblings, learned French, and at one point vowed to become a teacher when she "grew up." However, her scientific interests never waned. Her mathematical talent was apparent, and she eagerly read those articles in the Viennese newspapers on the famous French physicist Marie Curie and her research.

One evening Lise quietly approached her father and asked for his support to pursue her dream — "a scientific education" at the University of Vienna. During its 500-year history, several women had attended the university, but they had been privately tutored in working towards their doctorate, and had not attended regular classroom lectures. Nearby, in Germany, the first female Ph.D. in Physics had been awarded in 1895 to an American who studied under Kohlrausch at the University of Göttingen; the second Ph.D. in Physics would not be awarded to another woman for eight years (Olga Steindler, 1903).[12] Lise Meitner knew none of this. She only knew that at the university she could pursue her first love — science. Moreover, her father fully expected that once "justice prevailed" women would also be admitted to the medical school of the university.

Having studied to become an attorney, Philipp Meitner had no illusions about the Viennese university system. He thoroughly understood its discriminatory practices against Jews and women, and never dreamed that not only one but several of his daughters would wish to enter its domain! He was aware that entrance to the science departments would present obstacles. So it was natural for him (and his wife) to insist that Lise first learn a skill, such as teaching, "to support herself" before allowing her to undertake university preparation.[13]

There would be no more carriage rides with the children in the park. Lise dutifully turned her attentions towards obtaining her teaching credentials in French, eager to prove that she was capable of supporting herself. At the same time she studied late into the night for university entrance exams in the sciences, determined to pursue a "real course of study." Lise had already decided that it would be physics.

By the turn of the century science in Vienna was beginning to reach the proportions of a national obsession. Next to music, Austrians were spending as much time reading about such scientific advancements as Röntgen's new "X"-rays as they did about the praises of their current opera star. Philipp Meitner observed his daughter's sense of commitment and had to admit that she would be well-suited for an academic vocation. But the fact remained that only a few women had matriculated into the university before the turn of the century, and fewer still had entered the faculties of science. Philipp Meitner had eight children to support. Vienna of the 1890s was a world dominated by irrational social mores and prejudices in spite of intense cultural interest in science and technology.[14] All too often women found themselves denied goals they had been encouraged to pursue at an earlier age. Many wealthy families used to send their daughters for a university education to Switzerland, or to far-off America, but the Meitners could not afford this.

The tide seemed to change when, in 1899, the Austrian government decreed that universities had to admit women even without a *gymnasium* (high school) degree. After passing a rigorous state examination qualifying her to teach French, Lise began a year of trial teaching at a girls' high school in Vienna. She was then 20 years old. After this taxing pedagogical experience, her parents consented to her studying for the *Matura* certificate, which would qualify her for entrance into the university. The *Matura* test (unlike our modern college entrance exams) was nearly equivalent to an associates degree. Lise's sister Gisela had studied for two years to pass the *Matura*, which enabled her to enter medical school in 1900.

In-between household tasks, Lise became immersed in physics books and diagrams. She sharpened her calculus skills, and was drilled for the science exams by a private tutor. She also had to prepare for exams in psychology, German literature, Greek and Latin, zoology, mineralogy, and of course, religion. Lise managed to finish approximately eight years of study in two years. Her sisters used to tease her when she did *not* have a book in her hand: "Lise, you have just walked across the living room without studying!"[15]

Her life at this time seemed all about catching up; without proper high school training, private teaching was the best that could be done. Lise did not mind traveling across the city to the boy's *Akademische Gymnasium* on the Beethovenplatz where, along with two other young women, she met with her tutor. Arthur Szarvasy, her tutor, was a *Privatdozent*, an unsalaried academic instructor/lecturer whom Lise admired and eventually assisted in carrying out small tasks at the Physics Institute — a dream come true for a woman desperate to understand the intricacies of physics experiments. She recalled these long months of preparation with good humor:

> Dr. Szarvasy had a real gift for presenting the subject matter of mathematics and physics in an extraordinary stimulating manner. Sometimes he was able to show us apparatus in the University of Vienna Physics Institute, a rarity in private coaching — usually all one was given were figures and diagrams; today it amuses me to think of my astonishment when I saw certain apparatus for the first time.[16]

At last the fateful test date came. Out of the 14 women who took the exam, Lise Meitner was one of only four who passed. And in the autumn of 1901 many male faculty members and students may have been astonished to see a young *fräulein* enter the University of Vienna's crowded lecture hall, as astonished as Lise was when she was given the opportunity to see "real" physics apparatus and an experimental lab! At the age of 23, she was admitted to the University of Vienna, and took her place in the physics lectures with the male students.

From 1901 to 1905, she studied mathematics, physics, botany, and philosophy, the latter being a required subject in all university curricula. "No doubt," she observed retrospectively, "like many other young students, I began by attending too many lectures. Of course, it was very unusual for women to attend university lectures at all."[17] Yet in northern Germany there were signs that times were changing. One year earlier the Prussian Minister of Education had annulled the requirement that women needed "official state permission" to attend university lectures. But only several years earlier, for a variety of "academic" reasons, such permission had been generally refused. H. von Treitschke, the German historian, was particularly vehement concerning "surrendering universities to the invasion of women, and thereby falsifying [the universities'] entire character.... This is a shameful display of moral weakness; they are only giving way to the noisy demands of the press," he and other professors had protested.[18] Vienna was somewhat more liberal than Prussia in these matters, but individual professors in both Germany and the Habsburg Empire reserved the "right" to refuse women admission to their lectures, and did so in several disciplines, particularly medicine and anatomy. Hence, women were admitted and then denied access.

The issue of women's attendance, and later of postgraduates lecturing in the universities, is at the core of educational rights for women in the twentieth century.[19] Lise Meitner's experience as the first woman admitted to the University's physics department reflects the broader cultural shifts then taking place throughout academia.

Allegorical drawing of German Woman Student

In his book *Decline of the German Mandarins*, Fritz Ringer points out that within the tradition of German idealism and neo-humanism, the emphasis on *die Wissenschaft* (in the broadest sense of learning or scholarship) was tied irrevocably to the late nineteenth century "Germanic ideal" of education itself.[20] Austrian and German universities had been considered "national sanctuaries" of culture for centuries, an aura little tarnished with the advent of the twentieth century. Parallel to medieval monasteries, "inspired by German Idealism and dedicated to a Faustian search for 'pure' truth," Ringer observes, "the universities were carefully protected against premature demands for practical results. Like 'fortresses of the grail,' they were meant to have a spiritually ennobling rather than a narrow utilitarian influence upon the disciples of learning and upon the nation as a whole."[21] This philosophy held major repercussions, of course, for those seeking access to the "fortress," as women who struggled to earn their Ph.D.s in Germany and Austria soon discovered.

Male university students had traditionally been free to attend lectures of their choosing at any university once they were admitted into the system. Often they traveled from one town to another, listening to and participating in academic discussions inside and outside of the classroom. This freedom was seldom an option for the few matriculated women students. During the rigorous years of preparation for an undergraduate degree, all students paid fees to attend specific lectures (almost obligatory) in their chosen and required academic subject areas; hence, many gravitated towards a professor of their choice and remained in that department for several years. This remained problematic for women students.

Since German-speaking universities did not supervise the overall program of study or attendance of the student, it often rested on the *affinity* developed between a particular faculty member and the aspiring undergraduate to create the rapport for research and study. In Lise Meitner's case, the pursuit of a Ph.D. became a hope and then a reality when, during her second year of study, she found physicists who stimulated her interest in atomic structure and encouraged her to pursue experimental research as well as theoretical studies.

Certainly not all of those exceptional women entering universities in Germany and in the Austro-Hungarian Empire at the end of the nineteenth century found such faculty mentors. Those who did may have found that they were viewed not as independent students but rather as "disciples" of a figurehead professor. Ringer comments on this "mandarin" academic climate:

> If one thinks of education as a process in which a body of information and certain methods are communicated, one can arrive, at least in principle, at a means for evaluating

a student's receptivity. It is possible to assess his [or her] 'native' mental prowess, within certain limits, and to find out how much he already knows. Much more difficult problems arise, however, if one considers learning a *transfer* of cultural and spiritual values. If the pupil's whole personality has to be judged as a potential vehicle for the 'unfolding' of these values, it becomes quite impossible to make even moderately objective decisions about his aptitude.[22]

Such subjective educational dictates have major repercussions in the scientific sphere. This became a serious problem in the German pedagogic philosophy of "cultivation" since it tended to leave the selection of "candidates for higher learning" to the social prejudices and mores of the educated elite. Women entering this system found that "complete personal involvement" in their educational process was often fraught with personal and administrative difficulties, as well as many forms of social, sexual, or intellectual discrimination, including paternal patronage by the very elite providing instruction.

While Lise Meitner's first year at the University of Vienna did not entail major problems, the paternal attitudes she encountered did little to impress her:

> In my first term I studied differential and integral calculus with Professor Gregenbaur. In my second term, he asked me to detect an error in the work of an Italian mathematician. However, I needed his considerable assistance before I found the error, and when he kindly suggested to me that *I* might like to publish this work on my own, I felt it would be wrong to do so, and so unfortunately annoyed him forever.[23]

Had ethical considerations concerning joint research with such an esteemed professor entered into Lise's thoughts during her first year at the university? She would not publish results corrected by another, no matter how kind her professor's intentions. "This incident," she ruefully recalled, "made it clear that I wanted to become a physicist, not a mathematician."

However, mathematics is a fundamental tool for physicists, and Lise began to excel in its application, receiving excellent grades from professors who were influential and respected in their fields. She traversed the city daily from class to class; the university did not have one central location at that time, and lectures in various subjects were held in buildings of different sizes, shapes and locations. She retained vivid memories of her first laboratory: "I often thought, 'If a fire breaks out here, very few of us will get out alive.'"

Her first physics professor was Franz Exner who designed his course for pharmacy students and required much "hands on" work. Up the street from where Freud worked, between coffee houses and narrow apartments, this laboratory was directed by Anton Lampa. Lampa, who would later introduce Lise Meitner to Albert Einstein, had been skeptical of the atomic theory of matter at that time, [24] and although his lab was rather primitive, he struggled to provide Viennese students with new methods in science. Meitner later recalled:

> Lampa was an excellent experimentalist, but as an enthusiastic follower of Mach, was rather skeptical of the modern developments in physics. He was perhaps more interested in epistemological and philosophical questions than in pure physics, although he did write a manual on experimental physics that was really good for that time. An introductory course in practical work, it was certainly very well conducted, but the extremely primitive apparatus available limited the possibility of carrying out experiments. I remember I once asked him where to get ice for an experiment, and he replied in a rather scoffing way that I had only to go down into the yard and fetch some snow.[25]

Despite such drawbacks, experimental apparatus (primitive as it may have been) was at least available in the undergraduate laboratory, and Lise applied herself to mastering new lab techniques. By the start of her second year at the university, she began to see the results of her labors, and was convinced by 1903 that physics was to become her major focus of study.

The broader scientific climate of that period can also be observed through her choice of specialization. When early twentieth century academics complained about the dangers of students specializing, they were *not* referring to "experts" in different disciplines knowing less and less about each other's fields. Most of the professors took a lukewarm, indifferent, or actually hostile attitude toward the idea of interdisciplinary lectures or programs of study.[26] What seemed to worry them was not "the isolation of the disciplines from each other," but the "growing separation" between general scholarship and a certain kind of philosophy – "the search for 'total' knowledge, a vitality of *Wissenschaft* that exists only in relation to a whole – in essence, a philosophical *verité* from which a student's *Weltanschauung*, or overall worldview, would emerge."[27] Physics, with its questions of epistemological significance – literally exploring the question of how we know nature – challenged many academicians' fears as its realm of inquiry was expanded into philosophic domains. The issues surrounding any science gaining "total knowledge" about nature and its processes was at the heart of new twentieth century debates, and theorists as well as experimentalists joined with gusto in these duels against their more traditional peers in the Department of Philosophy.

The person who most influenced Lise Meitner concerning these issues at the University of Vienna was Ludwig Boltzmann. In 1900 Boltzmann had left Vienna to accept an academic position in Leipzig, where he had previously served as Dean of Faculty. Leipzig's new scientific research center was run by physio-chemist Wilhelm Ostwald, later given the Nobel Prize for his work on specific catalysts. The intense philosophical debates in Leipzig's physics department were overly taxing and Boltzmann's health began to fail.[28] Teaching physics, mentoring graduate students, he became depressed; greatly overworked, he succumbed to pressure from his wife and returned with his family to Vienna late in 1902. There he was greeted by a group of young enthusiastic physicists and the Dean who convinced him to begin his physics lectures again. From 1902 until 1906 Lise Meitner crowded into decaying lecture halls to attend his dramatic lectures. Boltzmann's influence on her future career was considerable. "In his opening lecture on the principle of mechanics," Meitner reflected,

> Boltzmann mentioned that there was no need for him to pay the usual compliments to his predecessor, since he was his own predecessor... [He] had no inhibitions whatever about showing his enthusiasm while he spoke, and this naturally carried his listeners along. He was also very fond of introducing remarks of an entirely personal nature into his lecture. I remember how, in describing the kinetic theory of gases, he told us how much difficulty and opposition he had encountered because he had been convinced of the real existence of atoms, and how he had been attacked from the philosophical side, without always understanding what the philosophers had against him.[29]

What a number of physicists and philosophers held against Boltzmann was an issue that had been simmering within scientific circles for centuries, and which broke out in full force during the last decade of the nineteenth century. Tensions arose over redefining the concept of *atomos*, the fundamental unit of matter postulated by Leucippus and Democritus in the fourth century BC.[30] Scientific debates in the nineteenth century over newly-proposed atomistic-based theories had their roots in tensions between the "idealist" and "materialist" conceptions of mind and matter. Those advocating the concept of *atomos* were labeled "materialists" by classical philosophers.[31]

Atomism, as debated during this period, must be distinguished from our modern day chemistry or atomic theories. In the nineteenth-century sense, it was defined as a "perennial" philosophy that sought to explain all change in terms of unchanging indivisible "ultimate units."[32] Some academics lumped together those advocating atomistic models with the "perennial" philosophical atomists. Progressive chemists and physicists in Boltzmann's time no longer believed that "ultimate" atomic units were "indivisible entities," but proposed that "divisible" atomic units could be used heuristically (as models or tools for thought and calculation, not as ultimate truths). However, this position was considered heresy in a number of scientific circles.[33] Chemists found themselves recipients of eighteenth-century models, and longed for more rigorous explanation systems. Certain religiously oriented philosophers and scientists thought that the new atomism represented the worst aspects of a "crass materialism" that challenged the "ultimate" power of a perfect God. In their view, a benevolent Lord had provided man with the powers of reason that could measure and quantify, but not divide the "indivisible." Atomism was therefore seen in some circles as an atheistic belief and outright blasphemy; how could man "divide," even by reasoning, what was given to us by God as "indivisible?" Deep divisions began to emerge between scientists in search of atomic explanations, while theologically oriented philosophers and positivists dug in to preserve their concepts of "ultimate truth," that is, that atoms were not really "real."

Nevertheless, the general premises of atomism continued to advance throughout the nineteenth century and into the twentieth. The systematic units used by chemists, going back to Dalton's insight into the structure of molecules,[34] had, over the course of 75 years, been related to atomic "units" used by physicists in their calculations. But toward the end of the nineteenth century, various shortcomings in theoretical physics, questions relating to specific heats, irreversibility, entropy, and other such physical phenomena were taken more seriously as "defects" in the concept of atomism and its philosophical implications.

With the publication of his famous 1872 paper containing a new derivation of the formula for the distribution of molecular velocities (which became known as the "Boltzmann factor"),[35] Boltzmann became a strident supporter of the *reality* of atoms. Opposition to his ideas was formidable. A counterargument was advanced by Wilhelm Ostwald in an 1895 lecture delivered in Lübeck entitled "The Superseding of Scientific Materialism." Ostwald supported his position with an admonition from the Ten Commandments:

> But, I hear you say, 'if the picture of moving atoms is denied us, what remains to form a picture of reality?' To such a question I would like to answer: Thou shalt not make thee an image or likeness! Our task is not to see the world in a more or less distorted or blurred mirror, but to see it as immediately as the nature of our mind will at all allow. To relate realities, demonstrable and measurable quantities, to each other so that, when some are given, the others may be deduced, that is the task of science. This task cannot be accomplished by hypostatizing some hypothetical picture, but only by proving relations of mutual interdependence between *measurable* quantities.[36]

In Ostwald's opinion, so-called atoms could not be measured. Even if Boltzmann had a new mathematical means to juggle pictures of matter, this "picture of reality" was *not* to be taken as truth. Ostwald continued to fight Boltzmann's atomistic concepts at every turn while his opponent was in Leipzig.

By 1900 Ostwald had developed an "energetics" theory of matter and motion within the physical chemistry department that he chaired — a theory that he propounded vigorously at conferences and in publications to counter Boltzmann's support of atomism.[37] During the late nineteenth century, Ernst Mach, who held the Chair of Natural Philosophy at the University of Vienna until 1902, also continued to propound *his* popular view based on positivism, in which he insisted that although the atomic hypothesis could be useful in science, atoms must not be considered to have a "real" existence. Max Planck, then a young Ph.D., recalled: "One just could not prevail against the authority of men like Ostwald or Mach."[38]

Mach's objections to the atomic theory were primarily of a philosophical nature. He outlined them in his 1882 lecture, "The Economical Nature of Physical Research,"

> It would not be proper for science to attribute to her variable economical instruments, created by herself, her molecules and atoms, realities behind appearances; to replace the animistic or metaphysical mythology by a mechanistic one, forgetful of the recently acquired wisdom of her bolder sister, philosophy, and thus to create pseudo-problems. The atom may remain a means, like mathematical functions, to represent phenomena. But gradually, as her intellectual education advances with the subject matter, science stops playing with pebbles to form mosaics, and seeks to understand the limits and forms of the bed in which the living stream of phenomena flows. She recognizes the most economical, simplest conceptual expression of the facts as her goal.[39]

At the famous Halle Conference of 1890, Ostwald backed Boltzmann into a corner, trying to convince him of the superiority of his "energetics" theory over atomism. Suddenly Boltzmann remarked, "I see no reason why *energy* shouldn't also be regarded as divided atomistically." Ostwald was thunderstruck by this suggestion, as was the young man next to him, Max Planck.

Five years later, at a controversial debate during the 1895 meeting of German natural scientists and physicians (*Gesellschaft für deutscher Naturforscher und Aertze*), Planck still hesitated to fully accept Boltzmann's views. But that year Boltzmann won over many other young physicists, including a student named Arnold Sommerfeld. Sommerfeld, who was to influence a generation of Germany's young physicists at the University of Munich, held vivid memories of this famous debate:

The report on energeticism was given by [the mathematician] Helm of Dresden. Behind him stood Ostwald, and behind both, the natural philosophy of the absent Ernst Mach. Boltzmann was the opponent, seconded by Felix Klein. The struggle between Boltzmann and Ostwald resembled, externally and internally, the struggle of the bull against the flexible matador. But this time, the bull won over the matador in spite of all his fighting skill. Boltzmann's arguments penetrated. We, all the then younger mathematicians, stood at Boltzmann's side.[40]

Some time later, Ostwald paid tribute to Boltzmann, calling him "the man who excelled all of us in acumen and clarity in his science."[41] Soon Planck would study Boltzmann's principles, apply them to radiation phenomena, and invent the quantum unit to account for energy not understood by physicists at that time.

When Boltzmann returned to the University of Vienna in 1902, the 58-year-old scientist was at the forefront of the atomic debates. He had firm empirical grounds upon which he based his convictions, and argued them vigorously against his opponents. His lectures were full of physics students as well as other curious academics who struggled with the various atomistic theories then proposed. Lise Meitner was inspired by this charismatic man and his ideas, and his influence upon her intellectual development cannot be overestimated. Indeed, Boltzmann remained Meitner's role model throughout her life: an independent thinker, a pioneer. Here was not only an Austrian physicist with superb theoretical abilities, but an individualist who withstood strong opposition and managed to win over a majority to his view.[42] Lise Meitner's classic introduction to the theories of atomism and its basic applications in physics and chemistry were fundamental to her university education:

> The internal fittings of Boltzmann's lecture room were, relatively speaking, very modern. He wrote down the main calculations in the middle of three large blackboards and the subsidiary ones on the boards on either side, so that it would almost have been possible to reconstruct the entire lecture. Indeed Boltzmann was not only a very fine scientist who opened up entirely new fields in thermodynamics and statistics but he was also a man who aroused admiration and affection.[43]

After Boltzmann's stimulating lectures there were hours of studying to be done. Lise Meitner did not make contact with many fellow students. It is interesting to note her response to a situation that developed with one of the brightest physics students, the charming Paul Ehrenfest. Lise later wrote the following in a private letter to Ehrenfest's biographer:

> I first met Ehrenfest in my student days in Vienna. He came from Göttingen to Vienna for the purpose of studying with Boltzmann. Somebody told him that I had carefully written down the whole series of Boltzmann's lectures. So one day Ehrenfest came to

see me and put forward the idea that we should study Boltzmann's lectures together as well as some papers written by Jacobi, which were perfectly new to me. This was an attractive offer for me, as Ehrenfest was by far much better acquainted with theoretical problems than I was and, moreover, he was an excellent and stimulating teacher. I am sure that working with him was a great help to my scientific development.

Nevertheless, I must confess that sometimes I was disturbed by his inclination to put questions about altogether personal things. At this time I was too young to comprehend that this habit of his was not the consequence of any absence of tact, but of his deep concern for every human being. Yet, it prevented me from coming into closer acquaintance with Ehrenfest.[44]

Various facets of Lise's personality are revealed through this memory: her meticulous study habits ("I had carefully written down the whole series of Boltzmann's lectures"); her consistent self-effacement ("Ehrenfest was by far much better acquainted with theoretical problems than I was"), and her willingness to work with others for mutual benefit although *she* had taken all the lecture notes! But characteristically, Lise shied away from "questions about altogether personal things" and kept their working friendship strictly "professional." Ehrenfest and Einstein would later become good friends. Both were charismatic "ladies men," and neither could understand Lise Meitner's pursuit of study, not flirtation.

The physics department was a lively but small community in the university's setting. Under the stimulus of its faculty and the research anomalies at hand, Lise began to write her doctoral thesis in 1905 under the direction of Franz Exner and his assistant Hans Bennedorf on thermal conductivity in non-homogeneous bodies. "Although Exner was an excellent experimentalist who did very valuable work in the most varied fields," Meitner recalled: "I cannot say I have a very lively recollection of his lectures on experimental physics. These were delivered almost *without* experiments between noon and one PM, when most of the students were already very tired. Sometimes I was really afraid that I would slip out of my seat."[45] Because of this dry atmosphere, Meitner focused on her own research and soon was deep into experiments about heat conductivity and its properties.

Exner proved to be a better dissertation advisor than lecturer, however, and under his guidance, Lise proved experimentally that Maxwell's formula for the conduction of electricity in homogeneous materials also held for thermal conduction in mercury. After many exhausting days and nights, her written dissertation, "Thermal Conduction in Non-homogeneous Bodies," was prepared, summarizing her research. It was formally published in the 1906 *Proceedings of the Vienna Academy of Sciences*, with due recognition by the university's physics department and proud members of

Boltzmann's Physics Institute.[46] In February, 1906, she was very formally awarded a Ph.D. in Physics, the second woman in the University's 500-year history, one of the first in the Austro-Hungarian Empire. Proud Lise was then 28 years old.

Her family was overjoyed. Lise's eldest sister had completed her medical training, and Lise had become a *Fräulein Doktor* in a very difficult field. But her elation would not last long; later that fall, the tragic news came that Ludwig Boltzmann, in a state of recurring depression and physical pain, had committed suicide.[47] The loss was felt by the physics community worldwide. Meitner took this news especially hard. She had spent most of her last year of research in the lab, but drew heavily on Boltzmann's influence to clarify her position and base her findings upon.

Without a theoretical physics mentor she returned to the old lab buildings, determined to keep informed of new developments. After completing her doctorate, Meitner intended to pursue research in general physics, but during her last year of study, Ehrenfest drew her attention to Lord Rayleigh's scientific papers. Both struggled to translate the English into German, but Lise tackled Rayleigh's complex formulas. The work was challenging, and she succeeded in proving a point that had puzzled Rayleigh about optical reflection. Lise was also able to predict other results that arose from her own experimental work in optics. Her first independent postdoctoral research was published in the *Proceedings of the Vienna Academy of Sciences* in 1906 under the title "Some Conclusions Derived from the Fresnel Reflection Formulas."[48] After its release she was strongly encouraged by her colleagues to consider a career in theoretical physics, but later recalled that it was the Curies ongoing experimental results with radioactivity that piqued her interest in continued lab research.

Stefan Meyer temporarily took over the chairmanship of the Physics Institute after Boltzmann's untimely death. Through him Lise Meitner became familiar with the emerging field of radioactivity. No longer regarded as one of the graduate students in the department, she was anxious to continue her work in experimental physics and to keep up her publishing success. Thus she shyly asked Meyer to let her remain in the field of his specialty – radioactivity. She recalled:

> I remember I had done work on radioactivity in Vienna on the scattering of alpha rays at small angles. This research emerged from a discussion between a physicist from Prague and one from Berlin – Cerniak, I think it was – who, if I remember rightly, then went to Giessen. One had maintained that there was no scattering at small angles, and the other said that there was. I then considered that one could prove this, and had even done so at Stefan Meyer's Institute. The Viennese were indeed very interested. Meyer and Schweidler had done a lot of work on radioactivity. I recall that when I had

my alpha-ray or electroscope there, they always came and held their hands out in the beam, in order to see whether they were 'radioactive.' I can still remember that! It was so at that time; one worked pretty carelessly [in terms of protection from radioactive materials].[49]

Four years earlier, in 1902, a New Zealand physicist Ernest Rutherford and his assistant Frederick Soddy had conducted research in England, providing quantitative evidence for the ideas of the Curies and others who believed that "radiation" came not from external sources, but *from within* the atoms themselves.

"Although alchemy had long been exorcised from scientific chemistry," one of Rutherford's biographers writes, "they declared that 'radioactivity is at once an atomic phenomenon and the accompaniment of a chemical change in which new kinds of matter are produced.' "[50] "The radioactive atoms decay," Rutherford and Soddy argued, "each decay signifying the transmutation of a parent into a daughter element, and each type of atom undergoing its transformation in a characteristic period."[51] These findings were to play a major role in Meitner's choice of specialization. Rutherford's pioneering research also opened the door to hundreds of new scientific avenues concerning research into the nature of the atom as well as radioactive decay.

After Boltzmann's death, university administrators began the search for a qualified successor. They invited 48-year-old Max Planck from Berlin to tour Vienna. Meitner had not yet heard of Planck's revolutionary concept of the "quantum" of action (represented by the symbol h), which he had formulated in 1900 to explain the spectral distribution of radiation phenomena.[52] In fact, in Boltzmann's final lectures, the quantum concept was not even mentioned. Few outside of Berlin had discussed its implications at all since it had been published. Commenting on the state of modern physics in these early years, Lise Meitner recalled:

> As is well known, in 1900 Planck developed the theory of thermal radiation, that an atom cannot take up or emit radiation in a continuous manner, but in quite specific, discrete 'quanta'; hence the name quantum theory. I have often wondered why Boltzmann never said a word to us about this. After all I was still attending his lectures five years after this discovery. It was, however, a very long time before quantum theory won general acceptance. Even so, Planck did not arrive at his theory until he had accepted Boltzmann's atomic theory as well as the use of statistics which Boltzmann had introduced.[53]

Planck took the long train trip to Vienna to interview for Boltzmann's position. His quantum "hypothesis" had not yet been accepted by the European physics community when an unknown Swiss Patent Office clerk named Albert Einstein, in the summer and fall of 1905, published a revolutionary triad of papers in *Annalen der Physik*. The first paper, "On a Heuristic Viewpoint concerning the Production and Transformation of Light," contained Einstein's bold suggestion that light could be considered as a collection of independent *particles* of energy called "light quanta."[54] Einstein had taken the revolutionary step of concluding that radiation must also consist of independent particles of energy, and this would be of extreme importance to the work of both Planck and, within a short time, Lise Meitner.

When he visited Vienna, Planck was already a full professor holding Gustav Kirchhoff's esteemed Chair of Theoretical Physics at the University of Berlin. At first he was skeptical of Einstein's application of his quantum hypothesis; however, he fully supported Einstein's relativity theory, which also appeared in the autumn 1905 issue of *Annalen der Physik*.[55] He discussed the implications of this revolutionary theory with many of his younger Viennese colleagues. However, Berlin still held his allegiance, and he left Vienna without accepting Boltzmann's position.

Planck had a solid circle of supporters in Berlin; one such physicist who was Meitner's age was Max von Laue. Von Laue, the son of an aristocratic Prussian judge, recollected coming home to Berlin after receiving his Ph.D. in Strasbourg. He quietly attended Max Planck's first lectures on Einstein's work before ever meeting the man, and became convinced of the revolutionary nature of quantum theory applied to atomic physics.[56]

Von Laue was soon thereafter hired as Planck's assistant in the Physics Department. In 1906, two semesters later, he spent his summer holiday on an important trip from Berlin to Berne, Switzerland, for a first meeting with Albert Einstein. Einstein had just received his doctorate from the University of Zurich, and was working in the Swiss patent office. Upon von Laue's enthusiastic return to Berlin, Einstein's work became the talk of the day among a small group of maturing physicists who had been equally influenced by Planck's "quantum constant." Von Laue and Einstein were the same age, born in 1879, and began a long-lasting friendship from their first meeting. Both physicists were just a year younger than another *Doktor* doing research in a similar field, Lise Meitner. Few foresaw that it would be the magnet of working with Max Planck that would draw each of them to Berlin to reside and work.

Max Planck's 1906 trip from Berlin to Vienna can be viewed more as a fitting memorial honoring Boltzmann's position than as a serious attempt to seek a new academic post. Planck's work and prestige rested fully in Berlin, the rising star in the constellation of twentieth-century physics centers.[57] However, Planck's interview trip held far-reaching consequences for more than one member of the Physics Institute in Vienna. After meeting Planck and discussing the possibility of further academic opportunities with her colleagues, Lise Meitner decided that the only place to continue her research on the physics of radioactive processes was Berlin. This would be the key to her independence, her way out of having to live at home and face limited or non-existent opportunities for research work in Vienna.

Hence, in the fall of 1907, the 29-year-old physicist held her breath and asked her father for financial support to make the long journey from Vienna to Berlin, "to study with Planck for a semester." Her determination to continue her research, her wish to attend advanced lectures on theoretical physics under Planck, and her courageous decision to live alone in a distant city after belonging to such a large family circle were remarkable qualities for someone raised in a bourgeois Viennese home. Her father, impressed with the tenacity of his third child, said that he would provide her with a small living stipend if she would report back to her mother and himself about her work. Being a single woman, Lise was counseled on suitable lodgings and travel procedures as well. Her original intention had been to study with Planck and spend just one season in Berlin. Few knew that her one-year visit to Prussia would become 31 years of commitment to work and life in the cosmopolitan center of German science and culture, and an opening prelude to research surrounding the central questions of twentieth-century physics.

CHAPTER II

Berlin: 1907–1909

In terms of 19th century stereotypes or rhetorical idealizations, a woman scientist was a contradiction in terms. Such a person was unlikely to exist, and if she did, she had to be "unnatural" in some way. Women scientists were thus caught between two almost mutually exclusive stereotypes: as scientists, they were atypical women; as women they were unusual scientists.

Margaret Rossiter
American Women in Science: Struggles and Strategies to 1940

BERLIN was modernizing rapidly at the dawn of the new century. Its population topped two million by 1900, and only five years later, horse-drawn omnibuses were replaced by subways and streetcars crisscrossing the sprawling city. Attitudes and value systems were changing more slowly here, however, than in the more progressive Viennese climate. Social change was generally not welcome and even adamantly opposed among German academics whose attitudes towards gender questions seemed stagnant. Debates about atomic theories, on the other hand, crossed national and cultural boundaries. The academic and social differences between Vienna and Berlin were underscored in Lise Meitner's first encounter with Prussian conservatism:

When I registered with Planck at the University of Berlin in order to attend his lectures, he received me very kindly and soon afterwards invited me to his home. The first time I visited him there, he said to me, 'But you are a Doctor already! What more do you want?' When I replied that I would like to gain some real understanding of physics, he just said a few friendly words and did not pursue the matter any further. Naturally,

I concluded that he could have no very high opinion of women students, and possibly that was true enough at the time.[1]

Planck's initial reaction to 30-year-old Dr. Meitner and her request to attend his lectures was typical of a German professor of his generation. Lise's assumption that Planck could have "no high opinion of women students" was probably close to the mark. However, since Planck was also a father of two grown daughters (twins) he must have had sympathy for Meitner's situation when he invited her to his home for dinner.

Official German state policy still barred university matriculation to *all* women, regardless of their citizenship, academic achievements, or social status. In 1900 there were over 100 women attending German university courses, yet *none* could matriculate as full-time students. Once the Reich Ministry of Education granted permission in 1901, after vociferous debates in Parliament, the University of Baden in Freiburg and Heidelberg immediately permitted women to matriculate in all of their departments; a small number of universities in other federal German states later followed suit.[2] True to its conservative reputation, however, it was not until 1909 – two years after Lise Meitner's arrival – that women's university education was officially sanctioned and regulated nationwide in Prussia, and that included the prestigious University of Berlin.

Discussions about the "women's issue" were alive in academia for decades before the turn of the century. In 1897, German professor Arthur Kirchhoff published a collection of statements by 104 university professors and lecturers entitled *Die Akademische Frau* [The Academic Woman]. Kirchhoff asked the following: What are the reasons in general, and from the special point of view of your discipline, for or against women students at universities? Which preliminary studies should young girls receive, and is it permissible for both sexes to study together at the university?[3] Planck was one of the experts quoted:

> In theoretical physics, the subject I represent, that question has not yet become so acute that it requires a special statement as far as I am concerned. . . . If a woman has a special gift for the tasks of theoretical physics, which does not happen often but it happens sometimes, and moreover she herself feels moved to develop her gift, I do not think it right, both personally and impersonally, to refuse her the chance and means of studying for reasons of principle; if it is at all compatible with academic order, I shall readily consent to women's admission on approval and always revocable to my lectures and my practical courses, and in that respect up to this point I have nothing but favorable experiences. On the other hand, I must keep to the fact that such a case must always be regarded just as an *exception*. Generally, it cannot be emphasized enough that nature

herself prescribes to a woman her function as mother and housewife, and that laws of nature cannot be ignored under any circumstances without grave danger, which in the case under discussion would especially manifest itself in the following generation.[4]

Along parallel lines Emil Warburg, Director of Medical Sciences at the University of Berlin, had earlier stated that although he did not doubt the capacities of women's mental achievements or interest in their studies, the "practical applications" of curricula for women had to be suggested with caution due to their "lack of experience" in applied subjects.[5]

Not so, wrote Wilhelm Ostwald in his reply in *Die Akademische Frau*. Ostwald was obstinately opposed to Boltzmann's atomic theory, and in pedagogical matters, he also held strong views. For him, the "right to knowledge" for women was not the question. Without doubt, lectures and laboratory work were essential to the training of a scientist, and the role of books and *Ratgeber* [mentors] for women students (above and beyond their university training) was also essential. There will always be those who will serve as guides for such *Studentinnen*, he argued, but like Planck, he reasoned that the number of women reaching such advanced levels within the university would always be few, and as such, they would always be the *exception*.

Unfortunately, this prejudice supported viewing bright women as "anomalies" to be tolerated but *not* encouraged. Ostwald's prejudices and those of other male academics became clear. Instead of stressing the merits of scientific achievement or the value of academic studies for female students, he speculated on the "suitable" (i.e., culturally proper) possibilities for women in the sciences. Given the many applications of chemistry to food preparation and medicinal usages, he foresaw opportunities for female chemistry *assistants* (note the emphasis on assisting male chemists) in the future. But in terms of *des Brotstudiums für Frauen* (wage-earning or vocationally oriented studies for women), he felt that wages in professions other than science "would always be greater," and hence speculated that as women gradually trained for professions such as law, they would find "practical applications" much more attractive than theoretical or scientific studies. Ostwald finished his conjectures by stating that motherhood provided a woman with one of the greatest sources of inner growth as well as a most satisfying sense of accomplishment . . . and hence "teaching and caring for the future generation should never be underestimated."[6]

What Ostwald implied is what Planck had also asserted, namely, that if a choice had to be made between childbearing and scientific training, a woman's allegiance should be directed towards her role as a mother. The general attitude of Germans seemed to be summed up by a deputy in the Reichstag who stated that little could be done to even *encourage* women's independent thought or action! The common Germanic phrase *Kinder, Kirche, und Küche* highlighted what the majority of Germans felt women should remain loyal to: children, church, and cooking. If, however, a woman did *not* marry (and there were many such women, often unfairly branded as "spinsters"), then her options were few: to work for meager wages or to stay at home all her life, caring for children, the elderly, or the sick.

Academics and politicians alike often support such stereotypes "scientifically" when necessary. In the book *The Study of Sociology*, first published in America and then in Europe twenty years after it was written, social Darwinist Herbert Spencer concluded that men and women were as unalike mentally as they were physically. For anyone to deny this "self-evident and inevitable truth," he stated, would be to suppose "that here alone in all Nature there is no adjustment of special powers to special functions." He further said that for all women, the "natural function" is to bear and raise children. Intellectual attributes are not necessary for this task and therefore these qualities had not been developed over the course of evolution. (Of course, he did not speculate on the need for "intellectual attributes" for fatherhood!)

Spencer went further: women's minds should *not* be developed. "Under special discipline," Spencer felt, the feminine intellect could equal or surpass the intellectual output of most men, but his point was that this would entail "decreased fulfillment of their maternal functions." He insisted that "only normally feminine mental energy" can coexist with producing and nursing the required number of healthy children. If powers of *mind* became "general" among the women in a given society he reasoned that it would lead to the disappearance of the society. Educating the intellect "is a power *not* to be included in an estimate of the *feminine nature* as a social factor."[7] With this kind of blatant prejudice supported in academic circles, it is no wonder that women throughout the world faced hardships entering the university.

Darwin thought he too had pinpointed this "unusual characteristic" of the "female intellect." In his *Origin of Species* he states: "A woman seems to differ from a man in mental disposition chiefly in her greater tenderness and less selfishness." These qualities would seem to favor women's social position, but Darwin did not think so:

> The chief distinction in the intellectual powers of the two sexes is shown in a man's attaining a higher eminence in whatever he takes up than can a woman – whether by virtue of requiring deep thought, reason, or imitation, or merely the use of the senses and hands.... We may also infer that if men are capable of a decided preeminence over women in many subjects, the average of mental powers in a man must be above that of a woman.[8]

Proponents of Darwin went on to defend his argument, giving examples of man's development of "higher mental faculties" in his fashioning of weapons, avoiding or attacking enemies, and so on. Obviously, long-held social values restricted women from entering into these kinds of debates in biology departments and medical schools. The double bind that blocked equal access to educational opportunities was not only denied but supported by "scientific facts."

Biologically-based social theories, together with views concerning woman's "true nature" and supposed natural role in the evolution of the species and society, abounded in twentieth century Germany; even university educated men regarded these theories as indisputable truisms. Adding fuel to the flames, the father of the new science of psychology, Sigmund Freud stated in his classic *Introductory Lectures to Psychoanalysis*: "Throughout history people have knocked their heads against the riddle of the nature of femininity. . . . Nor will any of you escape from worrying about this problem – those of you who are men. To those of you who are women, this will not apply; you are *yourselves* the problem."[9] Such outlooks and beliefs provided justification to those forming educational policies and engaged in the national debate. Accordingly, many argued that the "right" to a higher education never existed for women, and those females admitted to universities were always to be an exception to the rule, a "problem" but certainly *not* to ever be the "norm."

German feminists were thus uneasy with any discussion involving "natural and inalienable" educational rights, while their American counterparts, notes historian Amy Hackett, "easily slipped into natural rights language, the Declaration of Independence serving as a role model."[10] However, she continued:

> Historical theories of right superseded natural law in Germany, with consequences for both liberalism and feminism. Rights were contingent, not inalienable. Suffrage was a *civil* right. Enfranchisement should be decided, in the words of one feminist, 'merely according to consideration of political expediency.' It was 'contrary to duty' to give the vote to persons who might 'endanger the state or general well being.' German feminists skirted rights with anxiety or coupled their requests with duties or responsibilities.

The "duty" or responsibility for women in the twentieth century was whether to remain at home or assert the right to further their education. And the "right" to associate with colleagues or advanced peers in one's academic field was often coupled with an anonymous "duty" towards a department, professor, or institute as an assistant or unpaid aide. We see the educated woman being placed or placing herself in the role of the dutiful daughter or sacrificing wife and/or mother. All in all, as women around the world were granted their educational rights, historians note that women in Germany remained after puberty in a conservative, contradictory system fraught with hurdles and prejudice.[11]

Without question, it was an act of courage for 30-year-old Lise Meitner to move from Vienna into the unfriendly heart of the academic milieu of Berlin. She could have sought work as a French teacher, or as an anonymous professorial assistant in Vienna, but she elected to travel hundreds of miles to further her dreams: knowledge of physics and acceptance into the Berlin scientific community. Fortunately she had already published original scientific work by the time she made that decision. Keep in mind, however, that she was not only shy but very modest. While she had been attending lectures, taking notes and later joining in departmental discussions and debates, Planck and others did not learn of her achievements until many months after her arrival.

Lise Meitner was quickly catapulted into the heart of dramatic historical developments taking place in atomic physics. Planck's groundbreaking work was making Berlin attractive to many young physicists of Meitner's generation. The quantum constant was being discussed theoretically (Einstein elevated it to new heights in one of his papers in 1905) although Meitner was not yet familiar with the applications of Planck's ideas. But as her direct influences shifted away from the charismatic Viennese Boltzmann to the reserved Prussian Planck, one could begin to see a shift in her approach to physics. At age 86, she could still vividly recall her first meeting with the father of quantum theory:

> I officially met with Planck for the first time in the fall of 1907 when I called on him at the University of Berlin to register for his courses. A short while later I was invited to spend an evening at his home in the Wangenheimstrasse, where I met his first wife and his twin daughters [age 30], who were to become my close friends. Even during this first visit I was greatly impressed by the elegant simplicity both of the home and of the entire family. I must admit that to begin with I was a little disappointed in Planck's lectures because, despite their really classic clarity, they sometimes gave a rather colorless impression compared with Boltzmann's lectures, which were so strongly marked with feeling. But I soon saw that this was my mistake. I mention my initial error only because I think that others who never really knew Planck made the same error. They may have perceived his strong reserve as the arrogance of a person with high state honors, but there was no trace of arrogance in him. The purity of his character and of his inner single-mindedness were of a rare sort, parallel to his outer plainness and simplicity. For example, every day he rode to the university in the third-class carriage of the Berlin city railway; until he was quite old, he also rode third class on long journeys.[12]

Obviously Lise was impressed by Planck's personal traits, his "purity of character" and inner "single-mindness." By example, he showed her the way to approach physics not only as a research endeavor but as a "world view."[13] His strong spiritual views as he was the son of a Protestant minister also permeated his quiet realm, and his professional rapport with the young *Doktor* from Vienna enabled Lise Meitner to develop a broader sense of self in relation to her work. In her leisure, among other books she read Ovid's *Metamorphosis*, a revealing notation on her calendar, since her move to Berlin was a most marked metamorphosis in her own life.[14]

Since Lise was not a student of the University of Berlin, she was free to choose her own schedule. She immediately wanted to set up a research workplace for her continued study of radioactive processes. Her first goal was to work with the esteemed Heinrich Rubens, head of the university's Department of Experimental Physics, but she noted, "he did not have a place for me."[15] Since her results on Rayleigh's Fresnel reflection formulas had been published in the *Proceedings of the Vienna Academy of Sciences*[16] (Max von Laue, then Planck's university postdoctoral assistant, had told her that this was "quite a good piece of work in optics"), she knew that she had to continue her "track record" and make a name for herself.[17] Looking back, she recalled how she tried to disseminate this published work in Berlin in order to have professors like Rubens take note of her research. It happened that Rayleigh wondered about the fact that the yellow light had shown dispersion in the relative refractive index [*Brechungsexponent*]. She recalled:

> I had shown mathematically how to invert that, and had then, at the [Vienna] X-Ray Institute, conducted a calculation in optics, showing that a part of the *red* light must [also] be polarized according to whether the refractive index is greater or less than one. This was published in the Proceedings of the *Akademie der Wissenschaften* in Vienna. I had sent this to Planck before I went to Berlin, and also to Rubens, so that they could see that I had already done some good work.[18]

But whether or not Rubens reviewed her published results, he could *not* offer her a private workspace in his experimental physics department. Meitner ruefully recalled much later:

> He told me that the only space he had was in his own laboratory, where I could work under *his* direction, that is, to a certain extent, with him. While I was still considering how I could answer without offending him, Rubens added that a Dr. Otto Hahn had indicated that *he* would be interested in collaborating with me. Hahn himself came in a few minutes later. Hahn was of the same age as myself and very informal in manner, and I had the feeling that I would have no hesitation in asking him all I needed to know. Moreover, he had a very good reputation in the field of radioactivity, so I was convinced that he could teach me a great deal.[19]

Thus, in the winter of 1907, just days after being given permission to access the workshop, Meitner and Hahn cheerfully set up their experimental apparatus: three electroscopes modeled after Rutherford's to count alpha particles, beta radiation, and gamma rays. A picture was taken one year later of the two researchers in their best formal clothes. The lab portrait also shows how simple their apparatus was; yet for their time the two were considered pioneers in the study of radioactivity.

Personal factors, of course, also entered into Lise's adjustment to her new life. Working in an all-male environment at both the university and Chemistry Institute, she had few friends. Her budget was tight; in her notebooks of this time, we see her account for the purchase of one egg, one bus ticket, and, always, cigarettes. Hahn was careful to clarify his own early relations with her, in contrast to his rather jovial relations with his male colleagues in singing groups, beer halls, and hiking clubs:

> There was no question of any closer relationship between [Lise and me] outside the laboratory. Lise Meitner had a strict, ladylike upbringing and was very reserved, even shy. I used to lunch with my colleague Franz Fischer almost every day, and go to the café with him on Wednesdays, but for many years I never had a meal with Lise Meitner except on official occasions. Nor did we ever go for a walk together. Apart from the physics colloquia held at the university that we attended together, we met only at the carpenter's shop. There we generally worked until nearly eight in the evening, so that one or the other would have to go out to buy salami or cheese before the shops closed at that hour. We never ate our cold supper together there. Lise Meitner went home alone, and so did I. And yet we were really very close friends.[31]

Hahn's description seems to stress the fact that outside of his laboratory research there was "no question" of closeness to young Lise Meitner. In their later years Hahn reaffirmed his distance from a personal relationship in his early collaboration, a justification many male professionals seem to publicly address when working either in a team with a female or required to work with women on a one-to-one basis. However, the team did work well together. They would both conduct the often tedious alpha and beta ray readings and other sensitive measurements of radioactive substances long after others had gone home. In reality their *friendship* did blossom. During some of these evenings in the carpenter's shop Lise and Otto used to discuss music and then sing classical compositions to one another. In an 80th birthday tribute to her research partner, Lise wrote:

> When I think back on our more than 30-year collaboration, apart from the scientific experiences, my strongest and dearest remembrances are of Hahn's almost indestructible cheerfulness and serene disposition, his constant helpfulness, and his joy in music. Although he doesn't play an instrument, he is markedly gifted musically, with a very good

musical ear and an extraordinarily good musical memory. I remember he was given to singing or whistling the themes of all the movements of the complete Beethoven symphonies and some of the themes from Tschaikovsky's symphonies. If he was in an especially good mood, he would whistle large parts of Beethoven's violin concerto and would sometimes purposely change the rhythm of the last movement, only to laugh about my protest to it. When we worked in the woodworking shop (we still had no assistants) we would frequently sing Brahms duets, particularly when the work went well.[32]

Her exclusion from the Chemistry Institute may have also fostered sympathetic bonds between them. Yet Lise remained on formal terms, continuing to address him as "Herr Hahn." Later Hahn acknowledged that "at first, things were not easy for Lise." Lise noted that, when walking down the street near the Chemistry Institute with Otto, young members of the staff or other male workers and assistants used to address them with emphasis: "Good day, Herr *Hahn*!" Obviously, even recognition of the young Fräulein Doktor would have signified "acceptance" into their ranks, and Lise Meitner was pointedly *not* accepted by many in the Chemistry Institute. "Admittedly, the assistants in the Chemistry Institute had no particular love for women students," Lise later reflected ruefully,[33] but that did not stop her from traveling across the city from the university to their private workshop almost every evening, carrying her small dinner and hand-rolled cigarettes.

In the physics department of the university, however, her acceptance was almost immediate. Friendships formed between the younger members of the department over coffee and conversation, mathematics and music. From the outset, a major conduit for Lise into physics circles was Max Planck. Despite his important position at the university and in the Prussian Academy of Science, as well as numerous other scientific associations, Planck took a personal interest in a small number of his advanced students, and was quite open with them. Lise noted:

> In his outward behavior, Planck was very reserved, for all the affection he inspired. Some people mistakenly regarded this as a sign of conceit, but nothing could have been further from his character. He had a rare honesty of mind and an almost naive straightforwardness, well matched by his simplicity in externals. It was his expressed desire to enter into closer personal contact, at least with his advanced students, and he used to invite not only the research students, but also his own assistants and those of the Professor of Experimental Physics regularly to his home.[34]

In this way Lise became acquainted with a physicist her own age, Max von Laue, Planck's assistant since 1905.[35] When it came time for her to extend her stay in Berlin, it was von Laue who gave her the courage to persevere. Associations with her peers, married and unmarried, and visits to Planck's home in Grunewald, especially during the summer, left her with warm feelings for those early years in Berlin. She shared her heart and mind with fellow students and, always, with Planck. Her memories of her mentor are touching:

> He enjoyed cheerful company and his house was the center of good companionship. In the summer we ran races in the garden, and Planck joined in with an almost childlike eagerness and pleasure.
>
> He had a very good turn of speed, and was very happy when he caught one of us, as he often did. Planck once told us that Josef Joachim, with whom he often used to play chamber music, was such a wonderful man that when he went into a room, the air in the room became better. Exactly the same could be said of Planck. This was very strongly felt by the younger generation of physicists in Berlin, among whom I may include myself, and it undoubtedly made a very great impression on us.
>
> Planck also had friends who weren't scientists. He was very close with the art historian Burckhardt and his family in Basel. He also had a few friends among members of the Prussian Academy of Science. He was good friends with Mr. Erhard Schmidt.
>
> [Planck] was a very pleasant man; he liked to have people around his home, even when I was attending his courses. And as soon as we were in his home he played tag at least as eagerly as we young students: he tried to catch us while we ran, really he did. He could run very well – he had long legs – he was really a very agreeable man. He wasn't arrogant [with a government title] – a Prussian, the way one imagined him – not at all.[36]

The transition from Vienna to Berlin also meant a transition from Lise Meitner's immediate family to a new "extended" family, a circle of men and women of all ages who were to mean a great deal to her for the rest of her life. It was not by way of her previously published results, but only through slow and steady contact with colleagues in both the university and family settings that Lise came to feel accepted in the Berlin university circle. Her entry into Planck's inner circle of colleagues and students led to the transition from "visiting" Ph.D. to an accepted member of the physics department, and soon she was writing to her parents, shyly requesting that her father send her a further stipend so that she could remain in Berlin another year.

Lise's commitments to Planck's lectures and her joint radioactivity research at the Chemistry Institute led to a deeper understanding of the physics of alpha, beta, and gamma radiation. During her first semester in Berlin she published her own article "On the Dispersion of Alpha Rays,"[37] and later in the fall of 1907, a joint article with Hahn "On the Absorption of the Beta Rays of Several Radioelements" in the German journal *Physikalische Zeitschrift*.[38] Since Hahn owned an almost complete collection of known radioelements, he and Meitner then decided to systematically examine the beta radiation of *every* known element. Otto Robert Frisch (Lise's nephew, born in 1904) later wrote: "To Hahn, the discovery of new radioelements and the examination of their chemical properties was the most exciting part of the work. But Lise Meitner was more interested in disentangling the radiations."[39] The well-defined "range" of alpha rays made it possible to measure their energy quite accurately. With beta rays, however, no such simple method existed.

But they persisted. Nearly a year after their initial collaboration, Hahn and Meitner proudly announced in October of 1908 that they had actually discovered a *new* short-lived radioelement.[40] The product that they had found and verified was a short-lived product of the element actinium, which they named "actinium C." Frisch later commented:

> In a new field it is often good strategy to make simple assumptions even when there is no clear support for them. Hahn and Meitner adopted the assumption [proposed to Heinrich Willy Schmidt, 1906] that the beta rays of defined energy followed an exponential absorption law, and that each pure radioactive substance emitted electrons with defined energy. Thus a deviation from exponential absorption sometimes misled them into thinking that a preparation was a mixture when it was not; but on the whole those assumptions, wrong though they were, served them surprisingly well.[41]

It soon became clear to the research team that they needed a better method for analyzing beta rays. Magnetic deflection of the path of beta rays was a possibility, and with the assistance of Otto von Baeyer of the Physics Department, Hahn and Meitner were granted access to a suitable large magnet that was then taken to the shop and used in their readings. At the same time, Meitner developed a new method for analyzing radioactive processes that proved quite successful in the physics community at large. Hahn noted:

> In our carpenter's shop, work was making headway. The very low activity of actinium was [found to be] a result of the 'radioactive recoil.' While clearing up this problem, Lise Meitner and I found a new product of thorium, thorium D. This work, which came to be known as the 'recoil method,' turned out to be quite important, and Szilard and Chalmers (almost twenty years later) subsequently extended it by investigating the recoil in alpha-ray processes.[42]

What Hahn does not mention is that the recoil method was first introduced, and applied with great skill and precision, by Lise Meitner.[43] Methodological development was her forté, and the recoil process was *her* first major technique to be adopted by the scientific community. Meitner found that alpha particles, ejected from the radioactive atom with a great deal of kinetic energy, induced a recoil of any atom with a positive charge. Hahn later noted:

The actinium-X atom, at the moment it was produced by the radioactinium, carried a positive charge and experienced a recoil from the alpha particle of the radioactinium. This recoil frees it from its bonds and it is transported to the negative electrode. Rutherford [had] predicted this effect; now it had been confirmed by actual experiment.

At the time, I was so deeply involved with actinium that I hardly thought about any other element. But Lise Meitner, after reading my manuscript, said immediately: 'What you have observed there with actinium and with fairly thick layers of the preparation should be far easier to observe on alpha-emitting active deposits in infinitesimally thin layers.'[44]

Her observation was based on previous experience with beta-emitters, but more importantly, the measurements of 'recoil' and energy released were too minute for detection chemically: hence, her suggestion of the *physical* detection of alpha-emitters.

Their joint research, developed and articulated from the perspectives of radio-physics and radiochemistry, was read widely when published and duly noted by the scientific community.[45] They became known as a "team" although their individual research continued as well throughout the following decades. It was a most productive period: their night work was yielding monthly results, and when the New Year dawned, they were finishing up their fourth joint paper. Hahn recalled:

On January 16, 1909, [our] paper had been sent out for publication. Less than a week later, on January 22, 1909, Lise Meitner and I read a preliminary report to the Berlin Physical Society: 'A New Method for the Production of Radioactive Decay Products: Thorium-D, a Short-Lived Product of Thorium.' We reported on the making of radium-B by way of the recoil of the alpha rays of radium-A; we verified the existence of actinium-C (now called actinium-C''), which we had discovered in another manner by the recoil method; and we described a new beta-emitting product, thorium-D (now called thorium-C'), which we had found instead of the short-lived alpha emitter that we had expected to find.[46]

Despite such discoveries, for *Privatdozent* Hahn, there seemed little prospect of academic advancement based upon such research since the interdisciplinary field of radiochemistry fell outside the "established" departments of physics and chemistry at the University of Berlin. Not until late in 1908, when Ernest Rutherford was awarded the Nobel Prize in Chemistry for his research on the decay of radioactive elements, did those within the confines of standard scientific disciplines begin to take note of the tremendous growth of this new field.

After the Nobel festivities in Stockholm, Rutherford and his wife spent a few happy days in Berlin before returning to England. They received a cordial welcome in Germany from Otto Hahn and Lise Meitner as well as from a host of chemists and physicists. Hahn later recalled his former mentor, only eight years his senior:

> Rutherford told me that he was very pleased about the high distinction bestowed on him, but he smirked a little about the fact that he had been given the Nobel Prize for *chemistry*. Although his work was in fact of great importance to chemistry, Rutherford always regarded himself as a physicist; he had never done any chemical experiments, had never learned any 'proper chemistry.'[47]

Rutherford accepted the "pigeon-holing" of his interdisciplinary science by the Nobel committees. (Ironically, the Meitner–Hahn team would face the same challenges 30 years later when they were time and again nominated for the Nobel Prize in both physics and chemistry). Rutherford had looked forward for several years to meeting Hahn's collaborator; he had wondered who was the bright physicist with talent using recoil methods. When he was at last introduced to Lise Meitner, he said with great astonishment, "But I thought you were a man!"[48] However, they soon established a cordial relationship that was to continue through the many years Rutherford remained in contact with Hahn.

However, gender stereotypes persisted. While Hahn remembered Rutherford's 1908 visit fondly through the lens of his own professional development, proudly introducing the Nobel laureate to his colleagues, his memoirs illustrate how Lise Meitner was treated:

> While Mrs. Rutherford did her Christmas shopping, sometimes accompanied by Lise Meitner, Rutherford and I had long talks. He inquired about my prospects and was rather taken aback when I told him that as a *Privatdozent*, I was not [yet] on the permanent staff of the university. He thought that the fact that I had discovered a number of radioactive elements should have been recognized by some official appointment, preferably a chair.[49]

Roles were clear. While Meitner shopped with Mrs. Rutherford, the men discussed science. Lise was still considered a "visitor" to Berlin, but Otto Hahn was creating his career ladder. Through contacts with other academic chemists, Hahn became a recognized member of the faculty in the university's chemistry department a year later.[50] Employment was much more tenuous for Lise Meitner, however. Although women would soon be granted legal status to be hired in the university system, it was not until 1912, nearly five years after her arrival in Berlin, that she began to receive even a modest paycheck to support her internationally-known research. And it would be fifteen years before she was hired in the physics department of the university as a (*privatdozent*).[51]

Thus, when we compare the scientific careers of the research partners Hahn and Meitner, or many other male–female research teams, there is a recurrent theme. Scientific as well as academic recognition and salaried positions were awarded to Otto Hahn as he developed a reputation for his research in radiochemistry during the five years after his doctorate, whereas Meitner was denied paid academic positions at the University of Berlin. She struggled to maintain her "professional" status through her published results, and because she was denied professorial or teaching responsibilities, she focused on her research. A meager stipend from her father was her sole source of financial support in Berlin until she was 34 years old. Hence, she existed within a contradictory career status in the Berlin scientific community: excluded because of her gender from certain arenas, she excelled in realms that did not require her direct daily contact with students. And as her published results gained attention in both the Berlin "quantum circle" and international physics laboratories, her research compelled the Berlin scientific community to accept her as a representative of a newly emerging field.

Academia in Germany was in the throes of a generational watershed, a "paradigm shift" in models, mentors, and *Zeitgeist*. As an exception to the rule, Lise Meitner's life and work became a bellweather of changes yet to come. From the outset Meitner found the younger members of the physics community, both inside and outside of the university setting, to be supportive of her work. Many well-known physicists also became her acquaintances through the informal Wednesday colloquia at the physics department, which was initially led by Heinrich Rubens. In the beginning the colloquium consisted of a small group of at most 30 people. Meitner recalled some of the early participants: "the professors, of course," such as Planck, Nernst, and later, Einstein; Rubens' assistants Otto von Baeyer, James Franck, Gustav Hertz, Robert Pohl, Peter Pringsheim, and Erich Regener; and many other young physicists, all of whom were to have an impact on their field. Later, Otto Stern, Hans Geiger, and Hans Kampfermann became regular attendees, and Meitner would listen intently to their latest work, including technical details of the new "Geiger counter." She reminisced:

> By 1907 these colloquia were already an exceptional center of intellectual activity. All the new results that were then pouring out were presented and discussed. I remember lectures on astronomy, physics, chemistry – for example, a lecture on the stars of various ages given by Schwarzchild, a theoretical astronomer; another by James Franck on what were then called metastable states of atoms, or one on the connection between ionization energy and quantum theory. It was quite extraordinary what one could acquire there in the way of knowledge and learning.[52]

Through her participation at the informal weekly meetings of this outstanding group, Lise Meitner came to be accepted as a full-fledged member of the Berlin physics community. She was invited to share Sunday piano and violin concerts with Einstein and Planck, and daily coffee and cigarettes with her colleagues in the physics department. Hahn, as the extroverted member of their team, became the "public spokesperson" who delivered their joint research results and papers to colloquia. He related:

> Concepts such as radium C, thorium X, and radioactinium had now become, as it were, second nature to us, but for other people they still presented difficulties. I remember Professor Rubens once asked me: 'How do you manage to distinguish between all these names and remember all their chemical properties in the bargain? It's all so frightfully complicated!' And about fifty years later my friend Max von Laue confessed to me, after I had given a lecture covering the whole field: 'Now for the first time, my dear Hahn, I've really got some idea of what you do. Your work was always rather a mystery to me.' It just shows that one must never expect, much less take for granted, very much understanding from other people who are not familiar with one's field.[53]

Times were indeed changing in Berlin. In 1909, after university education for women was at last officially permitted and regulated, Emil Fischer at once gave Lise Meitner permission to enter the Chemistry Institute laboratories during working hours. This was a major victory. Afterwards, she remembered, he became much kinder towards her, supporting her in every respect, and later recommended her for various professional positions. Modestly, she later noted: "Although it naturally took some time for matters to proceed this far, this is not to say that I was in any sense isolated."[54] In many ways, Meitner's budding career mirrored the surrounding culture. She recalled:

> Radioactivity and atomic physics were making unbelievably fast progress. Nearly every month there was a wonderful, surprising new discovery in one of the laboratories working in these areas. When our own work was going well, we sang in two-part harmony, mostly Brahms; I could only hum along, although Hahn had a very good singing voice. Our relationship both personally and scientifically was very good with our younger colleagues at the nearby Physics Institute. They often came to visit us, and sometimes they just climbed in through the *window* of the shop, rather than coming in the usual way. In short, we were young, happy, and carefree — politically, maybe a little too carefree.[55]

These reflections in later years mirror Meitner's dawning social consciousness — a consciousness that was, in 1909, exposed daily to the Wilhelmian spirit of Prussian Berlin, which would, in five years, be challenged by a major world war. But during this period, with access at last granted to the Chemistry Institute, and research findings developing monthly, it was the structure of physical matter, not the political structure of German society, which kept Lise Meitner focused and challenged her scientific creativity.

CHAPTER III

New Explorations at the
Kaiser Wilhelm Institute: 1909–1914

The distinctive thing about real facts, however, is their individuality. One could say that the real picture consists of nothing but exceptions to the rule, and that, in consequence, absolute reality has predominantly the character of irregularity.

Carl Jung

''THE STRUCTURE OF MATTER preoccupied man long before the beginning of systematic natural science," wrote Lise Meitner in an article "The Nature of the Atom," published in 1946. "Such interest," she continued,

grew out of philosophical reflection, out of the need to introduce into a world of changing phenomena a principle of order. As early as the twelfth century BC, the Hindus developed a kind of atomic theory, according to which matter was composed of tiny particles separated from one another by empty space. These particles were supposed to attract one another strongly, thus explaining the resistance offered by any solid body to being divided. A considerable theoretical advance was made by Democritus and Leucippus, according to whom the particles themselves were indivisible (hence the name atom) and in rapid motion. The Greek philosophers even advanced the view that various substances differed from one another only in the number and arrangement of the atoms composing them. Although this Greek view seems quite close to present-day ideas, it was founded solely on philosophical considerations and led to no advance in factual knowledge.[1]

40

Meitner knew very well that theory often preceded experiment. Early Greek philosophers found that when amber was rubbed with cloth, it acquired the ability to "attract" bits of straw and other light material. They called this attraction *elektron*, from which the term electricity was derived one thousand years later. Up until the nineteenth century, it had been proposed repeatedly but never proven empirically that electrical attraction might be "atomistic," consisting of certain discrete "particles" of electric charges drawn together.

In the mid 1830s, Michael Faraday's studies on conducting electricity in liquids led him to the idea that electricity might be *transported* in the form of an electrical charge on the atoms themselves.[2] Thus if all atoms (regardless of type) had the same electrical charge, nineteenth-century atomists reasoned, the total "material" flow corresponding to a given "electrical" flow would depend only on the atomic weight of an element. Faraday's "charged atoms" were called ions, from the Greek meaning "wanderer." A new empirical field for studying the atom emerged, opening fresh vistas for chemists and physicists.

In 1897 the English physicist J.J. Thomson announced the discovery of a "new form of matter," which G.J. Stanley named the "electron."[3] Thomson's discovery was immediately hailed as the missing "atom of electricity" to which Faraday's work had pointed. Yet the position of the electron within the atomic structure puzzled Thomson: since it had been produced from the dissociation of very light gas atoms, he concluded that his electron must be an integral *component* of every atom regardless of the element's atomic weight.[4]

Since atoms in their normal condition are electrically neutral, it followed that they must have, in addition to their complement of negative electrons, an equal amount of *positive* charge to bring their "atomic charge" to zero. But because no "bits of positive charge" had ever been separated from atoms, it seemed to Thomson that the positive "part" must be all one piece. And since the mass of the negative electrons was found to be negligibly small compared to that of the atom itself, it followed that the positive part of an atom must represent nearly all of its mass, presumably most of the "bulk" of the atom itself.

In 1904, after years of gathering evidence, Thomson created a new model of the atom, noteworthy in the history of science as the first detailing a *quantitative* picture of atomic structure. He viewed the atom as a small sphere of positively charged matter with much smaller electrons embedded in it "like plums in a pudding" — the plums, or electrons, arranged in symmetrical order.[5]

Thomson's model of the atom generated great excitement. Rutherford, formerly a student of Thomson's, had learned a great deal about the properties of the newly-discovered electrons during his rise as a promising physicist. Building his own team in Canada, Rutherford challenged his young colleagues to find discrepancies in Thomson's model. He harnessed his own discoveries – alpha and beta rays – and set about using them as potential "probes" to test what might be *inside* the atoms.[6] The experiment is classic: Rutherford had a 19-year-old undergraduate Ernest Marsden check "alpha scattering" by producing a narrow beam of alpha particles – placing a bit of radium, an alpha-emitting element, inside a vacuum chamber tube. The alpha particles were then "aimed" at thin gold foil and those that emerged on the other side were counted on a "scintillation screen." With the assistance of Hans Geiger, another young German radiophysicist, Marsden began patiently recording the data.[7] Three years elapsed before empirical results from the McGill basement "counting room" were to justify an entirely new concept of atomic structure.

Hahn knew of Rutherford's experiments, and intently focused in his research with Meitner on examining the beta radiation of their nearly completed collection of radioelements. Their work presented its own complexities, so they gave little thought to the trickle of information coming from Rutherford's alpha research. Hahn, however, often wrote to Rutherford and asked for specific radioactive elements to be sent to him. A lively exchange of samples was carried out through the mail, without thought of "radioactive contamination." The lab assistants simply mailed the radioactive samples to one another in cardboard boxes! During her experiments Meitner used to pick up the substances with her bare hands because the radioelements she wanted to test decomposed rapidly; using gloves would have cut down her speed (the burned skin on her hands healed, she recounted in later years to a colleague.)

One joke that circulated in the Berlin workshop was about Meitner's "psychic abilities," which she used on the mailman one day when he entered the room. "Oh, you've brought me a package from Rutherford!" she announced to the startled postman, who had just come through the door. Even before the bag was opened, Meitner's sensitive equipment had picked up the radioactivity from the parcel, and the astonished mailman then delivered the package to her, wondering how she ever knew what was in his bag.[8]

In 1908 the Hahn–Meitner team proudly published three papers on radioactivity processes in *Physikalische Zeitschrift*, and by the end of 1909, five more. Meitner also wrote up her own results, "Radioactive Products of Radium," which were published in the respected journal of the German Physical Society *Verhandlungen der Deutschen Physikalischen Gesellschaft.*[9] These papers attracted the attention of the international scientific community and secured Meitner's and Hahn's places in Berlin's physics as well as chemistry circles. Soon they were asked to lecture on their research at various seminars and congresses. The results of Rutherford and Soddy, published in 1909, led to a radically new picture of the *composition* of the atom and stimulated debates by chemists and physicists. And, of course, the Nobel Prize in Chemistry to Rutherford further created a flurry of interest in radioactivity in laboratories across the globe.

In September of 1909, the German Scientific Society *Naturforscherversamm-lung* held its yearly conference in the beautiful alpine setting of Salzburg, Austria. Salzburg was not unknown to Lise Meitner. She was an avid Alpinist, accustomed to the high mountains, and during summers she used to hike in these regions with women friends, rucksacks on their backs. So she was overjoyed to plan a trip by train with Planck, von Laue, and Hahn. Meitner was one of the only women scientists to attend the conference which brought together for the first time the proponents of the "quantum revolution" (coming mainly from Berlin and northern Germany) and the "classicists," such as Wilhelm Ostwald.

Earlier that year, an honorary doctorate had been awarded at the 350th anniversary of the University of Geneva to a man literally unknown in academic physics only five years before — Albert Einstein. Einstein's reputation grew steadily after the mathematician Minkowski had praised Einstein's theory of relativity, which was presented in one of his papers within the "triad" published during 1905.[10] In 1908, Berlin physicist Rudolf Ladenburg visited Einstein to discuss his work. The encounter resulted in a formal invitation to Einstein, who was then 30 years old, to come and lecture about his work at the Salzburg conference. Only two months before his first "invited paper," Einstein had resigned from the Swiss Federal Department of Justice and Police, the department supervising the Berne Patent Office where he had been working as a clerk since 1902. His life would soon turn towards Berlin.[11]

After leaving Berne, Einstein traveled to Geneva to receive his honorary doctorate for his 1905 work. Praises were showered on him, Madame Curie, and on Ernest Solvay (whose profits from chemical discoveries served to fund the prestigious Solvay Conferences in physics). These festivities may not have made much of an impression upon Einstein, however; for soon after, in Salzburg, not Switzerland, he remarked that before *this* gathering, he had never met a "real" physicist![12] This fateful conference would bring together the younger generation – Einstein, von

Laue, Meitner, Hahn, Max Born, and others – face to face with the top physicists of Europe for an enriching exchange. In the cool climate of Salzburg, Einstein delivered his paper before an informed, critical audience. Planck, Wien, Rubens, and Sommerfeld were all there. Above all, Einstein respected Planck. And Planck became increasingly impressed by the charismatic personality of Einstein. The two had been corresponding since 1900, but they actually met for the first time at Salzburg.[13] Einstein was also quite impressed by another member of the physics department, someone who, like himself, was a young *Doktor* without a professorship — Lise Meitner.

Before a hushed audience, Einstein read his classic paper on "The Development of Our Views on the Nature and Constitution of Radiation,"[14] and set forth the radical idea that light demonstrated certain properties that could be more readily understood from the standpoint of Newtonian atomistic theory than from the classical wave theory of light. His proposition that such emitted "particles" of light (known later as photons) were literally quantum "packets of energy" heralded, in his own words, "the next phase of the development of theoretical physics which will bring us a theory of light that can be interpreted as a kind of fusion of wave and corpuscular theories."

More was to come after this paper. As they prepared their own presentation, Meitner explained to Hahn the details of quantum theory and the impact it could have in their own research. The following day, Einstein delivered another paper, this time on a subject even more fascinating than the one on quantum theory. Einstein called it "space-time." More than 50 years later Lise Meitner wrote:

> At that time I did not yet realize the full implications of the theory of relativity and the way in which it would contribute to a revolutionary transformation of our concepts of time and space. In the course of this lecture, he did, however, take the theory of relativity and from it derive the equation: energy = mass times the square of the velocity of light, and showed that to every radiation must be attributed an inert mass. These two facts were so overwhelmingly new and surprising that, to this day, I remember the lecture very well.[15]

The impact of $E = mc^2$ on the scientific world increased with time and further research. In Meitner's case, the equation was at the heart of her crowning achievement years later — in her interpretation of the tremendous energy released by uranium fission.

In terms of acceptance by an "informed, critical audience," Lise Meitner's presence at Salzburg was as essential to her career as Einstein's debut was to his. This

Congress was "altogether a very impressive experience" for her. At age 83 she recalled:

> It was attended by theoretical and experimental physicists from all over the world – Planck, von Laue, Born, Hasenhoerl (who succeeded Boltzmann), Schweidler, Stark, and R.W. Wood, the well-known specialist in optics from America. It was really something quite out of the ordinary, a most stimulating meeting. I reported on two minor pieces of work that Otto Hahn and I had carried out, in the course of which we had discovered and properly classified two new groups of beta emitters in the radium series.[16]

She was proud to present their joint research to an international audience. "I had, after all," she later exclaimed, "finally landed in work on radioactivity and the nuclear physics to which it was giving birth."

Several months after their return from the Salzburg conference, Emil Fischer consented to giving Meitner and Hahn an extra room to carry out their sensitive radiation measurements in his Berlin Chemistry Institute. Additionally, Alfred Stock, head of the inorganic chemistry department, also placed a private laboratory at Hahn's disposal, which Hahn gladly retained when Stock was succeeded by Franz Fischer. Hence, by 1910, only months after Lise Meitner had been granted "official" access to the laboratories of the Chemistry Institute, the "radioactive team" had three rooms that they put to use on a daily basis. By the end of 1910, their published joint research results included nine coauthored articles in journals ranging from *Physikalische Zeitschrift* to the prestigious *Verhandlungen der Deutschen Physikalischen Gesellschaft*, the journal of the German Physical Society. Meitner also published an article summarizing their cooperative research on beta rays and radioactive substances for *Naturwissenschaftliche Rundschau* in 1910 that gave a resume of their major discoveries together.[17] And at this point in her career, Meitner decided to make Berlin her permanent home.

Academically, however, Meitner existed in two very different "homes" — the laboratory and the lecture hall, the Chemistry Institute on one side of the busy city and the Physics Department within the quiet halls of the university. Her joint publications with Hahn and later with physicist Otto von Baeyer were discussed in many experimental circles by chemists and physicists alike. Both she and Hahn experienced the miscategorization of the young field of radioactivity: physicists labeled Hahn a chemist, while the chemists saw Meitner's interpretations of their research as "pure physics." Nevertheless, their joint work flourished and other researchers, with backgrounds in magnetic deflection, spectroscopy, and even Hans Geiger with his latest apparatus provided encouragement and insights.

In 1912, Max Planck formally asked Meitner to become his assistant at the university, succeeding von Laue. It was a great honor: also, it was Lise Meitner's first paid position since she had moved to Berlin five years earlier. She recalled:

> Not only did this give me a chance to work under such a wonderful man and eminent scientist as Planck, it was also the entrance to my scientific career. It was the passport to scientific activity in the eyes of most scientists and a great help in overcoming many current prejudices against academic women.[18]

Through her new role and responsibilities in the physics department, Meitner soon became a legitimate member of Planck's inner academic circle. The younger generation – James Franck, Gustav Hertz, Otto von Baeyer, Peter Pringsheim – had become her friends. She warmly recalled: "Not only were they brilliant scientists, they were also exceptionally nice people to know. Each was ready to help the other; each welcomed the other's success."[19]

At this time Planck was embroiled in a debate with the Viennese positivist/physicist Ernst Mach: again, the topic was the reality of atoms.[20] Meitner noted that Planck "rarely discussed" his own work, and that this "may have been precisely the reason that he had so few [advanced] students. But all of them were excellent." She commented that "he never felt the need to discuss his problems with other people," but that "he cared a lot about [von] Laue."[21] Von Laue was a friend of many Berlin physicists. In 1909 he was offered a Lecturer's position at the University of Munich's Institute for Theoretical Physics under the direction of Arnold Sommerfeld, and he moved to southern Germany. When Meitner started working as Planck's assistant in 1912, von Laue was completing his work in X-ray diffraction in Munich. Planck, Einstein, and others nominated him to the Swedish Academy of Sciences in 1913, and in 1914 he was awarded a Nobel Prize in Physics for this work.[22] Einstein considered von Laue's experimental work in crystallography as "one of the most beautiful in physics." Meitner recalled:

> I remember very well that [von Laue] sent a postcard with his diagrams to Pohl, who took the card to the colloquium. At the time the colloquium was still very small, just a few people, but everybody was very excited. I think everyone understood immediately that this would be a big thing.[23]

In addition to their own research and discussions about emerging findings, professors' assistants had pedagogical responsibilities. Meitner worked hard to keep up with Planck's student workload. In an interview at the age of 83, she described those years working as Planck's assistant:

I had to correct class assignments and tell Planck what errors had been made. I also had to select one student who had done well on the assignment and have him read his answer aloud the next day. Then Planck, based on what he noted I'd said about errors, discussed them. You see, it was a very minor job. But sometimes there were over 200 assignments, really quite a lot. They had to be turned in by Friday at the latest, and it was on Tuesday that I went to Planck to tell him what errors had been made, so it was not really all that much work. And of course I learned a lot in the process.[24]

With her usual modesty Meitner downplayed her own abilities. Yet it is clear that in her paid assistantship she bore a heavy academic load while carrying on her own (unpaid) research in the evenings. Years later Planck wrote to von Laue: "My maxim is always this: consider every step carefully in advance, but then, if you believe you can take responsibility for it, let nothing stop you."[25] Certainly, this philosophy made an impression on the Ph.D.s he had carefully chosen as his assistants. And in Meitner's case, her acceptance into Planck's family circle through her deepening friendship with his twin daughters and quiet, loving wife made her devotion to Planck's students one of loyalty as well as duty.

The Wednesday colloquium at the University of Berlin was a hub in which all physicists and their colleagues were invited. Not only was it a forum for discussion of physics, but a center for younger scientists to gather and discuss *their* latest research. Hahn and Meitner presented their new findings there, and Geiger discussed new applications of his "radiation counter." In 1911 Meitner published a paper with James Franck entitled "A Discussion of Radioactive Ions,"[26] which laid the groundwork for studying the radioactive properties of thorium. Between 1912 and 1913 she published seven papers in conjunction with Hahn on topics ranging from the magnetic spectrum of beta rays emitted from thorium to questions concerning the complex nature of radioactinium and its position in the periodic system. (See the list of Meitner publications in the Appendix.)

Meitner also participated in the many organizational developments then taking place in the Berlin scientific community. Although not directly involved in its meetings, her position as Planck's assistant exposed her to many facets of his role as the Standing Secretary of the Prussian Academy of Science. The Academy's four secretaries alternated the Executive Officer position every three months. Planck took his administrative role very seriously, and brought back a considerable amount of material for Meitner to read. Planck also maintained his involvement in the *Deutsche Physikalische Gesellschaft*, healing a split between the Berlin members and the many German physicists who then wished to form a separate organization outside of the Prussian capital.[27]

In 1913, Planck had been asked to serve in the creation of a newly-formed society funded by the Kaiser's government education ministry to support independent research in the sciences. As early as 1907, Ministerial Director Friedrich Schmidt-Ott (who later became Prussia's Minister of Culture) had been petitioned by Kaiser Wilhelm's aides to develop plans for an Institute that would include studies and research in physics, chemistry, physical chemistry, biology, anthropology, and medicine.[28] Of course, the Kaiser himself had not initially envisioned such a center, but when approached by Friedrich Althoff, Prussian Minister of Education, the Kaiser was said to have recognized the strategic importance of scientific research for "maintaining and increasing the leadership of Germany in science and industry, the basis of German power."

The Kaiser was eager to have Althoff's plan pursued, and met several times with physicist Walther Nernst. Greatly impressed with Nernst's outspoken way of presenting his views on the "necessity of scientific research for the Reich," he pressed his Ministerial Director to pledge direct government funding and land for the project.[29]

The relationship between German scientific organizations and industry had long been a close one, and was often based on service to a mutual political cause — the unification of the German states. In the late 1880s, the Society of German Natural Scientists and Physicians, a model multidisciplinary national organization, reaffirmed its position that Germany's importance in the world was due primarily to scientific achievements. "The vision of Germany as still predominantly a land of thinkers," wrote historian of science Russell McCormmach, "was the nostalgic vision of the traditional culture-bearers, and one that was to be increasingly challenged in the period following unification."[30] Walther Nernst, as respected as Planck among the older generation of scientists, had lectured before the Berlin Academy of Science in 1905 and stated that, whereas Germans used to look to "poetry and philosophy" for inspiration, they now were looking to "grainfields and factories." If an institution such as the Academy of Sciences did not adjust to that reality, he warned, it would lose its influence on the nation.[31]

The idea that Germany was a land of "scientific, technical thinkers" became the vision of the twentieth-century "modernists." After von Laue's departure for Munich, Planck and his university colleagues came under the influence of this vision as they witnessed the increasing association of their fields with technical and

industrial achievements. Many, like Hahn, became involved in professional organizations, addressing members of industry as well as academia. Other scientists,
however, viewed the alliance of science and industrial growth with troubled ambivalence. "They did not regard their national function," states historian Russell
McCormmach, "as economic in the first place but as *intellectual*, often priding
themselves on their independence from productive forces."[32] But the belief that
Germany's economic success depended on the *scientific* foundation of its industrial
technology grew, however, and many older scientists began to witness, through the
Kaiser's support of state-funded research institutes, the financial pendulum swing
towards the bridging of *techne* and *scientia*.

In 1909, Professor of Theology Adolf von Harnack, whose passion for science
was well known, was petitioned by the Kaiser's government to submit a detailed
memorandum on "the development of scientific research centers in Berlin." With
the assistance of leading scientists, Harnack produced a classic document that was
considered to be the basis on which the Kaiser Wilhelm Society was founded. This
"memorandum" not only presented a convincing case for the benefits of such research centers to German society in general, but also outlined fundamental principles
of the proposed Kaiser Wilhelm Society's organizational and financial structure.[33]

The document stressed the need for the cooperation of industry, science, and
government as well as complete academic independence and freedom in establishing
such research institutes *outside* of the universities. Harnack also pointed out the
dangers that might arise if only one of these entities (either the state or industry)
were in charge of the financial support of such research and teaching institutes.
Further, the memorandum outlined a concept for cooperation between the German
government and private citizens (including industrialists) in financing the proposed
scientific institutes.[34]

Leading industrialists and bankers, among them many prominent Jewish Berliners, contributed generously to the financial support of the independent institutes.
And in 1912, several years after the initial proposals for the foundation of the Kaiser
Wilhelm Society were developed, three large institutes of chemistry, physical chemistry, and biochemistry were constructed. Soon they were to officially open their
doors to working scientists and their students, changing the landscape of research in
Germany forever. Large institutes for experimental therapy, biology, coal research,
and physics were to open in coming years.[35]

The completion of the Kaiser Wilhelm Institute [KWI] for Chemistry in 1912 –
the first of the proposed institutes to be officially opened – was a major event in

the life of Lise Meitner. Radiophysics and a department of radioactive compounds were to be the first departments or "sections" established within the spacious KWI for Chemistry. The rural site of Berlin–Dahlem, approximately 15 miles from the center of the city, made an ideal location for this institute, its towers topped in the characteristic "helmet facade," symbol of the Kaiser's cavalry regiment. Lise happily recollected:

> Dahlem, which in those days wasn't built up very much, had an open view to Grunewald and was a charming area. The field between the Kaiser Wilhelm Institute for Chemistry and Fritz Haber's Institute for Physical Chemistry and Electrochemistry, which was opened at the same time, was especially beautiful after Willstätter planted it with tulips, chrysanthemums, dahlias, and other colorful flowers which he used in his experiments with dyes.[36]

Ernest Beckmann was appointed the Institute for Chemistry's first Director. He offered Otto Hahn the chairmanship of a small independent department for radioactivity research even before the Institute opened its doors. Hahn was delighted to accept the new position, and eager to take charge of the sparkling laboratories uncontaminated by radiation. As he later recalled, the years of research with radioactive products in their primitive carpenter's workshop were far from hygienic:

> I cannot help wondering how it was that Lise Meitner and I, working on preparations that were usually very strong, suffered no ill effects worth mentioning from radiation. The reason was probably that in the process of [chemical] fractionization of mesothorium, the 'radiothorium,' with the penetrating rays of its decay-products, was always separated. The same thing happened with the radium that was present in about twenty-five percent concentration. Hence the enriched fractions contained very little penetrating radiation.[37]

By contrast, the new Institute was designed with hygienic safety in mind. Hahn and Meitner spent long evenings discussing its organizational structure and potential. Procedures to prevent radioactive contamination were set in place. As Planck's assistant, Meitner was highly recommended for a position in the new Institute when the search began for appropriate staff. Thus, in the autumn of 1912, she was offered a post before the Institute's grand opening — a "guest" (unpaid) position in Otto Hahn's "radioactivity section."

Meitner was the first physicist on the staff of the Institute for Chemistry. Since her father's death in 1910, the Meitner family had struggled to provide her with a small stipend. Director Beckmann asked her to organize a group of students and laboratory assistants in the field of radiophysics, which she happily did without pay. Yet it was Otto Hahn who both socially and professionally was able to take an important step up in rank and salary to head the radioactive research department. Lise Meitner, his same age and research partner, remained contractually a "guest physicist" in an Institute planned and administered by chemists to whom she was subordinate in rank and administrative status. It would be another year, perhaps through the instigation of Planck, until Emil Fischer would advance her to a paid "Scientific Associate" position at the Institute, an equal to Hahn, and make the radioactivity section theirs.[38]

The official opening of the Kaiser Wilhelm Institute for Chemistry took place on October 23, 1912, in the presence of the Kaiser himself. The adjacent Kaiser Wilhelm Institute for Physical Chemistry, directed by Fritz Haber, a prominent Jewish pioneer in his field, was officially opened the same day, with pomp and ceremony, top hats, and dignitaries in carriages. Hahn noted:

> Of course there had to be exhibits for the Kaiser on that occasion, and I was asked to demonstrate a few radioactive substances. I used a preparation of mesothorium equivalent in radiation intensity to 300 milligrams of radium neatly mounted on a velvet cushion in a little box without any lead shielding whatever! I also had an emanating sample of radiothorium that produced in the dark very nice luminous moving shapes on the screen. But there was one entirely unforeseen difficulty. On the day before the opening, an adjutant [military officer] of the Kaiser's came in for a final dress rehearsal. When I guided him into the dark room to show him my radioactive preparations, he declared 'This is out of the question! We can't take His Majesty into a completely dark room!' There was a prolonged discussion; I called Emil Fischer to help me with my arguments. Finally we compromised on a tiny red bulb. But when the Kaiser arrived on the following day, he had no compunctions whatever about entering a dark room, and the demonstrations went off as planned. Lise Meitner tried to remain in the background, but she could not prevent the officials from presenting her to the Kaiser, who acknowledged the introduction with some friendly remarks.[39]

It is easy to picture four-foot ten-inch Lise Meitner's shy response to the Kaiser's "friendly remarks" when she was formally introduced as a well-known woman scientist and "Institute guest physicist." She was not political. Far from her Austrian home, meeting the Emperor of Germany was an honor that made a lifelong impression.

At last the days of work in the carpenter's shop were over. While the future for funding specific equipment and the hierarchy of administrative responsibilities in the new Institute were uncertain, the move to the suburb of Berlin–Dahlem was a pivotal moment in the professional and personal lives of Meitner and Hahn. Besides giving the team excellent facilities to continue their research, this institutional legitimacy added to their developing prestige in scientific circles. At this point Hahn took a major step: he proposed to his sweetheart, Edith Junghaus.

Lise Meitner now had dual responsibilities. At the new Institute she had even more opportunities to interact with chemists and radiophysicists. As one of the only "visible" women actively involved in physics research in the University of Berlin during this time, Meitner's isolated social position highlights the parallels between her status as a woman in science and her choice of a field such as the study of radioactive properties, then also considered a field of research "uncategorized" by traditional scientific disciplines. Hence, we can view Meitner's pioneering roles in the scientific community through the perspectives of both her professional status and that of her chosen specialty. Neither Meitner nor her field of study were fully "accepted" in traditional physics circles until the years after World War I when the need for radioactive materials in medicine and research increased dramatically.

From 1913 to 1914 Meitner and Hahn were introduced to new models of the atom that challenged the periodic table and were to have major implications. Their work – on the thorium decay series, uranium "Y" and "X 2," gamma rays, the complex nature of radioactinium, and the magnetic spectra of beta rays of radioactive products of uranium – produced new questions. During this period, a 26-year-old Danish physicist from Copenhagen, Niels Bohr, was spending a postdoctoral period at Cambridge, England, working in the Cavendish Laboratories of the aging J.J. Thomson. Bohr, described by Daniel Kelves as "having the habit of probing tenaciously for the core of theoretical conundrums, and of advancing ingenious ideas with quiet modesty and implacable insistence,"[40] soon went on to the University of Manchester to work under Rutherford, continuing his struggle with the then-current model of the atom. From data gathered by lab assistants Marsden and Geiger, Rutherford had, in the meantime, arrived at the idea that the positive charged "part" of an atom was not spread out through the whole (like Thomson's model of a "bowl of plum pudding"), but might be concentrated in a tiny core that would contain most of the atom's mass in a small central area, which he called the nucleus.[41]

Contrary to all expectations, Marsden and Geiger's findings demonstrated that out of the thousands of alpha particles that had been "tracked" when bombarded against a thin sheet of gold foil, a small number of them were deflected back from the sheet at more than a 90-degree angle — particles flying straight back from the target against which they had been bombarded! There must have been something *inside* the gold atoms off of which the alpha particles had bounced, Rutherford speculated, which was, in his words, "almost as incredible as if you fired a fifteen-inch shell at a piece of tissue paper and it came back and hit you."[42] Thus he was forced to postulate that the atom was not, as Thomson conceived, a gaseous cloud or "plum pudding," but rather that it must have a charged inner *core*. He boasted to Bohr and other younger researchers, "Now I know what the atom looks like!" And supporting Rutherford's model was empirical evidence emerging from laboratories around the world — including the Kaiser Wilhelm Institute for Chemistry in Berlin.

On the basis of his alpha scattering experiments, Rutherford was able to estimate the size of the nucleus, which he concluded would be ten thousand times smaller than the atom itself — in Rutherford's terms, "like a fly in a cathedral." Electrons with negative charge circled the nucleus like planets around the sun, he postulated, offsetting the positive electricity of the nucleus.[43] But to Bohr there was something puzzlingly wrong with a model that depicted the atom as a "miniature solar system." In the spring of 1913, just as Hahn and Meitner were beginning their move into the new KWI, Bohr returned to Denmark with his own formative ideas about atomic constitution.

In Copenhagen he chanced upon a formula named after Swiss schoolteacher Johann Jakob Balmer. Balmer's formula interpreted the spectrum of hydrogen empirically. Bohr recognized that it could also be written in a slightly different way, using "h," Planck's quantum constant, and hence introduced discrete quantized states into the analysis of physical spectra.[44] Bohr thus obtained a formula that described the spectrum of hydrogen in terms of energy quanta. Applying this formula, Bohr could also place the position of electrons in specific orbits, determined through their "quantum states." Bohr sent his meticulous calculations and results back to Rutherford. When he received a reply to "tighten his language" on a third draft, he set sail once more for England to discuss his calculations and their implications face-to-face. In order to match his theory to the empirical data of the hydrogen spectrum, he "restricted" classical laws using Planck's constant. Einstein was one of the first to note Bohr's papers.

Meanwhile, members of the KWI and the Prussian Academy of Science were locked in negotiations regarding Albert Einstein's position. Many wished that he would become Director of the soon-to-be-completed Kaiser Wilhelm Institute of Physics. While still deciding upon offers of positions in Berlin, Einstein was informed by Hungarian physicist George von Hevesy of evidence supporting Bohr's new atomic theory. Hevesy is reported to have said that on that day "the big eyes of Einstein looked still bigger" reflecting on the newest application of quantum theory. "Then," Einstein was to have exclaimed, "it is one of the greatest discoveries!"[45]

Since Einstein had been struggling with his own theories for many years, and had been involved in heated debates against opponents of relativity, he immediately sensed the significance of Bohr's work in applying the quantum theory as well as its overall applications. Among Einstein's generation of young physicists who were pioneering new theories about quantized matter, light, or radiation, no one was more than 35 years old.

In 1905 Einstein had made the innovative proposal, also utilizing Planck's quantum constant, that radiant energy itself could be "quantized." He visualized a beam of light consisting of waves as a stream of speeding "particles" of energy, later to be called photons.[46] This was a daring theoretical step, and one of the greatest discoveries of the twentieth century. But during the year in which Bohr's papers appeared, Lise Meitner later recalled, many physicists in Berlin "still did not want to believe in [Einstein's] photon." There were many debates about Einstein's theories until 1913, adding further fuel to the fires that Bohr's ideas about "quantized" atomic structure had set ablaze. Meitner later recalled that she too was at the heart of these debates then taking place at the university. Nernst, Planck, and a select few senior physicists were the first to point out that while the "idea of the photon was still a fantasy," they had again and again demonstrated that "great risks are involved if one wants to do something extraordinary."[47]

Not all physicists were easily convinced of Einstein's ideas *or* of Bohr's new model of the atom. The *younger* generation seemed to embrace the concept of the "quantized light packets" enthusiastically. "They gave [Einstein] a chance for his ideas about the photon in that they believed in it at that time,"[48] but they felt that Bohr's model would have to be tested, recalled Meitner. Working daily with questions of atomic structure and radioactive decay, Meitner also gave Bohr's quantum explanation of the hydrogen spectrum "a chance." She felt Bohr's work to be "an incredible success" and noted that in certain Berlin circles "it was immediately embraced with enthusiasm." When questioned if there had been people who did not accept the Bohr atomic model, she said that it was the consequences of the theory that became the talk of the day in Berlin scientific circles:

> [We felt it] was fairly curious that there were select ways in which the [electron] can 'move' without radiating, and not fall into the nucleus. This was discussed and pointed out again and again. But many slowly believed [Bohr's theory] since experiments proved it.... You could see that there was a model which explained how one atom emits so many [spectral] lines, while up to this time, one asked: 'How can an indivisible atom send out so many lines?' This was very curious. Riecke always used to point out how curious this was in his experimental [research]. As soon as it was published, I knew

it [that Bohr's theory was correct] and of course, it was such a marvel that all these complicated spectral lines could be explained. . . . Naturally, [von] Laue as well as Stern saw exactly the problems with Bohr's model, the half classical and halfway quantum type of explanation. Even Bohr himself knew that this was only an 'attempt' to do something [with the conflicting experimental data]. But from 1911, the Rutherford atomic model had already been acknowledged with the greatest interest — at least by us 'radioactives.'[49]

Emerging new theories and their experimental confirmation continued to generate enormous changes and opportunities within the physical sciences. Lise Meitner's experiences can be viewed against these dramatic developments. Her acceptance as a Viennese Jewish woman scientist in the Berlin scientific community also illuminates shifting social mores. Once her presence had been acknowledged through institutional recognition and legitimacy, she was treated as a participating member of the rising young generation. Even Planck, who had once viewed the possibility of women entering the field of physics in the abstract, witnessed Meitner's work and progress in concrete, day-to-day situations. During the period in which she served as his assistant (1912–1915), their friendship was solidified. Years later, Planck wrote to W. Wien: "I value her very highly not only scientifically but also personally; she is a true friend."[50]

She visited her widowed mother regularly in Austria when time permitted, and her large family was extremely proud of her accomplishments. Two of her sisters also entered into their own Ph.D. research in Vienna during this time.[51] Lise also had made friends with several woman scientists in Berlin: Eva von Bahr-Bergius from Sweden and Elisabeth Schiemann, a lecturer in biology and researcher in the university's Institute for Plant Research. They shared long walks through the city's wooded areas, and often met for picnics or other recreation with Planck's twin daughters.

These first years at the new Institute for Chemistry were among the happiest of Lise Meitner's life. She had completed research with James Franck on gas-phase radioactive ions as well as electrochemical classification schemes during her last months in the carpenter's woodshop; also she had developed six new articles with Otto Hahn between 1913 and 1914. But underlying these personal and institutional victories was the shadow of a Berlin culture still based heavily on rank, bureaucratic procedure, and the slow, distrustful acceptance of changes in social mores and roles. Despite the academic community's acceptance of Lise Meitner, many younger scientists as well as established academics remained aloof from the *Fräulein Doktor*.

Women's "rights" were contingent upon many hidden social factors in Germany. The gifted woman physicist remained within a sometimes hostile milieu. But in those six years since her move from Vienna, she had landed in the midst of the paradigm shift of the quantum revolution. She wrote to a dear friend: "I love physics with all my heart. I can hardly imagine it not being a part of my life. It is a kind of personal love, as one has for a person to whom one is grateful for many things."[52]

World War I and Its Consequences: 1914–1920

It must be granted that many aspects of the intellectual life of that era showed energy and grandeur. We moderns explain its concomitant uncertainty and falseness as a symptom of the horror which seized men when, at the end of an era of apparent victory and success, they found themselves suddenly confronting a void: great material scarcity, a period of political and military crises, and an accelerating distrust of the intellect itself, of its own virtue and dignity and even of its own existence.

Hermann Hesse
The Glass Bead Game

From 1914 to 1919, the close of the Wilhelmian age and the beginning of the Weimar, science in Germany shifted from the academic arena into factories, battlefields, and the world theater of war. Countless issues were raised during this period concerning the role of scientists in wartime and their governments' application of scientific research to wartime purposes,[1] viewed in this chapter through the eyes of Lise Meitner and the experiences of her colleagues.

On June 28, 1914, the foreboding headlines of the Berlin newspapers reported the assassination of Archduke Franz Ferdinand in Sarajevo. It did not take long for Austria to retaliate by seizing Serbia. Germany, as the protectorate of Franz-Josef's Austro-Hungarian Empire, was immediately drawn into the conflict.[2] In universities across Europe, students and professors were making plans for summer holidays, unaware that the "war to end all wars" had begun.

Six weeks later, Germany's declaration of war in early August threw the Berlin scientific community into an uproar. Labs and lecture halls were quickly emptied of both their youngest members and mid-career scientists, all anxious to enlist and "demonstrate" their love for their German nation. Men urged others into the defense of Austria, confident that the national pride was at stake. The quiet calm of a normal summer became a patriotic circus. Buildings previously used as warehouses or classrooms were transformed into hospitals and mobilization centers; soldiers were outfitted overnight, and sent off with flowers, kisses, and hopes for their return by fall.

By late summer 1914 Otto Hahn had been drafted. It was a shock to both him and his family. Still living like a newlywed eighteen months after his marriage to Edith Junghaus, a beautiful, talented art student and daughter of the Mayor of Stettin, Hahn had to face separation from his wife and work. He was ordered to register and then report to an infantry regiment at Wittenberg.[3] Suddenly Lise Meitner and Edith Hahn were both without a partner. Little did anyone in Berlin foresee that the war would drag on for four devastating years. The government had no plans for a war of more than six months duration.

An early casualty of the war from the Berlin circle was one of Hahn and Meitner's young assistants, Ph.D. candidate Martin Rothenbach, who had been working on the half-life of beta decay associated with rubidium (later found very useful for dating the geological age of specific minerals). He was killed in France in November 1914. Lise Meitner took over a month to break the news to Hahn in a letter. She recalled that Rothenbach was "as industrious an assistant as he was agreeable a person"[4] and mourned his passing.

The war began to take an immediate toll on university students in Germany and Austria-Hungary, as well as throughout the rest of Europe. Soon women were also called on through songs, posters, and public speeches to contribute their full efforts to the "national family." Very few registered for fall semester studies that year.

Meitner became active in her own style: during the fall of 1914, she expanded her research to include investigation and training in the latest *medical* applications of X-rays. She worked at the city hospital in Lichterfelde in between her duties to Planck, studied human anatomy and "röntgenology": X-ray technology.[5] She also continued to write articles on the latest questions in physics which were published in the journal *Naturwissenschaftliche Rundschau*, edited by a Prof. Sklarek whom she had known in Vienna. Her contributions aroused the interest of the editor of the Brockhaus Encyclopedia who wanted this "prolific physics writer" to develop articles for the Encyclopedia. He contacted Sklarek and asked for the address of "Herr" Meitner, whereupon he was informed that the expert was a female. The editor quickly replied that he would not think of publishing an article by a woman![6] Such was the mood of the times: men continued to be valued for their contributions, women as supporters of the cause.

War fever had reached a high pitch in Austria. Lise visited her mother there, recalling the view from her apartment window of the train station, full of recruits and dramatic goodbyes.[7] In August her brothers enlisted, and the family discussed their situation and what they could do during the war. Lise's own career advancement beckoned. Before the first news of casualties had reached them, Lise had unexpectedly been offered a permanent academic position from her former lab instructor Anton Lampa at the University of Prague. Prague was still a major center within the Austro-Hungarian Empire. To have a university position offered to a woman scientist, albeit during wartime, demonstrated that the long-sought academic acceptance of her accomplishments had been achieved, and that her former Viennese professors still highly valued her work. She debated the offer with her relatives, uncertain as to what to do.

The offer also opened the door for further salary negotiations with the Institute for Chemistry in Berlin. Evidently, the war was bringing dramatic changes in the leadership and staff within all of the Kaiser Wilhelm Institutes. Einstein, a pacifist, did not get on the war "bandwagon" and did not enlist. With the sudden "emptying" of academic and research posts, leadership was sorely needed. Max Planck pressed Emil Fischer, then Director of the Kaiser Wilhelm Society, to counter the Prague offer and keep Lise Meitner in Berlin. By July, a counter offer was made for a salary and lodgings. In late August 1914, Lise Meitner sent her acceptance of the Berlin offer, as the drums of war began to roll.[8]

When Meitner returned to her research at the KWI for Chemistry, she was lonely for Hahn, their assistants, and the former bustle of the labs. The administration stated their intention to reevaluate her contract as "guest" physicist. Final documents show that she was granted a yearly salary of 1000 marks after the war began (not even close to Hahn's 3000 marks per year, which also included a marriage stipend). Other factors in Fischer's decision to offer Meitner a paid job was her strong influence on younger members of the Institute. When the section officially called "Laboratorium Hahn–Meitner" had opened a year before, most chemists had gone to *her* with questions about radioactive processes as well as physics. Meitner loved the Institute enough to move back to Berlin–Dahlem, close to Planck in the tree-lined suburbs near Berlin's lakes and forests. She *chose* to give her allegiance to the KWI for Chemistry despite the low pay since she felt this place was certainly her "home base." It would be two years, until 1916, before her salary would be raised another 3000 marks, reaching parity with Hahn's.[9]

The intrigues in the German capital during World War I grew deeper. Most Germans felt that by expanding into Belgium, marching towards France, their armies would be triumphant in their "noble mission." Hahn noted in his memoirs that, as a loyal soldier, " . . . hardly anyone had any doubt about our winning this just war." And throughout 1914, *Wissenschaftspolitik* – the politics of science – was as difficult to follow as the orders being issued from the German Army's Central Headquarters.[10] War fervor was replaced with Germany's determination to expand its boundaries, its culture, and its army.

Philosophically, all this was backed by certain intellectuals who were intent on rescuing their national honor. In the late fall of 1914, a document called "A Manifesto to the Civilized World" was circulated among Berlin academics and caused an outcry in the international scientific community. Later known as the "Manifesto of 93" for the 93 notables who signed it, including Ernst Haeckel (evolutionist), Paul Ehrlich (biologist), and Wilhelm Röntgen (discoverer of X-rays), the text declared that "were it not for German militarism, German culture would have been wiped off the face of the earth." The signers expressed a "pained expression of surprise" that the world objected to Germany's invasion of Belgium, noting that it would have been suicide to do anything else since Germany, "like no other country, had been ravaged by invasion for centuries." The text concluded:

> We cannot wrest from our enemies' hand the venomous weapon of the lie. We can only cry out to the whole world that they bear false witness against us. To you who know us, who have hitherto stood with us in safeguarding mankind's most precious heritage — to you we cry out: Have faith in us! Have faith in us when we say that we shall wage this fight to the very end as a civilized nation, a nation that holds the legacy of Goethe, Beethoven, and Kant no less sacred than hearth and home. In token whereof we pledge our names and our honor![11]

One of the signatures on the Manifesto appeared to be Max Planck's. Years later Lise Meitner vehemently defended her aging mentor:

> Truthfulness was so ingrained in him that even an innocent offense against such would quite depress him. That was why he took an official position with regard to the well-known appeal of 93 German scholars and artists in 1914 to which his name had been added *without* his instigation.[12]

Planck, since 1913 Rector of the entire University of Berlin, had misgivings about the Manifesto, and denied ever having signed the document. He wrote a letter in 1914 to Nobel Laureate A.H. Lorentz in which he asked his Dutch colleague to make public a response to the manifesto's false assertions. Planck affirmed to Lorentz that: "... several of the appeal's assertions restricted its logic. What I would like to stress with special vigor, face to face, is the firm conviction never to be deeply altered by the events of the current war, in that there are domains of the spiritual and moral world which lie beyond the battlefields of nations."[13] Planck may have voiced the sentiments of many academics of his generation who were making difficult moral choices regarding their actions and "service to the Fatherland," but that fateful year, for the majority of Germans, wartime frenzy was mounting.

Planck's denial of having signed the Manifesto does not alter the fact that most of the German public agreed with its contents. 1915 marked a New Year few in Berlin would forget: the streets were nearly empty, and there seemed little to celebrate even though "advances" into France were reported close at hand. Lonely for her former research tempo, and the students who had once filled the new Institute labs, Lise Meitner fretted away hours with her work, barely making progress. Hertz, Geiger, and other physicists had also been called up from the reserves; however, most still felt the war would be over in a few months. Meitner persevered and completed research on the end product of the uranium decay series, as well as on magnetic research that had been started with Otto von Baeyer before the war. He had been supportive and continued to be a close colleague for the next decade.

A year after accepting her post at the Institute, and after completing her final responsibilities as Planck's paid assistant, Lise could no longer resist the call of duty to her homeland. In July of 1915 she volunteered for service as a certified X-ray nurse/technician for the Austrian Army. Her enlistment changed her life. She soon found herself shoved aboard crowded trains, and then, after weeks of diagnostic training with primitive X-rays, was sent behind the lines of the Austrian front where the battle was raging. She had been given specialized medical training, and thereafter dedicated herself to helping the wounded, working all day and many long, exhausting nights. She later recalled the gruesome 20-hour shifts.[14] She was often grouped with the physicians, not nurses; her expertise was valued and put to use.

Meitner was placed in one of the first hospital X-ray units, and experienced firsthand the carnage of war. Broken bodies, shattered skulls — in observing every type of presurgery condition, she came to understand how X-rays saved lives. But after several months she became very depressed. The war had robbed her of her first love — science. She wrote to Hahn on October 14, 1915 from Lemberg:

> You can hardly imagine my way of life. That there is physics, that I used to work in physics, and that I will again, seems as much out of reach as if it had never happened and never will again in the future. At night, in bed, when I cannot immediately fall asleep, sometimes homesick thoughts come to me. During the day, I think only about my patients.
>
> I have completed over 200 X-ray films, but that still leaves me much time to help in the surgery room. Work is not at all straining this week, since we did not have many transports of wounded soldiers – and the convenient side of my job is that it is not purely mechanical, so that I am not tempted to forget the personal element when dealing with the wounded. One always gets the opportunity to fulfill little desires or spend extra effort, especially for someone who is seriously hurt. The gratitude these people feel and show always makes me feel a little ashamed. Often times these people suffer so terribly and cannot be helped at all.
>
> You can imagine that we follow events on the Western front with much interest. I am personally completely convinced that the French will not accomplish a real breakthrough. But when is the war going to be over? The Russians could already retreat, couldn't they? What else do they want to lose? We have many Russian captives here, some wounded, some not wounded, who do jobs as male nurses: all are huge fellows, big, strong, and very mellow. Few of them know why they had to go into the war and all are very happy to be with us.[15]

In the meantime Hahn and other scientists from Berlin were sent into the trenches. Intense fighting on both the Eastern and Western fronts occurred in the summer and fall of 1915. Meitner began to see a steady stream of Eastern European casualties in her wards. She felt lonely and isolated when she turned 37 that November, although she had been granted occasional leave to visit her family in Vienna. A small triumph was news that Willstätter, who had planted the lovely flowers in fields around the Dahlem Institutes while researching innovative plant dyes, had been awarded the Nobel Prize in Chemistry for his research on chlorophyll.[16] But, for the most part, "it was a harrowing time," she later wrote her nephew Otto Robert Frisch (then being raised in Vienna). She was "working up to 20 hours a day with inadequate equipment, coping with large numbers of soldiers who had every kind of injury without knowing their language."[17]

Her "homesickness" for Berlin continued. In her first six months of service she earned a furlough every ten months and could get to Vienna, but she was too far away to travel by train to northern Germany. Hahn was shuttling back and forth with the newly-formed chemical warfare unit and also served in the trenches near Belgium. Very ill, he urged Meitner by letter to return to the Institute for Chemistry, fearing that their years of carefully assembled research apparatus would be "disassembled" by the army. Her earlier 1915 research on thorium, radium, and pitchblende seemed like a distant memory, and she wrote to Hahn in detail about the administrative reports she had filed on their behalf from Vienna:

Dear Herr Hahn!

On Christmas we of course had a celebration, distributing presents for the wounded and workers, and since our department has about one thousand people there was much to prepare. It was a nice job 'on the side' and the wounded became as excited as children. There was a formal dinner for physicians and officers (I am counted in the first because I am independently heading the X-ray room). It was like it always is among people who are more or less – mostly more – strangers to each other.

My brothers and sisters and I let New Year's Eve pass [in Vienna] without any party. We did not even stay together until midnight. My brother-in-law had just lost his only sister, completely unforeseen. She died of influenza, leaving a three-year-old child and a tubercular husband, who had been staying at Davos at the time of her illness. . . .

We are not aware of the Russian offensive that started recently. We are now so far behind the front that we do not get the really serious wounds anymore. Happily, the frost exposures have decreased significantly too. We have a lot of success in healing with a new treatment method of our professor's, consisting of immobilizing the closest joint with casts. In about 150 frost exposure cases, we had to take off limbs only twice.

Beckmann wrote me a letter regarding the [Institute for Chemistry] report about three weeks ago. You are completely right. I did in fact write the previous year's [1914] Annual Institute Report too. So I prepared this one, as you told me, naming title and contents of written work with a remark up front saying that you have been engaged in war service since August 1914, that I have been gone since August 1915, and that therefore the department has been closed down since that time. I hope it is all right with you as it is written.

I am glad you were able to spend Christmas with your wife. But now in return, you have to tell me a little something about Berlin. You must have met all our acquaintances and picked up some academic and otherwise interesting information. I sometimes still have homesick thoughts and, furthermore, the feeling that I do not even know what physics means anymore. We have positive news from my brother and other brother-in-law out on the front. I am of course always very fine healthwise, although I do not know if I, in fact, even weigh 100 lbs right now; but after all, I am not here to conduct a fattening treatment. I do not get around to reading much – in the daytime, not at all, and evenings, it is an accomplishment to get through the Berlin newspaper that I ordered sent here. The whole day I am stuck in the hospital. If I am not taking X-ray photographs, I am helping in the surgery room. I even performed anesthetization, but I do not like it. I am also 'hospital mechanic,' fixing bad electrical wiring and equipment, producing T-pipes, catheter plugs and other useful things. However, there is still enough time to take care of the personal worries of our patients, who are really thankful for each little gesture of friendliness.[18]

Meitner's combined professional, maternal, and continuing Institute roles are well illustrated in this revealing letter. Always hard-working, included with the physicians, Berlin administrators, and other circles before the age of 40, Lise Meitner applied herself wholeheartedly to every job she undertook. At last she earned enough leave to travel north to Prussia. And when she finally arrived in the familiar Berlin train station in January, she found the German capital full of uniformed soldiers, many of them wounded or disabled. Food shortages were common throughout the country. While she maintained her former friendships with her (Lutheran) minister's family, women friends at the university, and Planck's daughters, Lise had to fend for herself. By November 1916, two weeks before her 38th birthday, she wrote in a somewhat depressed state to Hahn:

> It really is very friendly of you to even spend time thinking of my diet; you imagine things [the shortage of food] to be worse than they really are. I am muddling my way through. My minister's daughters do not want to feed me and since it is impossibly dirty and ugly in the train station restaurant, I often have lunch in the Steglitz Rathskeller. But the way down there is fairly far and takes up my entire lunch hour. Frau Meisenheimer [Lise's landlady] kindly offered me 'lunch board' the other day.
>
> You know, I am in love with my individual freedom and I respect that of others, and so being a daily guest in other peoples' houses does not seem appropriate to me. But maybe I will take the offer for a few days a week. Last night I was at Plancks'. They played two gorgeous trios, Schubert and Beethoven. Einstein played violin and on the other side volunteered some naive and strange opinions about politics and war. Just the fact that there is an educated person around who in this time does not touch any newspapers is certainly curious.[19]

Those who "kept in touch" knew that the Kaiser was *not* about to withdraw from occupied territories. By 1916 his industries were creating new weapons: chemical warfare research was underway, and factories were supplying soldiers with the latest in automatic machine guns and modern supplies. The Berlin War Office found increasing uses for the facilities of the Kaiser Wilhelm Institute for Chemistry for military research on chemical gases, unknown to the majority of citizens, or to the world.

Hahn and Meitner were deeply troubled by the military "appropriation" of their laboratories and offices. After the years of preparation and joyous opening of the Institutes in Dahlem, the chemistry hub had now become a busy place for military research, as were the other developing Institutes in the so-called "German Oxford." What was once hailed as the new intellectual center now looked like a military installation, with uniformed officers coming and going at all hours of the day and night. Lise Meitner remembered these years with ironic clarity:

> The Kaiser Wilhelm Institute for Chemistry at this time evolved more and more into a workplace for military purposes. Professor Willstätter went to Munich as successor to Adolf von Baeyer. His rooms were relinquished to the research department of the aerial photography pilot commandos. Accordingly, Professor Stock (successor of Willstätter) continued in the Chemistry Institute of the University for the duration of the war. With the exception of our small section, all the remaining rooms of the Institute for Chemistry were placed at the disposal of Professor Haber and his group for the use of the army.[20]

Meitner knew Fritz Haber from earlier encounters with him in Dahlem during the formation of the Kaiser Wilhelm Institutes. A friend of Einstein as well as of influential Berlin industrialists, he had been regarded as the best choice before the war to lead the adjacent Institute for Physical Chemistry. Haber was a scientist who combined a knowledge of his field with shrewd business contacts, and his synthesis of ammonia would provide an invaluable basis for producing synthetic nitrates and fertilizers during the war.[21] At age 47, he was not eligible for the usual military service, and because of his Jewish heritage, he was ineligible for an officer's position under Reich law. Yet his fame and pride in Germany catapulted him to committing his Institute to dangerous war research: chemical irritants and poison gases. In shock Lise Meitner learned that, under Haber's leadership, all *but* the section she and Hahn had created for radioactive research in the KWI for Chemistry would also be turned over to the military for further chemical warfare research. And ominous signs appeared to their scientific "family" when, before the first offensive, Haber's wife – a Ph.D., humanist, and adamant opponent of her husband's war research – committed suicide.

Haber, like many Jewish Germans, insisted that his loyalty lay with the state. War research continued and the German army advanced across Europe. Meitner had mixed feelings about Haber. She later commented on Haber's administration of the military group that was to work with existing staff at the spacious three-story Institute: a top-secret staff that developed and organized the offensive and defensive preparations for the first gas warfare attacks. She reflected:

> In his charming and lighthearted manner, by his own telling, [Haber felt that he] committed many amusing blunders in his encounters with customary military procedures. [But] because his joking was unmistakably based on his fair inner attitude, he found equal favor with his superiors as well as with his colleagues and subordinates.[22]

What Meitner did not note, of course, is that as power accumulated in the hands of the military, more deadly weapons were being invented by the very scientists she once held in great esteem. The more Haber assimilated into the upper echelons of the military, the more he and his work would face grave repercussions *after* the war. Lise Meitner also fretted privately that all of their work setting up their apparatus, counters, and carefully designed experiments would be in vain. She wrote to Hahn about Haber (nicknamed "The Lion"), sinking his "claws" into their modest physics laboratoy. She also yearned to repeat Chadwick's work, that is, to measure the spectrum of beta radiation in a magnetic field.[23]

By the summer of 1916 Lise Meitner had requested either a transfer to Berlin or an immediate release from active duty. After her 1914 promotion of the Institute of Chemistry, she was frustrated at not being able to be part of the "inner circles" of the Institute who were making important decisions about the fate of her own workplace and about the applications of science to war research. Advising government officials every day about scientific policies, such military personnel were forming a strong bloc inside the Reich's decision-making and funding process, and each of Dahlem's Institutes would witness this tension throughout the coming decades.

Meitner struggled to train other young X-ray technicians in the Austrian army during 1916, and worried a great deal that her prewar research efforts would sooner or later be destroyed by those now in charge of her former area of responsibility. Personal needs seemed far from most people's minds. However traumatic as these times were, Meitner received a flowery proposal of marriage from a Greek professor who must have been completely smitten by her. The proposal was found among her private papers at her death.[24]

Her true commitment, however, remained to science, as she recalled:

> Every two or three months I was able receive a leave to carry out the necessary research and measurements in Dahlem, and was in continuing contact with Hahn. In the middle of [1916], Hahn wrote me that our section would also be utilized for military ends if I didn't return there permanently. In our research into protactinium as a matrix of actinium, we made the necessary very exact reproducible measurements with precisely-calibrated apparatus. The seizure of our section would have ruined our work of several years. For that reason, I returned permanently to Dahlem in September 1917 to conclude our project. This didn't occur without opposition on the part of a few gentlemen, who wanted to occupy our section for military research. In spite of Hahn's warning to return, I felt uncertain for just how long, in those times, we could justify using our room only for scientific work. So it was a great help, for practical as well as personal considerations, that Max Planck, whom I asked for advice, declared explicitly that it *was* right to support the scientific work.[25]

Petitioning Planck's aid, knowing his influence and concern for the "right of research," Lise Meitner appealed to the principles and sentiments of the tradition-
bound older generation of Berlin. She knew that this segment of the population
emphatically affirmed that science, art, and German *Kultur* must survive, despite
the "necessities" and shifted priorities brought about by war. Scientific research had
to be maintained not to *hinder* State policies but to *serve* the Fatherland, Planck
would continue to affirm, and he supported Meitner's release from military service
for her to continue her research. Too many of Europe's finest physicists already lay
dead, casualties of a war that changed Europe forever.[26] Planck was overjoyed at
Meitner's return to Berlin.

We should note that it was Lise Meitner who struggled to maintain the "home
front" at the Institute upon her permanent return. A salary increase did little to
counter the rampant inflation, or check the growth of military appropriation of
materials and supplies. Throughout 1918 she coexisted uneasily with the army's
administration in a workplace emptied of its students and professors, and managed
to complete her radioactive readings during long nights alone. While Hahn worked
on poison gas development and its applications in the field, sending long letters
back to his wife, Meitner *resented* the militarization of science. She would later
be shocked that there would be no "unification" of the tattered Austro-Hungarian
empire with Germany; she felt a similar sense of betrayal in the militarization of
the Berlin chemistry labs that were serving industry and not defending Austria. No
doubt there was tension between the wartime roles of Meitner and Hahn during
those years; the conflict surfaced again in a much more ominous form 30 years
later when their research was harnessed, unknown to both, to create the first atomic
bomb.

During their brief coordinated leaves, the Hahn–Meitner team was still hot on the
trail of Frederick Soddy's work. Soddy's concept of isotopy in 1913 had led to the
reorganization of radioactive elements in the periodic table. Several months before
Hahn was drafted, he and Meitner had discovered a new method for separating the
tantalum group from pitchblende, thereby minimizing radioactive contamination
from other substances.[27] The government's priorities were directed towards gas and
chemical research, not the search for radioactive isotopes; yet Meitner and Hahn
continued to stress the importance of their work for the future, perhaps even for
cancer research.

Shortly after the New Year of 1915, Otto Hahn had been ordered to attend a
meeting at the German Army Headquarters in Brussels where Haber informed him

(and other officers) of a proposal to use poisonous gas against the French. Since he had been officially serving in the German Gas Warfare Corps under Haber, Hahn had only conducted chemical *research*. Now the sobering news came that their noxious substances were to be used against soldiers. The General Staff was eager to achieve a breakthrough on the western front. Hahn recalled that he "protested, noting that such a proposal violated the Hague Convention," but that he was assured by Haber that the French had already used gas "in a small way" and that many lives might in the end be saved. This same argument would be used again with the application of nuclear fission to the development of the atomic bomb.[28] In both instances Otto Hahn seems not to have had the nerve to stand on a higher ethical ground. His moral guilt would only emerge *after* the war.

Hahn and the others were briefed on the formation of a new regiment that was to include many well-known chemists, including some of his colleagues from Berlin. Hence, when Hahn was posted as a "Gas Pioneer" to the 126th Infantry Regiment near Gheluvelt, Belgium to prepare for the mounting of the first gas attack using cylinders of chlorine, he knew what horrors lay ahead. Ironically, plans for this attack were canceled in favor of an attack on the small Belgian town of Ypres.

Hahn had the good fortune not to witness the gruesome gassing of British, French and other soldiers near Ypres. However, a young Austrian corporal named Adolf Hitler was one of the many overcome by the fumes that day, and recorded the cruelty of the event in his memoirs.[29] Later on, in Galicia, Hahn witnessed the horrible effects of chlorine and phosgene gas on soldiers. In May of 1916 he was stationed at Verdun preparing for a gas attack to be launched from Fort Douaumont, but heavy machine gun defense prevented him from entering the fort, and again he was spared being near the poison gases.

During 1916 and 1917 Hahn was able to time his leaves of absence in order to continue ongoing radiochemical research and often tried to get home to Berlin. On February 22, 1917, Meitner, alone in the laboratory and impatient for news from the front, wrote to him:

> Dear Herr Hahn!
>
> The pitchblende experiment is of course important and interesting but you must not be angry if I cannot do it right now. I have ordered the vessels for our actinium experiments, will get them in a few days and begin right away . . . be well and please *don't be angry* about delays with the pitchblende. Believe me, it is not because of lack of will, but because of lack of time. I can't very well do as much work as the three of us did together. Yesterday I bought 3 meters of rubber tubing for 22M! I was quite shocked when I saw the bill.[30]

By the end of the cold February of 1917, Meitner began the chemical separations of pitchblende and monitored the radioactivity and alpha activity of the new preparation she and Hahn had initiated years before. Alone, she began to write up results that occurred exactly as she had expected, and noted that "the original substance possesses a short-range alpha radiation, and decays to products with faster alpha radiation...whose intensity climbed to double the initial intensity in several weeks." The untreated 1.5g SiO_2 showed exactly the same behavior, "proof that the SiOx residue of pitchblende contained [a] new substance and was indeed a suitable starting material for its extraction."[31] We must note that it was *Meitner*, not Hahn, who had these insights; her interpretation of the decay curves from the "faster alpha radiation" demonstrated that an exciting "new substance" was present and radioactive.

Since Kasimir Fajans in 1913 and then Soddy had suggested the existence of an empty "position" between thorium and uranium in the radioactive elements of the periodic table, and noted that the yet undiscovered element probably resembled tantalum in its chemical behavior, Meitner had been cautiously looking over her shoulder for other elements, eager to make a new discovery. She felt triumphant that their "team" was close to a breakthrough. After testing silica residue from uranium ore and verifying the radiations coming from "something new," Hahn was able to work with Meitner during a short leave to at last characterize chemically this "mother" substance. Without a name as yet, they estimated its half-life, measured the range of its alpha particles and began to write up their discovery. Meitner surely did not think of claiming this work for herself despite Hahn's long absences. (But, in the reverse case, when Meitner was absent from the team during wartime separation forty years later, Hahn accepted the Nobel Prize in his name alone.) The Berlin editor of *Physikalische Zeitschrift* was delighted to receive their paper and published their discovery of the mother substance of actinium in 1918.[32]

The German military pressed on with research: Haber had nearly 1,500 men under his unit, and despite serious accidents and deaths in the testing procedures, he pressed on in developing lethal gases. Geiger had developed an innovative gas mask that saved countless lives; others applied their chemical knowledge to specific catalysts and compounds. Moral scruples remained silenced. Forty years later, a German journalist questioned Hahn's role during those years in the following interview:

Interviewer: You have told us that you had qualms about helping to organize gas warfare, and that you spoke to Geheimrat Haber about your scruples.

Hahn: I knew that the Hague Convention prohibited the use of poison in war. I didn't know the details of the terms of the Convention, but I did know of that prohibition. Haber told me the French already had rifle bullets filled with gas, which indicated that we were not the only ones intending to wage war by that means. He also explained to me that using gas was the best way of bringing the war to an end quickly.

I: And you found those arguments convincing?

H: You might say that Haber put my mind at rest. I was still against the use of poison gas, but after Geheimrat Haber had put his case to me and explained what was at stake, I let myself be converted. I then threw myself into the work wholeheartedly . . . .

I: . . . You have told us that you several times saw with your own eyes the effect of poison gases on enemy soldiers. And you also say that what you saw made a very deep impression upon you.

H: Yes, that is true. I felt profoundly ashamed, was very much upset. First, we attacked the Russian soldiers with our gases, and then when we saw the poor fellows lying there, dying slowly, we tried to make breathing easier for them by using life-saving devices on them. It made us realize the utter senselessness of war. First you do your utmost to finish off the stranger over there in the enemy trench, and then when you're face to face with him, you can't bear the sight of what you've done and you try to help him. But we couldn't save those poor fellows.[33]

Other atrocities of war lay ahead. The bombings of Belgium by inflatable Zeppelins pulled England into the war, and the British retaliated by bombing hangers in Düsseldorf. But these military maneuvers paled in comparison to a new mode of warfare: airplane attacks on civilian populations. While Dover, England was used as a port for allied soldiers getting ready to go to France, on May 25, 1917, the town exploded into fire and death as German Gotha biplane bombers, heading for London, attacked. Strategic bombing and more atrocities followed.[34]

All such news was censored in Vienna and Berlin. Meitner missed all of these stories, as well as the close contact with her colleague during the war years. As she returned that fall to Berlin, her letters fluctuated between the personal and professional. She was proud of her growing confidence in speaking, teaching and lecturing to those unaware of her field of specialty. Although she was still on formal terms with Hahn, she confided to him on March 22, 1917:

Dear Herr Hahn!

Yesterday I lectured in the colloquium [at the University]. I thought about you and I spoke loudly, looking at the participants rather than the board, although sometimes it is more pleasant to look at the board than at some of the people.[35]

Several months later, one of her letters revealed a rare personal holiday:

I had a really good time in Vienna and was utterly lazy, having a bad conscience but nevertheless enjoying great pleasure. My sister married on the third. It was quite a cozy, tiny celebration, a real war wedding, with only closest family attending, which in our case is quite numerous; and so it was truly lovely and intimate. I left the same evening, while the newlyweds had already taken off to Salzkammergut in the afternoon. From the card I received yesterday from your wife, I realized that you had photographed it (which I had assumed!) The telegram is probably already in Vienna by now. Never mind, the telegrams of families Schiemann and Franck have had the same destiny. Tell me, please, did you get the watch from Felsing? I hope you will write soon, and as thoroughly as I did. Fraulein Mueller was just here [in the Institute for Chemistry] and mentioned in the conversation that Knoeffler's produces mesothorium in somewhat sizable quantities for luminous paint. What do you think about that?

Take care, lots of love,

Lise

P.S. It really hurts that Kurz got killed. Did you know that he was supposed to stay here in Berlin and that he requested his transfer back to the front? The women that know their husbands and sons are on the western front now surely need a great deal of hope.[36]

By that summer, she grew frustrated over Hahn's silence. On July 27, 1917, she scolded him:

Dear Herr Hahn!

Much respect for the consistency you are showing in *not* writing. Anything else would not be 'timely,' would it? We save on everything, why not on a friendly word? Unfortunately, as you can see, I am not quite up to date, since I am writing you. However, I can throw in as an argument in my favor that my words are not only intended to send you a friendly 'hello' but also some factual information. Not to damage the utilitarian spirit of the time too much, I will immediately get down to business. In my high spirits, I have tried to process 43 grams of Giesel's substance. It seemed to be going quite well, but the end results were less than satisfying. I am writing you the exact procedure of the treatment so that you will be able to recognize the source of my errors.[37]

Lise's sarcasm was, by now, evident. Sick of the war, she wanted Hahn back at the Institute. A few weeks later, as the hot August sun tempted even the most stoic workers outdoors, she wrote to Hahn once more, with family as well as "utilitarian" news in mind:

I got thus far in writing you when your wife visited me and lured me into going swimming with her in Krumme Lake. When I came back home in the evening, I found two of your letters; I imagine the first sentence of my letter is still correct, because you only wrote for radioactive reasons! And a telegram arrived from my brother, announcing his arrival the next morning. Since he had to leave Saturday evening, I spent all of the

3 1/ 2 days with him and did not do anything in the Institute. My brother would have been very glad to see you and honestly felt sorry that it was not possible. Of course, we went to many places, so that he could get an idea of Berlin and its surrounding area. Early Saturday we both went to Dresden and spent the day there. Since the train to Vienna left so late that I could not get back to Berlin, I stayed the night in Dresden and traveled back yesterday. So now this short pleasure is over and the work day shall once again get its due.[38]

Lonely and yet loved by family and friends, Lise carried on.

In November 1917, Meitner traveled to Braunschweig to meet with an industrial chemist (one of the first to produce radium) to obtain radium "residues" for her research. This was an important step for her: she was able to lobby for the Institute for Chemistry, and gained invaluable experience in doing so. Then, continuing on to Vienna, she took the time to call on her former graduate advisor Stefan Meyer and renewed ties with him. She caught up on news of the Vienna Institute, which had become stripped of its finest young students, and they animatedly discussed the burgeoning field of radioactivity.

For many years researchers interested in radium had been searching for the parent substance of actinium. In studying radioactive substances, measurements have to be taken at fairly long intervals to let activities build up or unwanted ones decay. During one of Hahn's leaves, he and Meitner had taken the final measurements of the insoluble radioactive residues from their treatment of pitchblende and agreed to announce their discovery to the world in the spring of 1918: the parent substance of actinium. They deliberated together about the choice of a name for their newly-found radioelement, settling upon "protactinium." When Lise wrote to her teacher Meyer, he teased her that it should be called "Liseottonium"! In the spring of 1918, she again wrote to Meyer and asked for an appropriate symbol for their newly-discovered element. He wrote to her in June 1918, when the war was winding down:

> In your letter you pose a terribly difficult question about Protactinium. I would prefer the names Lisonium, Lisottonium, etc. and I therefore propose the symbol Lo, but unfortunately these are unsuitable if one wants general acceptance. In Pn a completely unimportant letter is brought to the fore, which leaves me most sympathetic to Pa. Although I still prefer Lisotto, it is much more significant to have discovered Pa or Pn than to come up with the most beautiful name....[39]

Under Meitner's prompting, the Berlin team proudly wrote up their findings as "The Mother Substance of Actinium: A New Radioelement of Long Half-Life" and the announcement of their discovery of protactinium was published in *Physikalische*

Zeitschrift during November 1918.[40] The naming controversy and priority credit for their newly-found radioactive element, however, did not simmer down until after the war.

At the same time that their independent research was progressing, Hahn and Meitner were awarded new contracts with an increase in salary as "Permanent Scientific Members" of the Institute for Chemistry. With the financial support of I.G. Farben, the major German chemical firm funding the Institute during the final years of the war, Meitner was given the responsibility of developing a full-fledged physics section, and closely supervised the necessary conversions, purchase of apparatus, and hiring of staff for her second-floor laboratory.[41] On her 40th birthday, celebrated in the new lab, she was as excited as a young child! She wasn't as mindful of scientific work then published in England, which had been almost impossible to get under wartime censorship and the blockade of trade. Hence, busy with her new "department" – the entire physics laboratory – she had not been aware of Soddy's work on the same topic, which occupied her for so many years during the war: the mother substance of actinium. Over a year later, rivalries would erupt.

In the latter part of the nineteenth century, the rise of science in German *Technische Hochschulen* and universities had laid the foundations for *industry* to train scientists and develop its own privately funded research institutions. In the twentieth century, the burgeoning industrial base, supported by government funding, emphasized the relationship between Germany's political and scientific "great power." By 1917, large industries such as I.G. Farben and their scientific laboratories played a key role in the German economy, and became even stronger as new markets were provided by the military.[42] Despite food shortages, disabled veterans unemployed, and growing unrest across Germany, private research was funded by key industries whose profits had exploded during war. Many of these capitalistic firms had their roots in scientific innovations and discoveries that received public recognition; in turn, industrialists continued to support scientific as well as government expansion, stressing their own private sector importance "in helping Germany attain worldwide scientific leadership."[43] Hence, applied research was on the funders' minds.

In Berlin, the research activities of the scientific academies and the Kaiser Wilhelm Society had been expressly developed for the *benefit* of German industry as well as the government, illustrated by wartime research priorities. At the close of the war, Max Planck became a visible spokesperson for German science in his role as Secretary of the Berlin Academy's Mathematical–Physical section (one of the most influential positions in this politicized body). Yet his views – mirrored by Lise

Meitner – differed dramatically from the German military. He had convinced Albert Einstein to join the Academy, despite Einstein's pacifist views. Planck attempted, as the war wound down, to rise above the nationalistic fervor of his colleagues. Not all of the younger generation agreed with his views. Years before, Planck had written about the philosophic nature of the physicist's work:

> If any thought strengthens and elevates us during the patient and often modest detailed work that demands our minds and bodies, it is that in physics we labor not for the day, not for momentary success, but as it were, for eternity.[44]

But many had lost this idealistic spirit as the war dragged on; so many had been killed, and so few remained to teach or carry on research duties. Yet somehow, Lise Meitner's dedication to her work seemed unflagging, based more on Planck's idealism concerning laboring "not for the day, or for momentary success." She had great empathy for Planck, and resonated to his deep harmony, inner fortitude, and strength in the face of personal tragedy. Years later, she recounted:

> Planck could be a very alert and compassionate, a warm-hearted observer of his fellow man. I personally experienced this quite frequently with great gratitude. What human insight and what assistance I received from him! His genuine outward restraint was to a large extent due to his great conscientiousness. He was to be fully answerable for everything he said. When I inquired about something to do with physics during the time of my assistantship, how often would he say to me, 'I will have your answer tomorrow.'[45]

Others were equally as reflective during a time in which war's intensity stretched most of them to their limits. Between facing urban life's difficulties and resettling into her own small apartment, Meitner corresponded with several scientists, including Einstein, about experimental problems in physics. Einstein even sent her a "thought experiment" that they both eagerly approached, only to learn later that Hans Geiger in Berlin and Meyer in Vienna had been tackling similar issues relating to gamma radiation *before* the war. In the fall of 1918, Einstein wrote:

> Dear Miss Meitner,
>
> This afternoon I mentioned our problem to Mr. Hopf. During the conversation, I realized that accomplishing our variation experiment would really be of exceptional importance because Meyer's result seems to be *confirmed* and it is viewed as proof for the continuous character of the absorption process. [Einstein then included a diagram and detailed description of his idea.]
>
> To reach this goal, we do not need to be too precise. So we should well be able to approach it. What do you think? Why don't you call me to give me your opinion.

Even though in *our* minds, there cannot be any doubt about the results, nearly all *other* physicists think differently.

Best regards,

Albert Einstein[46]

Clearly Einstein valued Meitner's views, recognizing that "nearly all *other* physicists think differently." The two physicists would again find themselves in a "minority" position about views of war and peace many years later. During 1918, it was Max Planck who kept them both assured that it was fine *and* proper to think about physics.

At last, in November 1918, the war ground to a bloody halt. The people of Berlin on Armistice Day appeared depressed, sullen, quiet. The war had ended, but their Kaiser had abdicated and then fled. With Berlin trying to recuperate from its wounds, even Einstein's enthusiasm flagged. He wrote to Meitner on October 29:

Dear Ms. Meitner:

... Meyer wrote me that he is already busy revising his variation work. Good that *we* did not start yet. Heartfelt greetings,

Yours,

Einstein.[47]

But Einstein and Meitner had more in common than physics at the end of a long four year war. Planck meant a great deal to both of them, and they often crossed paths in this son of a Protestant minister's home. But tragedy seemed to cast a shadow on Planck's life during the war years: his beloved twin daughters would both die within a year of one another.[48] They had been Lise's close friends since her arrival in Berlin. Grete had married a professor in Heidelberg and died suddenly during childbirth in 1917. Emma, her sister, came to care for Grete's infant, and in January, 1919, married the widower. But Emma too died in childbirth at the end of 1919. More sorrow was to come. Planck's eldest son Karl had been killed at the front. The family held out hope for their youngest son Erwin, who barely survived the war as a prisoner in France. Lise Meitner was embraced both as friend and as a source of solace and comfort by the Plancks in these years, as was Einstein and the very few other scientists not in the military. When Erwin was released from the prisoner of war camp at the end of the war, he was bitter and wan. He slowly began to confide in friends of his age and generation, such as Lise — all were now veterans who had witnessed horrors on the front lines.

The Versailles Treaty was, to say the least, an uneasy peace negotiated under conditions new to the world. When the German soldiers finally returned home, tired and hungry, there were months of riots and chaos on campus and in the government. General strikes broke out across the city; food was so scarce that mobs often ransacked small stores. Lise often remarked to Planck's daughter Emma, before her tragic death, that she felt lucky to have returned to a job and place to sleep; others did not fare as well during the socialist revolution in November 1918 in Berlin. The political confusion that ensued after the Kaiser abdicated was bloody. The subsequent provisional government, without police support, left the streets of Berlin in an uproar; it took some time for the new Weimar constitution to take hold.[49]

World War I in Germany ended not with a bang but a whimper. There were political disturbances and strikes for months after the armistice, as trainloads of weary soldiers returned. Some grew angry and restless, resorting to acts of violence to vent their frustrations against the new democratic Republic of Germany; some joined right-wing political parties or communist organizations. Others returned to their former activities, including scientific laboratories, but were forever altered by the conflict.

Hahn was not released from the German Army until months after the end of the war. In looking back at 1918–1919, Lise Meitner recalled that the governance of the Institute was left to the discretion of several junior staff members, with many "comic rather than tragic" consequences:

> The provisional Worker's and Soldier's Council Advisor Beckmann invited Professor Liesche and me to a meeting in order to choose the permanent Worker's and Soldier's Privy Councillor (as we were the only academics then extant at the Institute). We said, 'Of course, nominations must immediately be made for the persons to be chosen.' Thereupon, we received the answer that the selection would be secret, and therefore no such nominations could be made. We succeeded only with some effort to bring some order into this confusion of ideas.[50]

Meitner's attitude towards the disorganized Council's model of worker self-governance demonstrates her frustration at having to conduct work "around" the newly-empowered socialists of Germany. But were such Councils truly detrimental to Institute governance? Politically, Lise seems to have yearned for the prewar days of calm and order in the Institute structure, when she did not have to deal with the "business" of administration on a day-to-day basis. With her growing organizational responsibilities and seniority at the Institute after the war, creating "order out of confusion" was a role she seemed to take on only out of institutional "necessity" and her sense of performing honest administration.

Her attitude towards the workers' governance councils, as seen through her later recollections, sheds light on the feelings of many academics about the young Weimar Republic: namely, that its early disorganization forecast its doom.[51] Few academics supported what they deemed "further chaos" after a war that, to many, had dethroned reason throughout Europe. Hence many Councils set up in post-World War I Germany lacked the key ingredient they had been formed to provide: the support and leadership of experienced administrators and intellectuals.

Although many heated debates arose at the close of the war concerning the readmittance of German scientists to international congresses and associations (the French demanded that all Germans be banned due to the use of gas warfare),[52] international scientific communication was gradually reestablished. Scientific journals, unattainable in Berlin for nearly four years, began to circulate once more. Wartime obstacles had been responsible for Hahn's and Meitner's lack of awareness of the published results of Soddy in England. In February 1918, Soddy and a coworker had declared that *they* were the discoverers of the mother substance of actinium.[53]

The Berlin team began to examine Soddy's article, having already published *their* "discovery" in March of 1918, without knowing the existence of Soddy's work. The Soddy team described how a "heat treatment of pitchblende yielded a sublimate which gradually generated increasing quantities of actinium;" hence, they had not obtained the *chemical* results of the characteristics of the new radioelement that Hahn and Meitner had labored over. Later on, Hahn recalled that the field became even more crowded when it came time to agree on an official name *internationally* for the new radioelement:

> Fajans and Gohring, the original discoverers of the element, had had the right to name it; they called it brevium because of its short half-life. But the International Atomic Commission could not bring itself to use the name brevium for the element *we* had discovered, since it had a half-life of many thousands of years. Hence, a fully correct designation would be 'protactinium (Pa),' the long-lived isotope of brevium.[54]

It took an international commission to referee the priority dispute. Thus, through one of the first decisions of the newly-convened International Atomic Energy Commission of 1919, Lise Meitner and Otto Hahn became the "official" discoverers of protactinium. This official recognition led the way for other German scientists to be recognized once again in the international community.

As research settled into reorganization, Hahn, Meitner, and their new teams settled back into their labs. The year 1919 began with good news. It was with pride and pleasure that Lise Meitner learned that the 1918 Nobel Prize in Physics had been awarded to Max Planck, then 60 years old. His fundamental interpretation (in 1900) of the quantum concept applied to black body radiation phenomena had laid the groundwork for twentieth century atomic and nuclear physics.[55] It was with great pride that the Berlin community supported Planck's trip to Stockholm to receive the coveted Nobel Prize. Other good news appeared on the horizon for the Berlin physics community. Max von Laue had returned to Berlin in 1918, exchanging a more prestigious full professorship at Frankfurt for Max Born's *Extradordinariat* [associate professorship] position in Berlin.[56] Lise Meitner could not have been happier. At last, the university's physics department, emptied of so many of its finest students and professors during the war, was beginning to regain the air of authority and adventure that, with its impressive weekly colloquia, had drawn her to Berlin ten years earlier. The prestige of German science had been enhanced by the impact of Einstein's general relativity theory on scientists throughout the world. The Dutch astronomer De Sitter had taught the theory to Eddington in England, who, in turn, mastered Einstein's difficult techniques and published proof of the theory at the end of the war.[57]

And there was more "scientific diplomacy" to come. In January 1919, despite international animosities still simmering in Europe, Planck traveled to Stockholm to receive the Nobel Prize. The ceremonies in Stockholm that year were indeed sobering for the entire international community of science. By presenting Planck with the Prize, the Swedish Academy of Science affirmed that the world community still held the German scientific establishment in esteem. Ironically, one year later, the Nobel Prize for Physics was awarded to German experimentalist Johannes Stark who, under the Third Reich, would lead the movement in Germany to establish an "Aryan physics."[58]

In the wake of the First World War, the spirit of scientific "internationalism" was again emerging in Germany. Classical scholars of the Kaiser's era had, in one historian's words:

> endeavored to broaden the age-old formula providing national *prestige* from scientific achievement to an equation between political and scientific great-power status, contending that a decline in one of these sources of 'power' must adversely affect the other. In the early Weimar period, however, the *Gelehrten* [scholars] shifted this rhetoric subtly and significantly to the proposition that scientific great power status could function as

a substitute for political great-power status — a notion that found wide appeal across the German political spectrum.[59]

Germany had lost political power but not its scientific preeminence. Planck himself represented the German Academy of Sciences and was its visible leader in many state functions after the war. Yet the *Akademie* became increasingly utilized as a political forum as well, and mirrored the tensions of the post-war period.[60] While Meitner shied away from organizational politics (and preferred to correspond with colleagues in other countries about their work, after her disputes with the British about protactinium), Planck fostered a spirit of scientific cooperation among his German colleagues, attempting to heal the splits caused by the war. This spirit of cooperation was fostered during the Weimar Republic, and it produced mixed results.

Because of the widespread assumption that Germany's economic success was based on its industrial technology and scientific "great power status," both government officials and citizens of the Weimar respected the work of physicists, chemists, and other scientists. Such intellectuals now formed a new modernistic "vanguard" of a resurrected Germany.[61] In their suburban surroundings, the Kaiser Wilhelm Institutes were often viewed as "removed" from university and government politics, and their members were again regarded as dedicated to "pure" science, the perfect cultural role models in a society yearning for idealism. In reality, German and Austrian scientists themselves continued to mirror the divisions and misgivings apparent in the young Republic. For example, as Einstein's theories were incorporated into the new physics, opposition arose to his ideas from right-wing nationalists bent on smearing all "Jewish scientists." While on an autumn trip to a scientific congress, Lise Meitner wrote back to Berlin–Dahlem:

> Dear Herr Hahn!
>
> Many thanks for your first letter. By now, your lecture will probably be over and will surely have received much interest and applause.... According to my feelings, the anti-Einstein lectures with anti-Semitic background do not give any honor to us Germans — you could really speak of barbarism. Shall the Holy Inquisition rise again with Herr Gehrcke as the Grand Inquisitor? (... Your wife could do me a very big favor if she could get Mrs. Kasedorf to give you the ration-coupons for bread....)
>
> Many cordial greetings,
>
> Lise Meitner[62]

Social theorist/critic Max Horkheimer observed the following regarding the role of science: "Insofar as science is available as a means of creating social values, that is, insofar as it takes shape in methods of production, it constitutes a means of production."[63] The control and influence of that "means of production" still rested heavily in State hands during the dawn of the "roaring twenties," despite the new vanguard of intellectuals in the university and scientific research institutes – as well as right-wing radicals who denounced science and its practitioners.

Lise Meitner was critical of the anti-Semitic undercurrents she observed in many German circles, but did not feel threatened herself. Like most of her male colleagues, now that she was salaried she continued to work long solitary hours *after* her daytime responsibilities in her laboratories. By the middle of the decade, she was even able to employ domestic help to cook, clean, and do laundry, but she seldom had an assistant to conduct the often tiring, elaborate physical analysis of her radiochemical research. She did not complain.

Fifty years later, Hahn, in his old age, reflected upon his position concerning the military application of science during the war. He regretted his involvements with gas warfare, and wrote with ironic clarity:

> As a result of continuous work with these highly toxic substances [used for gas warfare by his Chemistry Division], our minds were so *numbed* that we no longer had any scruples about the whole thing. Anyway, our enemies had by now adopted our methods, and as they became increasingly successful in this mode of warfare, we were no longer exclusively the aggressors but found ourselves more and more at the receiving end.[64]

Meitner, in a sense, also found herself in a new position in the game of institutional politics. It was to her benefit that she was now regarded as a war veteran, on contract and free from university demands. The Weimar Republic, with its liberal policies and fresh perspectives, ushered in new winds of change. The peace, however, was an uneasy one, and a vocal few sought to revive a "stable" German climate by reviving anti-Semitism, using the appeals of divisive rhetoric to attack even the sacred domains of physics and Einstein's theory of relativity.

CHAPTER V

Shadows Lengthen: The Struggle
Out of the Causal Chain, 1920–1932

On the Remains of a Past Age

Still for instance the moon
hangs above the new buildings at night
Of the things made of copper
It is
The most useless. Already
Mothers tell stories of animals
That drew carts – called horses.
True, in the conversation of continents
These no longer occur,
nor their names;
The great aerials
Know nothing now
of a past age.

Bertolt Brecht, 1925

LIFE BECAME dramatically different after World War I for the generation raised with horse drawn carriages. Automobiles, airplanes, and telephones burst upon the scene for those who could afford the luxuries; the ordinary person was fortunate to have even a radio. Research in science and technology led to new ways of making electricity, heat, and matter benefit humanity, and in theoretical physics attention turned to the "quantized atom."

81

Lise Meitner was actively forming new friendships with physicists who had not entered Germany or attended scientific congresses because of the war. A paradigm shift was in the making, and she felt herself in the heart of it. She recalled:

> During the First World War, physics had been placed on an entirely different basis, both from the experimental and theoretical points of view. The main credit for that rests with Niels Bohr and his work on the structure of the atom. Not only did this work have a very decisive effect on physics, it also had repercussions in astronomy, chemistry, and even biology. I do not think that any scientist has had such a worldwide influence on at least two generations of physicists as did Niels Bohr... not even Rutherford, despite his immense genius. Certainly, Rutherford had a great many pupils, but I do not think that [fact], or the influence [these pupils] exerted, can be compared with the [accomplishments of] Niels Bohr.[1]

Bohr's atomic model (1913) had revolutionized atomic physics. Subsequent applications and developments after the war in other fields earned him the respect of scientists on a global scale. His Copenhagen school attracted young physicists, and an international atmosphere, especially at his summer colloquia, was created between his advanced students and visiting physicists.[2] Meitner's group pressed to invite Bohr to Berlin:

> I first met Bohr in 1920 when he was lecturing at the Physics Society in Berlin. In his lecture, he stressed the importance of a series of spectral lines and, for their interpretation, introduced the correspondence principle for the first time. I must confess that when James Franck, Gustav Hertz, and I came out of the lecture, we were somewhat depressed because we had the feeling that we had understood very little. In this half-depressed and half-playful spirit, we decided to invite Bohr to spend a day at Berlin–Dahlem, but not to include any physicists who were already professors in the party. This meant that I had to go to Planck and explain to him that we wanted to invite Bohr, who was staying with Planck, but not Planck himself. In the same way, Franck had to go to Professor Haber – because if we were going to have Bohr in Dahlem for discussions the whole day, we [naturally] wanted to give him something to eat – and ask Haber for the use of his clubhouse 'without the bigwigs' (*bonzenfrei*). Franck too had to stress that we did not wish to have Haber himself present since he was a professor. But Haber was not the least put out. Instead he invited us all to his villa. You must remember that this was in the very difficult period after Germany had lost the war, and to get something to eat was rather difficult in Dahlem.
>
> Haber only asked our permission to invite Einstein as well. We then spent several hours firing questions at Bohr, who was always full of generous good humor, and at lunch Haber tried to explain to Bohr the meaning of the word *bonze*.
>
> I did not really get to know Bohr personally until a year later, in 1921, when I was invited to give a lecture in Copenhagen on beta and gamma radiation and had the good fortune to spend many hours with him and his wife. The great difficulty for Germans

during this period was that they had been excluded from all scientific congresses, and Bohr put himself to a great deal of trouble to get them readmitted. He also told me a great deal about the war and his experiences in England. In a word, we spoke about everything under the sun, whether grave or gay. Even today, I can still feel the magic of this, our first meeting, a magic which was only enhanced in subsequent years when it was often my privilege to take part in his famous conferences. I have often thought how fortunate it is that there are such people and that [a chance] is given to a person to make a close acquaintance.[3]

Although she was rather shy at first in the noisy gatherings at Bohr's Institute, Lise relished going to Copenhagen in the summer of 1921 during which she formed a lifelong friendship with the Bohr family.[4] She fondly recalled the Bohr's five boys, and his wife Margrethe's hospitality. She also enjoyed the heady, brisk debates on the latest in quantum theory. Bohr used to organize small groups in the cafeteria or conference rooms of the Institute for Theoretical Physics at a moment's notice, and Meitner participated in these animated dialogues long into the Scandinavian twilight.

Later that summer she accepted an invitation to Sweden and traveled by ferry and train from Copenhagen to the University of Lund. There she spent a month conducting experimental work under the Swedish experimentalist Manne Siegbahn. She was quite happy in the land of the "midnight sun," so different from the atmosphere of inflation-ravaged Berlin, and enjoyed meeting other experimental physicists with interests similar to her own. She also met active women in the physics department, who invited her for hikes and concerts. Most fortuitous that summer was her meeting with one of Ehrenfest's students from Holland, Dirk Coster. Coster and Siegbahn were to play a major role in Lise Meitner's life some twenty-five years later.

Her ties with the Scandinavian physics community were cemented in those early 1920s, and her trips left a strong impression on her each time she returned to Berlin. Contact with new colleagues served to broaden her "international" attitude towards scientific exchange and communication. Her research on radioactive substances and their properties was now firmly situated in the mainstream of scientific debate. In 1922 Bohr received the Nobel Prize, which placed him and his Institute at the center of quantum theory research and the "new physics." In her usual modest style, Lise wrote Bohr from Berlin:

Dear Professor Bohr,

When I think about the congratulatory letters that have been showered upon you recently, it seems almost unjust that I too should write. But I'm so very happy that you received the Nobel Prize that I have to express my happiness in spite of my hesitations. Please permit me therefore to congratulate you and your wife most heartily; I can assure you that the Nobel Committee's decision has rarely been received with such agreement and enthusiasm as their last one has been.

I still think with joy and a kind of homesickness of the lovely time in Copenhagen. I hope that you'll come again, with your wife of course, to Berlin. This would be enjoyable in every way, and would also enable me to have a few physics problems solved that have been causing me much hard study.

Please ask your wife not to be angry that I haven't yet answered her last letter....[5]

In Berlin, as elsewhere in Germany, the spirit of international scientific cooperation lay under a xenophobic shadow. The importance of "German science" and its "essential contributions to the Fatherland" were becoming the subject of speeches and public debate. Immediately after the war Max Planck addressed the Prussian Academy of Sciences on this theme:

> If the enemy has taken from our fatherland all defense and power, if severe domestic crises have broken in on us and perhaps still more severe crises lay before us, there is one thing which no foreign or domestic enemy has yet taken from us: that is the position which German science occupies in the world. Moreover, it is the mission of our Academy above all, as the most distinguished scientific agency of the state, to maintain this position and, if the need should arise, to defend it with every available means.[6]

What position did "German science" and scientists actually occupy in the world during the early 1920s? Planck, von Laue, Hevesy and others had made Berlin a mecca for the younger generation even before the war, and Einstein's work was central to their thinking. The University of Göttingen was enjoying a rebirth as an intellectual center[7] with physicist Max Born and mathematicians such as Emmy Noether, Felix Klein and David Hilbert, all of whom crossed national boundaries and intellectual disciplines. In general, physicists were eager for new tools, and Klein's mathematics was applied to quantum mechanics with startling results. In 1924, the first volume of *Methods of Mathematical Physics* was published by Richard Courant. Courant's work heavily reflected the ideas of Hilbert and also exerted a strong influence on physicists. The new mathematical ideas and techniques stimulated many a young person, like Werner Heisenberg, to be drawn to what was soon called the "new physics" school.

Respected educators and politicians in Germany noted the impact of quantum physics on the liberal Weimar Republic, particularly within the Berlin group. "People complain," theologian Adolf von Harnack wrote before the war, "that our generation has no philosopher. Unjustly: they [the philosophers] now belong to other faculties. Their names are Max Planck and Albert Einstein." Interestingly, in his later years Planck wrote:

> ...a philosophy becomes untenable if it conflicts with inanimate nature. I need not here refer to the considerable number of religious dogmas to which physical science has dealt a fatal blow.[8]

Unfortunately, right-wing ultra-nationalistic views were also a part of the "post-religious" philosophies propounded in Germany during the 1920s. Ideas of "racial hygiene" were presented to the world even by Nobel laureates. Philipp Lenard and Johannes Stark championed their *völkisch* ideas in published journals and at congresses. Most of their colleagues, however, did *not* share their ideas and considered them renegades.[9] Scientists seemed willing to serve the State but not the political aims of the Social Democrats or the ultra-right. Historian Alan Beyerchen notes:

> They regarded parliamentary politics as social and factional, but did not realize that their own stance, which was allegedly 'above politics,' was just as divisive as that of the parties they abhorred.[10]

The economic maneuverings occurring in the Berlin scientific community underscored the political ties that bound science, state, and industry to each other. These ties enabled funds to flow to certain centralized research centers, which included the Kaiser Wilhelm Institutes and the more "established" scientific organizations. For the most part, these organizations tended toward "state conservatism." The need to maintain scientific research in Germany during the economic depression that followed World War I was a high priority for many, including the industrialists, who found numerous civilian uses for the breakthroughs made during the war in chemistry, metallurgy, and so on.

For Otto Hahn, scientific goals did not preclude the possibility of an industrial career. He found himself happily in the midst of new colleagues in industrial business circles; also he was doing well financially. Patents based on his earlier discoveries gave him income from industry. While his wife was content in their large new house, Hahn spent a considerable amount of time expanding the research laboratories in the three-story Kaiser Wilhelm Institute for Chemistry, courting industrial contributors when he could.

During these years, Lise Meitner was comfortably settled in an apartment within the Harnack House (property of the Kaiser Wilhelm Society) across the street from the Institute. She and Hahn continued to press the Society for their economic advancement. From records in the post war archives of the Institute, we see that

85

Meitner and Hahn had to "lobby" to advance their own causes with the Kaiser Wilhelm Society administration. In a letter dated February 14, 1919 addressed to von Harnack, president of the Kaiser Wilhelm Society, Hahn and Meitner wrote:

> ... the requests, which we permit ourselves to present today, are basically of three kinds: 1. A salary increase. 2. An increase in our budget. 3. Long-term contracts....[11]

For Lise Meitner, "political involvement" during the turmoil following the war meant obtaining financial as well as professional security for herself and her Institute.

The *Notgemeinschaft der deutschen Wissenschaft* (Emergency Association of German Science), a foundation created to support research and scholarship in Germany after the war, became a major source of funds, although most of the money supporting scientific research and publications came from the central government. The Berlin physicists, however, "never lost the inside track of the *Notgemeinschaft*," and were able to administer and benefit from funding coming from foreign benefactors; for example, the elusive Japanese industrialist Jaiami Hoshi gave a major contribution to the Society to aid German chemistry.[12] In an administrative coup, physicist Fritz Haber managed to divert a large amount of this contribution for the support of atomic physics in the Dahlem Institutes.[13] In 1922–1923, Haber designated Max Planck, Otto Hahn, and himself to represent half the disbursing committee of the *Notgemeinschaft's* Finance Council. As funds for research and expenditures came under the control of Berlin's physical scientists, the city became both honored and resented as the "stronghold" of German scientific research.

Lise Meitner, whom Einstein called "our German Madame Curie,"[14] remained outside of the many economic decisions made during this time. At the organizational level, however, she was a visible leader in Institute affairs, and she still traveled across town (when time permitted) to the university, where she attended the colloquia and stimulated students and professors with reports of her ongoing research. Hahn maintained a more administrative position, accorded to him when the Institute was first opened.

In late 1921 Planck approached Meitner to discuss the possibility of her teaching in the physics department of the university. The offer came as a pleasant surprise. Lise had *never* had a female instructor in her entire university education. Even after the war, women in Germany and Austria could not "officially" lecture in the university, let alone become professors. But such prohibitions were diminishing. Doors were beginning to open for women in many sectors of society, scientific education included. In 1922 Lise Meitner qualified as one of the first women in Prussia to become a *Dozent*, a lecturer, and, four years later, assistant professor. It was a milestone celebrated in her "Berlin family" that summer.

Her *Habilitation*, the oral examination to qualify her for an assistant professor lectureship, was far more formal than the similar procedure Hahn had joked about when he first arrived in Berlin and qualified for the same position. Max von Laue and Heinrich Rubens (the same person who years ago had no place for her in his laboratory) served as Meitner's reviewing committee; a written thesis was waived. In her oral presentation, some sixteen years after attending Planck's lectures, she demonstrated her thorough knowledge of physics. By the end of the examination, Rubens warmly wished Meitner well in her new University Lectureship position.

The *venia legendi* certificate was formally presented by her second reviewer, colleague and friend Max von Laue (then a full professor in the Department of Physics).[15] The *venia legendi* qualified Dr. Lise Meitner to give lectures and conduct courses in the Physics Department each semester. And she did so with dedication for the next ten years.[16] On August 12, 1922 she wrote to Planck, formally thanking him in his official capacity as Chairman of the University's Physics Department:

> Highly esteemed Geheimrat!
>
> Please permit me to thank you and the faculty for the *venia legendi*, awarded without a trial lecture, and give my sincere, heartfelt thanks to the colloquium. I highly value this showing of trust by the faculty, and am fully aware of the responsibility which has been awarded me.
>
> With preferential esteem,
>
> Your deeply devoted,
>
> Lise Meitner[17]

Every new lecturer had to give a public "inaugural" university presentation; the subject of Lise Meitner's was cosmic physics. She had become interested in the radiation studies of astrophysicists, and spoke about "The Significance of Radioactivity for Cosmic Processes."[18] However, a misinformed Berlin academic press published the lecture by the female *Dozent* as the significance of radioactivity for "*cosmetic* physics!" The press seldom if ever made such blunders when men were hired by the university.

Chauvinism and biases notwithstanding, Lise Meitner achieved many milestones in the next several years. Her scientific research on radioactive processes broke new ground experimentally and provided evidence for theoretical speculation. The study of the physical and chemical properties of radioactive substances assured

Hahn and Meitner continued prominence in the field. During these years they also conducted separate projects on topics of mutual interest. At the Physics Department of the university, Lise Meitner served as an example of the pressing need for female educators and role models in higher education. She was promoted (in 1926) to the adjunct position of *nichtbeamteter ausserordentlicher Professor*. Her friendships with other women scientists such as Elisabeth Schiemann, Eva von Bahr-Bergius in Sweden, and Hertha Sponer (who would later marry James Franck) led to personal and professional connections to the disciplines of physics, biology, and botany.

Meitner continued to pay close attention to the developments at Bohr's Institute in Copenhagen, Hilbert's mathematics department in Göttingen, and other significant scientific institutions around the world. She herself worked on the relation of beta and gamma radiation. Hahn, meanwhile, was conducting independent research at the large radiochemistry section of the Institute, concentrating on applied chemical analyses. Meitner often worked late, returning to her nearby apartment only after the Geiger counters were shut off and the last assistants departed, well after midnight. Together, she and Hahn were known as the "mother and father" of the Institute for Chemistry.

Between 1922 and 1925 questions of atomic structure, beta, and then gamma radiation (as well as spectral analysis) intrigued her, and she wrote up her findings in 16 separate articles. The "observation and identification of an *anomaly* in the absorption and scattering of gamma rays began as a test of relativistic quantum mechanics," noted historians Brown and Moyer.[19] It was this anomalous absorption rate that led Meitner and her assistants to delve into the complex physical problems surrounding gamma radiation. As coauthor of the Kaiser Wilhelm Institute for Chemistry's 1921 "Yearbook," Meitner looked back at the rapid growth of quantum physics. In 1924, she observed in findings published by the *Zeitschrift für Physik* that the "Compton effect" of recoiling electrons was being heralded in America: her challenge was to colleagues in Europe to apply the same techniques that she did with original results.[20]

Lise Meitner did not suffer from a "mid-life crisis" in her forties. Far from it. She was an extremely productive woman in a growing discipline. She stayed in close contact with friends and family during holidays and free Sundays, often boating on the lakes of Berlin, or picnicking with the Hahns and other Institute colleagues. But she was still an anomaly with respect to her university teaching position. In 1914, there had been 770 women attending the University of Berlin, but it was not until 1922 that *any* woman in Germany was granted the right to lecture to university students. Meitner's pay was far lower than that of a male assistant professor. Most of the early women assistants at the Institute had never worked with any female mentor during their university career before they met Lise Meitner. So Meitner was a role model both at the Institute where she carried out her research, and at the University where she was able to encourage the few young women who had broken with tradition and entered graduate-level science courses. At the war's end she had written to Hahn:

June 23, 1918

Did I write to you that I lectured recently about our work at the Colloquium and that afterwards Planck, Einstein, and Rubens each separately said to me how substantial and elegant our work is? From this you can see that I did present quite reasonably, although I was again in a silly way very embarrassed, much worse than at the Bunsen Society. I was glad you were not there. I must surely have maligned you! So I escaped with a friendly jest by Planck and very benevolent psychological understanding from Einstein about my shyness, and with this, the two both gave me consolation that I had lectured clearly and well.[21]

Four years later, full of confidence, she stood in front of rows of students and carried on as if she had always been "Fräulein Doktor Professor." Quantum physics had a woman spokesperson, and Meitner was soon busy with lesson plans and students as well as Institute for Chemistry research.

The curricula of girls' school were under revision throughout Europe during the 1920s. Germany finally instituted academic programs for girls that were equivalent to the twelve-year *Gymnasium* course of study for boys in preparation for a university education.[22] Women were granted the right to vote for the first time in Austria in February 1919, and in the German parliament during the first years of the Weimar Republic, women in the Social Democratic and Communist parties triumphantly obtained many seats.[23] In the words of a leading sociologist of the time, "who so urges the mental inferiority of women in bar of their claim to equal rights with men may be met in various ways." And "met" may have been an understatement — women in Europe were at last asserting their rights, with support from around the world.

Britons and Americans had begun to sponsor research for scientists of both genders in *applied* radioactivity. National centers were funded throughout the United Kingdom by a Medical Research Council, which granted diplomas in radiology to women as well as men. But there was no "storming of the medical citadel" by women medical students in Germany, even though women pioneers such as Lise Meitner had trained numerous male veterans in radiological techniques.

As X-rays began to be applied by untrained personnel, an epidemic of radiation burns and illnesses broke out in medical facilities. With pressure from the military, chemical industries, and physicists worldwide, new sources of radium and radioisotopes began to be explored and safety precautions instituted. Faced with rising costs, the medical community looked towards state-supported research centers, such as the Kaiser Wilhelm Institutes, for answers to a number of puzzling questions concerning the characteristics of radioactive elements, their safe handling, and less expensive radiochemical sources for their manufacture and use.

Medical researchers were now aware of Hahn's and Meitner's work, and throughout the 1920s the team was asked to speak at congresses and international conferences with increasing frequency. Despite these demands on her time, numerous unpaid travel invitations, and friendly international correspondence, Meitner's focus continued to be on her first floor Physics Section in the Institute for Chemistry. Franco Rasetti, a graduate student, who visited from Fermi's laboratories in Rome, recalled:

> Miss Meitner...almost daily checked what her pupils were doing, being extremely courteous and pleasant even with the less gifted ones. It is true that she was very strict in enforcing, quite rightly, precautions to avoid radioactive contamination of the laboratories. I believe that there was no better place anywhere to learn clean handling of radioactive elements, and from this I profitted when I had occasion to introduce these techniques in [Fermi's lab] in Rome.[24]

Having used radioactive materials throughout her career, Lise Meitner knew their dangers, in contrast to many throughout the international scientific community who did not take the safe handling of radioactive materials seriously. It would be years before the development of systematic procedures for *measuring* contamination; and of course, many scientists were exposed to extremely high levels of radiation without fully understanding the consequences until decades later.

But social issues of health and welfare and the state's role in health education were receiving greater attention. In Berlin, numerous new political parties were formed, each putting forth proposals to solve the contradictions of its predecessors concerning social ills. It was an era of innovations in technology, art, music and theater. In this city of great cultural diversity, Lise Meitner spent her leisure hours, enjoying music, and participating in animated discussions about social questions in the liberal Weimar Republic. One of her close lifelong friends, Elisabeth Schiemann, said:

> [Meitner's] native country and her home were fertile ground for the love and understanding she devoted to literature and art; they accompanied her through life. Most of all, she enjoyed exquisite music. The Viennese 'student' did not miss a single chance to thoroughly learn about great composers and to enjoy their music with or without a full score. In those years, Berlin offered many such opportunities, and provided her with a broad and profound knowledge of music literature. This inclination further intensified her friendship with the Planck family and soon made her a regular guest in my family, where she enjoyed – though not practicing herself – chamber music played with love and serenity.[25]

The decade between 1923 and 1933 was a major turning point in world history. *Kunst und Kultur* marched alongside a creeping nationalism that assumed the forms of economic boycotts, scientific chauvinism, and overt racism in Europe. Nazi groups organized in Germany and Austria as early as 1923, while in Italy Mussolini formed a small cell of the Fascist government in Rome in 1922. Economic and political upheavals caused by the First World War lay the foundation for the events that culminated in the Second World War. But most people, including the Berlin physicists, could not foresee eventual trouble: the Roaring Twenties were full of innovation and hope after the devastations of World War I.

Within the Kaiser Wilhelm Institutes seeds of future conflict were visible. In a group photo taken during this time of this small influential urban group, (see photo in center of book), Einstein stands to the left with an all-encompassing smile, as physicist Hertha Sponer, one of Lise's close friends, and Ingrid Franck, James Franck's wife, look on. It is a jovial, almost intimate, group portrait of the great figures of this "younger generation" then in Berlin. Physicists Grotrian, Westphal, von Baeyer, Pringsheim, and Hertz stand confidently behind, and Haber is posed easily in their midst. Hahn leans in to participate, and in the center of the group, Franck sits self-assuredly next to a relaxed, almost carefree, Lise Meitner. Generally regarded as shy, we have perhaps a rare glimpse of her confident, self-assured side, at ease with peers and colleagues. All of them are in their mid-forties, mid career. Many considered one another as very close friends, and more than half were Jewish.

*   *   *   *   *

If I were asked what was Christopher Columbus' greatest achievement in discovering America, my answer would not be that he took advantage of the spherical shape of the earth to get to India by the western route – this idea had occurred to others before him – or that he prepared his expedition meticulously and rigged his ships most expertly – that, too, others could have done equally well. His most remarkable feat was the decision to leave the known regions of the world and to sail westward, far beyond the point from which provisions could have gotten him back home again. In science too it is impossible to open up new territory unless one is prepared to leave the safe anchorage of established doctrine and run the risk of a hazardous leap forward. . . . However, when it comes to entering new territory, the very structure of scientific thought may have to be changed, and that is far more than most men are prepared to do.

Werner Heisenberg, *Physics and Beyond*

When the general public was admitted to Einstein's lectures on relativity in the early 1920s, controversies once confined to the university rippled through newspapers and were often perverted by anti-Semitic groups. At first, Einstein joked with characteristic modesty about the right wing "study groups" that denounced relativity as a "vast Semitic plot." But it was soon apparent that these politically-motivated groups, which had earlier backed an unsuccessful coup at the end of the war, were at the forefront of a growing anti-Einstein movement.[26] New territory in physics, as Heisenberg noted in his later years, always runs the risk of a "hazardous leap forward." Many in the Berlin scientific community vowed to take no notice of the public spectacles staged by the anti-Semites, but most physicists were shaken by their forcefulness. As Einstein and his work were defamed in public, private conversations about the fate of Germany grew more serious.

Werner Heisenberg, a student in theoretical physics at the University of Munich in the summer of 1922, witnessed the anti-relativity movement firsthand. He recalled the first public lecture by Einstein that he attended:

The lecture theater was a large hall with doors on all sides. As I was about to enter, a young man – I learned later that he was an assistant or pupil of a well-known professor of physics in a south German university – pressed a red handbill into my hand, warning me against Einstein and relativity. The whole theory was said to be nothing but wild speculation, blown up by the Jewish press and entirely alien to the German spirit. At first I thought the whole thing was the work of some lunatic, for madmen are wont to turn up at all big meetings. However, when I was told that the author was a man renowned for his experimental work, to whom Sommerfeld had often referred in his lectures, I felt as if part of my world was collapsing. All along, I had been firmly convinced that science, at least, was above the kind of political strife that had led to the civil war in Munich, and of which I wished to have no further part. And now I made the sad discovery that men of weak or pathological character can inject their twisted political passions even into scientific life.[27]

The son of a classics professor, young Heisenberg could not accept the simplistic drawings in his physics textbooks of the atomic structure of a gas connected by "hooks and eyes" to represent chemical bonds. "The whole of modern physics is false," he stated resolutely at the age of 17, and arrived at the University of Munich the next year to begin his mathematical training. He was bent on discovering a deeper truth to the "reality" of atoms. His formative years were about to be spent in a society undergoing radical changes in its definition of science, culture, and politics, as well as in its struggle to find new heroes to replace the old.

Still wearing the khaki shorts of the German youth movement, Heisenberg took the train at the Munich station and headed for the University of Göttingen.[28] Bohr later invited him to spend a summer at his Institute and a lasting bond formed between them. Heisenberg was only 23 when he formalized the notion of matrix mechanics, which arose from Bohr's earlier work; it was a key in the door to the quantum interpretation of the atomic line spectra, and much more. He was now in the forefront of a new generation of German-born physicists.

Lise Meitner was also encountering bright new doctoral candidates daily in her travels on the trains between the university and the physics section of the Institute. Just as she herself had once rushed to hear the latest physics seminars, many of her assistants now flocked to the university whenever a colloquium was held or a visiting professor came to lecture. Throughout the 1920s, the front bench of the Wednesday Physics Colloquium at the University of Berlin held an impressive array of five Nobel laureates, all of whom argued the finer points of theories in the making. The chairman was Max von Laue. Planck, Einstein, Haber, and Walther Nernst, the feisty Berliner whose "Nernst lamp" and other patents had brought him wealth and fame before the war, challenged participants. The heated debates over the latest theories of atomic structure, the role of causality and the new quantum mechanics left many feeling defensive. Some expressed the notion that a relaxation of complete determinism was taking place, while others saw the new quantum physics as based upon a total abandonment of causality itself.[29]

On June 29, 1922 Planck objected to the causality "controversy," stating that when his quantum "hypothesis" had been developed sufficiently so that one could properly speak of a "quantum theory," then and only then would it be time "to consider consequences for scientific-causal thought."[30] Members of the Berlin circle, especially Einstein, were continually at odds over the acceptance of the newly-proposed quantum mechanics of the Copenhagen school. Lise Meitner participated in these debates. At the same time, she guarded her practical views and worked to *apply* theoretical developments to her field. Einstein was later to say: "She knows her way in the family of radioactive substances better than I do in my own."[31]

As quantum mechanics and its newly found applications emerged, so did controversies outside of science over such "abstractions." In a time of recession and scarcity in Germany, intellectual debates were viewed as beyond the country's postwar "utilitarian needs." Departmental budgets were slashed at the university. Inflation was out of control. Financial difficulties reached a climax, when, on January 10, 1923, Germany was declared in default in its reparation payments and the mark lost its monetary value.[32] Within days, Franco-Belgian forces occupied the Ruhr valley to "enforce" reparations, and the German government panicked. Passive resistance began that same week; in Munich, more violent means were adopted to bring "order" to the mounting chaos. Some months later there was an abortive *putsch* led by the self-appointed leader of the new Nazi party, Adolf Hitler.

Although the mark was stabilized (a short-lived achievement of the Stresemann government) by November 20, 1923, the public had lost confidence in its leaders. This "loss of confidence" was also mirrored intellectually in the growing debates over the nature of causality in physics.[33] Amidst this ferment, Lise Meitner continued to work and recollected those times somewhat ironically:

> It is of course known that the economic conditions of the Kaiser Wilhelm Institutes were very unfavorable after the First World War; to be specific, the complete reorganization of the financial foundation for the individual Institutes had to be undertaken. Given inflation, there were times when our primary occupation was securing sufficient cash to cover the rising demand for currency. All the employees [with cash in hand] would run off to take care of their purchases as soon as possible, since the exchange value of the money could markedly diminish in a few hours.[34]

Whether male or female, politically left or right, the economic situation affected every person and organization in German society. Early in 1923 the US dollar was worth 18,000 marks; by August of that year, it was worth 4,600,000 marks, and during the month of Hitler's *putsch*, the US dollar reached the astronomical equivalent of four billion marks.[35] Two thousand printing presses worked night and day to supply the Reichsbank with what would soon become useless currency; the need was so great, it was often printed on just one side. By the time the mark stabilized in mid-1924, there had been several Reichstag elections. Five political parties were represented in an uneasy government: 100 were declared Social Democrats, 95 Nationalists, 62 Communists, and 32 Nazis held parliamentary seats. Hitler had been sentenced to five years' imprisonment in April of 1924, along with countless other agitators throughout Germany, but was released from prison by the end of the year.

Lise's mother, widowed for many years, passed away in Vienna in 1924, leaving a traditional "Last Will and Testament" later copied by each of her daughters.[36] Lise Meitner traveled to Vienna to be with her brothers and sisters in the large family home for the funeral. She also had some time to reunite with the "educated circles" in Vienna, and strengthen her ties in a country that, like Germany, was suffering from high inflation. She met with academic friends and Institute assistants — "minority women" in professions filled with men. Two of Lise's younger sisters, supported by the family, were working hard towards their Ph.D.s at this time. It was a comfort to support one another, and Lise's career served as encouragement to all.

Upon returning to Berlin, Lise realized that few *German* families were as willing or able to support their daughters at a university. For her own type of research, there were still few, if any, academic positions available. Meitner was in an enviable position: her publication record, her position as a full academic member (*Wissenschaftliches Mitglied*) of the Institute, and her recently acquired University Lecturer status placed her in a preeminent position that was not easily attainable by other German women scientists until the end of the decade. Despite the liberalization of the culture, professional realities were still very harsh for women in science. Lise maintained a determination to guard her salaried position in spite of cutbacks within all of the Kaiser Wilhelm Institutes. She also was aware of Hahn's intrigues in wooing interested industrialists in their joint work.

It was a rather traumatic period for her. Lise recalled that as Hahn became more and more involved with Institute administration and politics, their joint research was put on hold. Yet, they met daily to discuss projects of joint interest even after the Institute was reorganized into separate Physics and Chemistry sections. She later said:

> As a result of this division of the Institute into two sections [radiophysics and radiochemistry], Hahn and I no longer worked together from about 1920. In the chemistry department, Hahn and his colleagues did very important work on applied radiochemistry. Hahn also found the first examples of nuclear isomerism, uranium-Z, which he found to be isomeric with uranium-X2.
>
> Our work was naturally directed more toward physics; for example, we investigated the line spectra of beta radiation and were able to establish its relationship to gamma radiation. We also checked the theory of Klein and Nishina with regard to the passage of gamma radiation through different materials and, in this connection, incidentally discovered pair formation – not that we recognized it as pair formation, but we did report

the presence of some previously unknown effect of the atomic nucleus. Despite this break in direct collaboration between Hahn and myself, there was still very close indirect collaboration. Indeed, it was only natural that the chemistry assistants should help and advise the physicists on all chemical problems. They also made up any preparations we needed for our experiments, while the physicists, in turn, built auxiliary apparatus such as amplifiers or counters for the chemists.[37]

By the mid 1920s, Meitner's publication list grew to nearly ten articles per year on topics ranging from radioactive processes to cosmic rays. This visibility brought its rewards. In 1924, she was honored with two scientific awards based on her achievements as a woman scientist. The American Association to Aid Women in Science (known formerly as the Naples Table Association) had become dissatisfied with awarding honors to women recently awarded the Ph.D. for initial research efforts. Why continue to honor beginners for one piece of good work, they questioned, when they could be rewarding the *best* among older woman scientists, many of whom were as yet unrecognized by their professional colleagues? Basing its new rules on the Nobel Foundation's, the AAAWS attempted to upgrade its award to a "woman's Nobel Prize." Hence, its first award was presented jointly to chemist Pauline Ramart-Lucas of the University of Paris and radiophysicist Lise Meitner of the University of Berlin.[38]

More awards would be announced. In 1924, Max von Laue nominated Lise Meitner for the coveted Leibniz Prize, a nomination endorsed and signed by Max Planck and Albert Einstein. Their formal nomination read, in part:

> As a woman, she is excluded from many kinds of academic recognition. Exactly for this reason, it seems fitting to express such recognition by conferring the Leibniz Medal.[39]

The award of the second place Silver Medal to Lise Meitner in 1924 was the first Leibniz Prize granted to a woman. She was indeed highly honored by this recognition from her own colleagues in Berlin, and once again, her independent research was honored and recognized outside of the Institute for Chemistry circles. More accolades would follow. In May 1925, Meitner was awarded the Ignaz Lieber Prize in Vienna by the Academy of Sciences for her work during 1922 to 1924 with beta and gamma rays and radioactive substances.[40]

The scientific community had begun to recognize the research of Hahn and Meitner. In 1923 Otto Hahn (alone) had been nominated for the Nobel Prize in Chemistry by Max Planck. Hahn's discovery of the process of nuclear isomerism had clearly demonstrated his expertise in the field of applied radiochemistry and contributed an essential "missing piece" to the puzzle of nuclear constitution. In 1924 the research *team* of Otto Hahn and Lise Meitner was nominated for the Nobel Prize in Chemistry by H. Goldschmidt and M. Bergmann. Kurt Fajans and H. Goldschmidt nominated the Hahn–Meitner team for the Prize again in 1925 for their joint research in radioactive processes. And from 1929 to 1934, Max Planck nominated them both yearly for the Nobel Prize in Chemistry.[41] This is an important point to note: the *team* of Hahn and Meitner were nominated for ten years.

Physicists as well as chemists recognized the team's work on an international scale. In their day-to-day professional lives, however, Hahn and Meitner did not focus on international honors or recognition. Both were dedicated to their work at the Institute for Chemistry. In the gradual reorganization of the Institute, Meitner's research responsibilities increased, while Hahn, with the retirement of Institute directors, accepted more administrative duties. In 1921, the Kaiser Wilhelm Society's annual report detailed the budgets of 19 scientific institutes, and the employment of 371 persons.[42] But only after inflation had been brought under control could Hahn and Meitner seek defined budgets for their sections, or could they request materials, assistants, and experimental equipment to continue their research on radioactive substances.

Hahn was appointed "Deputy Director" in 1924 in order to prevent him from accepting an offer to become chemistry chairman at the Technische Hochschule of Hannover. When chemist Alfred Stock retired at the end of 1927, Hahn was nominated and appointed Executive Director of the Institute; his 1928 inaugural party was a gala event. He later jokingly characterized the evolution of the large third-floor radiochemistry lab and the first-floor physics lab as follows:[43]

1. the Hahn radiochemical section with Lise Meitner as scientific associate;

2. the Hahn–Meitner section;

3. the Hahn and Meitner section;

4. the Hahn radiochemical section and the Meitner radiophysics section.

Obviously, Meitner's partnership was invaluable. As Hahn assumed the Institute's Directorship, Meitner's responsibilities also increased.

By the late 1920s, chemists' work became more complex with the emerging theories and data concerning the discovery of new elements, properties of radioactive emissions, and improved models of atomic constitution. Physicists were increasingly needed to interpret chemical developments. In turn, findings in radiochemistry sparked further developments in physics, which then often led to the discovery of new radioelements.

97

Lise Meitner was particularly occupied with research in X, beta, and gamma radiation and related nuclear processes. Her studies on secondary beta radiation are an outstanding example of pioneering research in the field. Her nephew Otto Robert later noted:

> In 1925, she decided the controversial question concerning the sequence of fast nuclear processes by proving that the secondary beta radiation is derived from the electron shell of the atom formed by the transformation. So gamma radiation had to succeed beta transformation. She also succeeded in taking the first picture of traces of positrons excited by gamma radiation, and she was one of the first to demonstrate pair production (electron/positron) from high-energy gamma radiation.[44]

Meitner's independent scientific research was certainly maturing. She introduced C.T.R. Wilson's cloud chamber (scarcely used since its invention in 1911) and was one of the first physicists ever to use it experimentally. Together with graduate students and her colleague Kurt Freitag, Meitner used the cloud chamber to study electrons at greatly reduced speeds years before others entered this field.[45] Meitner also correctly interpreted a physical process called the "Auger effect," which she described in papers published in 1922 and 1923 in the *Zeitschrift für Physik*. In 1925, she and Hahn jointly published two papers on the beta ray spectrum of radioactinium and the gamma rays of the actinium series,[46] and in 1927, she and von Laue published a series of papers on the scattering effects observed in cloud chamber photographs.[47]

The silver Leibniz Medal, the Austrian Lieber Prize, and continued Nobel Prize nominations assured Meitner's recognition in international circles. She also had a large group of young graduate assistants to draw upon from her university classes. Quantum physics and its theoretical intricacies were her domain at the Institute, and at the weekly university colloquia, she participated with the eagerness of one demonstrating a new language in the interpretation of physical phenomena. The staff at the Institute for Chemistry recalled that she often called out to Otto Hahn: "Hähnchen, leave that to me – you know *nothing* of physics!" – and they would all joke that she knew very well that *she* was the Institute's authority!

But political storm clouds darkened the horizon. Many in the academic and industrial communities felt threatened by the new political order. While others endorsed the open policies and reorganized currency of the new coalition government, entrenched academics and industrialists shared analogous views of *die Wirtschaft* (the economy) and *die Wissenschaft* (learning) as, in the words of Paul Forman, "autonomous realms to be governed by strong personalities whose *authority* within their own factories and institutes must not be limited, and in whose hands lay the true interests of the German people." Each group offered the other both moral and political support in common resistance to "interference" from the parliamentary-democratic government.[48] Hahn handled his sensitive role in the dispensing of Association for German Science (*Notgemeinschaft*) funds with tact and discretion. The esteem many held for him, together with his good humor, direct manner, and evident administrative abilities placed him in the forefront of scientists who directly influenced the reconstruction and funding of scientific research.

Many other German scientists – for example, the physicist Philipp Lenard – who were members of the same scientific organizations as Hahn, went to remarkable lengths to demonstrate their contempt for the Weimar government and its Berlin leaders.[49] Hahn's position was a conciliatory one, whether in his dealings with the affairs of the Institute or in debates in scientific organizations. He had many channels from which he could obtain funds, which reassured the KWI administration.

Hence Hahn became the "paternal" provider for the Institute during the 1920s, and in 1922, a family provider as well when his wife Edith gave birth to their only son Hanno. Lise Meitner was named godmother at the christening, a role she enjoyed the rest of her life. They were two caring figures: Otto Hahn, the "father figure," and Lise Meitner, more "maternal." Lise Meitner's door was always open to all who came to her with their problems as well as academic questions. This high-strung, petite woman was a whirlwind of efficiency, a chain smoker working hard to maintain radiation contamination-free facilities, and set high standards for all in her domain, including Hahn. A later assistant to both Hahn and Meitner, Fritz Strassmann, observed:

> It turned out that she deserved just as much respect in her taking care of, and in the training and guidance of, the rising scientific generation as in her character as a human and a scientist. Herself inspired by an indefatigable interest in work, she ... knew that in the long run it better served the end of research and researchers not to force one's working speed on a coworker; and what is more remarkable, she had the self-control to act on that judgment.[50]

Thomas Mann, in his essay on Freud and Einstein, maintained that everyone has a role model on which to build his or her own goals.[51] For many women during this era, such models were hard to find. Lise's drive and determination are difficult to "characterize." Since there were no female role models in her scientific education, we sense that she worked herself particularly hard to "bridge the gap" between theoretical and experimental realms, physics and chemistry, the University and the government-supported Institute.

Hahn's first major mentor and perhaps role model had been Rutherford. Eventually knighted by the Queen of England, Lord Rutherford was politically astute, solicited industrial support for his research facilities, and maintained close ties with university administrators. He also worked daily with a host of research students, staff, and primary scientific investigators, and remained "headmaster" of his laboratory, calling his research staff "my boys" all his life, although many were senior to him in age. Niels Bohr had been one of Rutherford's pupils, and consciously or unconsciously, both Hahn and Bohr may have modeled their congenial interactions and family-style relationships with students and staff at their own Institutes after those of their influential mentor. Democratic, not authoritarian, leadership was their style.

Bohr's Institute in Copenhagen mirrored the sense of "family" at the KWI for Chemistry in Berlin–Dahlem. The influence of the Copenhagen interpretation of quantum physics on the Berlin circle was extremely important in those years. Lise Meitner fit into both groups. She looked forward to spending summer vacations with Bohr's family, and "mentoring" his brightest postdoctoral students, such as Heisenberg and Gamow. In Copenhagen she later discussed Heisenberg's "uncertainty principle" and other "theories in the making." Bohr also introduced her to Pauli, Born, and other German-speaking colleagues. We can appreciate the "invisible college" phenomenon at work: intellectuals drawn together working with other scientists from diverse backgrounds and various nations on similar problems in physics.[52]

Bohr's leadership of the Copenhagen Institute, unlike Hahn's gradual administrative rise in authority in Dahlem, was unequivocal. The Carlsberg Foundation, based on its brewery company profits, had granted land (and later family residence in a stately mansion) to the Nobel Prize recipient for the development of a scientific institute.[53] The Carlsberg donation was a feather in the cap of Danish national pride as well as a tribute to the soft-spoken physicist whose commitment to educating the new generation of scientists was as great as his devotion to understanding complex atomic and nuclear processes.

In 1925 a new form of quantum theory, known as matrix mechanics, was being developed by Max Born, Pascal Jordan, and 24-year-old Werner Heisenberg. In 1926 another form of quantum mechanics, which came to be known as wave mechanics, was proposed by the Austrian Erwin Schrödinger.[54] "The incontrovertibility of these two forms of quantum mechanics was shown very quickly," recalled Max Delbrück, later an assistant of Meitner's (and twenty years afterward, famous for his applications of quantum theory to molecular biology). Physicists in the Copenhagen Institute relished these debates, and Bohr's institute became a central gathering place for younger physicists to debate the fine points of a new quantum mechanics. Delbrück recalled:

> Heisenberg formulated the uncertainty principle as the real root of meaning of the quantum of action, and Bohr, in a lecture at Como, Italy, gave his version of its deeper meaning and formulated what was called the 'complementary argument.' The essence of this argument was that for any situation in atomic physics, it is impossible to describe all aspects of reality in one consistent space-time-causal picture. The various experimental approaches that you use will reveal one or another aspect of reality, but these various experimental approaches are *mutually exclusive*; this means they are such that you cannot get information that you get out of one arrangement, and simultaneously use the other arrangement to get other information. So these various experimental arrangements stand in a mutually exclusive relationship. The nature of the formalism of quantum mechanics is to permit you to derive the predictions for the outcome of the arrangement (if they are done successfully); these predictions are of a statistical, probabilistic nature. This feature of atomic physics, expressed in the way Bohr expressed it, or in the more popular way that Heisenberg expressed it (as an 'uncertainty relation') was, of course, a total shock to everyone concerned; in fact, so much a shock that Einstein never got over it. During the rest of his life, Einstein tried somehow to get back to the classical picture in which reality is just one reality, and if you can't get at the full reality with present methods, then presumably there must be other methods to get at it; on the other hand, Bohr was insistent on saying that this limitation to the classical picture of reality was not a preliminary stage to be replaced by a return to classical notions, but was an advance over classical notions — that we now had arrived at a new dialectical method to cope with a feature of reality that was totally unexpected....[55] Bohr continued to elaborate and restate his position year in and year out ... through innumerable lectures.[56]

Lise Meitner, after attending such lectures in Copenhagen, would then return to Berlin with these new concepts and participate in the heated debates over the modern role of causality. Atomic physicists were, in the words of Paul Forman, becoming convinced " ... of the fundamental inadequacy of the existing quantum theory of the atom, which supposed classical mechanics to be valid for motions within the stationary states...."

The younger generation, led by Heisenberg, were "beginning to doubt the reality of the visualizable atomic models to which that theory had been applied."[57] As in Boltzmann's lectures decades ago, Lise again found herself in the midst of fundamental heuristic arguments over the applicability of atomic models. Some began to draw parallels between the crises in the governments worldwide and the paradoxes debated by physicists. However, Forman notes that the Weimar's intellectual milieu

" . . . at the very least *facilitated* the precipitation of a generalized conviction of a crisis of the old quantum theory."

Meitner, practical to the core, focused on what her own intuition pointed to as a window into this dilemma: the activity of the electron. She wrote to Hahn in 1927:

> Franck just called and told me that he had met Hertz and that I should eat with them in the city. So I drove into Berlin and met them; then Pringsheim and Max Born came, and we had a nice get-together. We talked about all kinds of personal things and also about some physics. Born told me that he had, for the time being, only a qualitatively elaborated ionization theory, into which *my* conclusions for the primary beta-spectrum fit well. He promised that he would allow me to calculate more precisely the occurrence when the nuclear electron runs through the K-shell.[58]

Meitner found herself between several schools of thought throughout the late 1920s and early 1930s. By then she had become a "senior" visiting physicist at Copenhagen. The information she accumulated was introduced into her lectures at the University of Berlin and other physics departments across Europe. She also encouraged many postgraduate students to go to Copenhagen whenever possible; their philosophic questions and enthusiasm sparked her interest and determination to apply the emerging quantum mechanics to her research results in radiophysics. Max Delbrück recalled:

> Anybody who was at all interested in the result of the questions couldn't help but be fascinated. It also motivated me to look at the writings of Kant on causality to see how Kant, who was so clever and thoughtful, could have overlooked this possibility. So, for the first time, and with a real motivation, I looked at Kant, and it was very clear that this situation was just utterly removed from anything that Kant had thought of — so there was no doubt that the physicists had been pushed into an epistemological situation that nobody had dreamed of before.[59]

*   *   *   *   *

The uncertainties, ambivalence, and contradictions in philosophy, art, and physics were mirrored in the political realm. Internationalism was ushered in with optimism and foreboding. Germany was admitted to the League of Nations in September of 1926 as fascism spread to Italy and Austria. The liberal Bauhaus movement in the arts did not penetrate academia, where rising nationalism spread to international scientific congresses. Many scientists refused to even recognize the elected Weimar government as the "legitimate" voice of the German nation! Scientific academies and professional associations which had always been international in scope now became "authoritative" with respect to *their* "international relations." Still, as Paul Forman notes, "...below the level of formal international relations [in science], there was a very extensive development of personal scientific contacts and of recruitment of individual German scientists into international collaboration."[60]

The younger generation at the Institute had not experienced the nationalistic hatreds of World War I. Many took international scientific relations and communications for granted, and were at ease with the utilization and application of theories emerging from all parts of the globe. Delbrück formed a study group with other physicists and biologists from the Soviet Union, England, and Denmark to discuss the applications of their physics to cellular structure and biology.[61] Hungarian-born Leo Szilard also received his advanced training in quantum physics in Berlin, and through Meitner's sponsorship at the University in 1930 and 1933, he lectured as a *Privatdozent* in Meitner's classes on questions of atomic physics and chemistry.[62] Meitner's own lectures each semester from 1923 through 1933 at the Physics Department continued to be attended by an international blend of scholars, scientists, and students.

Otto Robert Frisch had moved to Berlin after his graduate work in physics in Hamburg. Frisch retained fond memories of both his aunt and Hahn during this period:

> Lise Meitner helped me find lodgings; she lived in a tiny flat and couldn't put me up, but often asked me to dinner. Through her I met Otto Hahn, with whom she had worked for twenty years. I always feel a glow of pleasure when I think of Hahn, with his bluff Rhineland accent and his self-deprecating sense of humor. He liked to whistle the last movement of Beethoven's violin concerto in an oddly syncopated manner; 'Doesn't it go like that?' he would say with assumed innocence when challenged.
>
> I changed landladies several times but always stayed near where my aunt lived in Dahlem.[63]

Frisch always credited his aunt with his early interest and entry into a career in physics. Ten years later he and Meitner would coin the term "nuclear fission."

Meitner had been skeptical about Chadwick's contention that certain spectral lines arose from "secondary electrons," and followed her intuitive sense that primary electrons must form a "group" with well-defined quantitative energies. Her measurement techniques, focusing electrons through deflecting them, were tedious but necessary. A later graduate assistant, Hendrik Casimir, recalled that she:

> practiced all the manipulations required like [playing] a difficult sonata before doing the experiment...the equipment consisted mainly of gold-leaf electrometers, string electrometers, ionization chambers, and Geiger counters...today's physicists may have difficulty in imagining the ways particles were counted in those days.[64]

Isotope separation was also of interest to Gustav Hertz, who was awarded the Nobel Prize in Physics along with James Franck in 1925. Collaborations in radiophysics and the interpretation of the disintegration processes of radioactive elements were making rapid progress, and the Berlin circle continued to lead the way. When Hahn became the Executive Director of the entire Kaiser Wilhelm Institute for Chemistry, new opportunities opened up for Meitner as well. She undertook a study of all the known elements, analyzing their radiation properties. In 1929, the Italian physicist Corbino concluded an optimistic speech as follows:

> Only technical and financial difficulties, not insurmountable in principle, oppose the realization of this great project [the nuclear 'disintegration' and restructuring of elements]. The object is not only the artificial transmutation of elements in appreciable quantities, but the study of the tremendously *energetic* phenomena that would occur in some cases of disintegration or recombination of atomic nuclei . . . although it is improbable that experimental physics will make great progress in its ordinary domain, there are many possibilities in the attack on the atomic nucleus. This is a true field for the physics of tomorrow.[65]

Note that in 1929, even a professional physicist was still skeptical about experimental physics " . . . making great progress in its ordinary domain." The bullet necessary to pierce the atomic nucleus core – the neutron – had not yet been discovered; and although the "release of atomic energy" had not been accomplished, it was discussed in many circles. Writers and journalists alike speculated that "atom power" might drive submarines, power spaceships, and someday light entire cities. H.G. Wells and Hans Dominik fueled the imaginations of young and old readers with their "science fiction."[66] In a 1933 newspaper interview with Einstein, it was reported that Einstein felt the attempts at "loosening the energy of the atom were fruitless."[67] And Rutherford himself quipped (1933): "Anyone who expects a source of power from the transformation of atoms is talking moonshine."[68]

But the transformation of radioelements was exactly the process Lise Meitner and her staff were pursuing. Frisch recalled:

> At that time, her chief worry was the continuous spectrum of the primary beta rays. At first she thought this was a secondary effect, perhaps due to the random energy loss by radiation as the beta electron leaves the nucleus; but that explanation was refuted by the well-known experiment of Ellis and Wooster (1927), confirmed by Lise Meitner and one of her students, Wilhelm Orthmann, in 1929.[69] It seemed incredible that the transition between two well-defined long-lived nuclei should lead to the emission of electrons with a continuous energy distribution. Peter Debye is said to have called it 'a topic best not talked about, like new taxes.'[70]

One physicist, Wolfgang Pauli, would not let go of the energy distribution question. In 1930, he wrote a now famous letter to Lise Meitner and Hans Geiger in which he proposed a new neutral particle, later to be called the neutrino, which would be emitted with the electron, and share with it the "available energy at random." However, Pauli stated that such a neutral particle might be "too elusive" to be detected by the available means.[71] But he was soon convinced, and bold enough to lecture on this topic, opening his remarks on a humorous note: "Dear Radioactive Ladies and Gentlemen!" It is important to note that Pauli kept careful watch on the results coming from Meitner's second-floor laboratory in Berlin and included her in his correspondence on the subject, which soon led to the famous formulation of "Pauli's neutron" (after the discovery of Chadwick's neutron, later called the "neutrino" by Fermi) .

As usual, the Kaiser Wilhelm Institute kept Meitner more than busy. Looking back, she wrote:

> There were 25 scientific workers in our two sections by 1932, and several foreigners from the most varied countries among them. A good spirit and a happy mood predominated in this working community, reflecting Hahn's personality [as Institute Director], which not only had a very favorable impact on the work, but was also expressed time and again, at Christmas and birthday festivities.... There existed a strong feeling of belonging together, the basis of which was mutual trust.[72]

Although Meitner now concentrated her research on projects with her own staff and laboratory, and no longer collaborated with Hahn, her relations with him were solid, and as they built up their "family," young physicists visited from the university as well.

In the early 1930s, Meitner published a series of independent papers on the beta-ray controversies, then turned her attention to the theory of Klein and Nishina on the passage of gamma radiation through matter, and empirically proved that secondary beta radiation comes from the electron shell of the new, not the original, atom.[73] Thus, gamma radiation was found to *succeed* beta decay. Meitner took the first known pictures of positron traces excited by gamma radiation utilizing the Wilson cloud chamber.[74] She also observed and correctly interpreted the radiationless transitions "in which an electron jumps down to a lower energy level, transferring its energy to another electron which is then emitted," now called Auger electrons.[75] Focused upon her work, she had little time for politics.

Historian Peter Gay reflects that Berlin politics mirrored the fate of the liberal Weimar Republic:

If Berlin tasted of the future, the taste of Berlin was cruelly deceptive; there was little future left. Life would not leave art alone. Beginning in 1929, the Republic suffered a series of traumatic blows from which neither the Republic nor its culture was to ever recover. Gustav Stresemann [Chancellor of the Weimar Republic] died [of natural causes] on October 3, 1929. Count Kessler was in Paris at the time; on October 4, he wrote in his diary: 'All Parisian morning papers are reporting the news of Stresemann's death in the largest possible type. It is almost as though the greatest French statesman had died.'[76]

The French had good reason to mourn the passing of the leader. The London Times stated: "Stresemann did inestimable service to the German Republic; his work for Europe as a whole was almost as great." Then came "Black Friday" 1929 — a tumbling stock market brought on a global economic crisis. The impact on Germany was immediate. International loans were suddenly withdrawn, and the Weimar government collapsed without a leader or an economic framework. Peter Gay describes these events:

> Then came the Depression, unemployment, continual political crises culminating in the elections of September 1930, which decimated the bourgeois parties, gave the Nazis six and a half million votes and 107 deputies in the Reichstag, and led to Bruning's semi-dictatorship, and to government by emergency decree. 'The German intellectual's state of mind,' Franz Neumann later recalled, '... was, long before 1933, one of skepticism and despair, bordering on cynicism.'[77]

A global economic depression struck in the fall and winter of 1929. Several right-wing German physicists attempted to become "international representatives of the Fatherland" at scientific congresses and in government and industry circles by 1930, with repercussions for all. Other members of the scientific community joined the ever-increasing, vociferous Nationalists, dragging anti-Semitic elements into their denunciation of scientists like Einstein, and attacking physical theories such as relativity as a "Jewish theoretical construction," the tabloids screamed. In 1932, Schrödinger wrote:

> In a word we are all members of our cultural milieu. As soon as orientation of our interest plays any role whatsoever in any matter, the milieu, the cultural complex, the *Zeitgeist*, or whatever one wishes to call it, must exert its influence. In all areas of a culture, there will exist common features deriving from the world view and, much more numerous still, common stylistic features — in politics, in art, in science.[78]

Max Planck witnessed these "common features" influencing his cherished domain, and feared for the safety of Einstein and even Meitner. Three years after the Great Depression, Planck, a Nobel laureate with the privilege of nominating others, held up the contributions of the Hahn–Meitner team to the international community for commendation and nominated them for the 1933 Nobel Prize in Chemistry.[79]

Such "cultural features" as scientific achievement or Nobel Prize nominations, however, would soon be proselytized under the Nazi banner, which would also banish or exterminate those deemed "superfluous" to its "Aryan milieu." Meitner and Hahn were not awarded the 1933 Nobel Prize. Many began to fear the loss of free scientific inquiry. The carefree times were coming to an end. Because of growing political unrest, friends and colleagues of Lise Meitner became extremely concerned for her safety. Lise, with a sense of foreboding, watched uniformed Nazi guards take the place of students in the train stations and university hallways of an increasingly troubled Berlin.

Science in Nazi Germany: 1933–1936

Science can only ascertain what is, but not what should be, and outside its domain, value judgments of all kinds remain necessary.

Albert Einstein

National Socialism is not unscientific, only hostile to theories.

Bernhard Rust
Nazi Reichsminister for Science, Art, and Education

Our national policies will not be revoked or modified, even for scientists. If the dismissal of Jewish scientists means the annihilation of contemporary German science, then we shall do without science for a few years.

Adolf Hitler

LISE MEITNER invited Erwin Schrödinger's wife Annemarie to her apartment on January 30, 1933. The two women listened pensively to the swearing in ceremonies on radio of Germany's new Chancellor, Adolf Hitler. Only a month earlier President von Hindenburg told one of his generals that he hoped the German people would not suppose that he "would elevate the lance-corporal from Bohemia"[1] to Führer. However, von Hindenburg finally conceded to political pressures. Hitler's NSDAP party never obtained an absolute majority in the Parliament, but they now had a platform and leader whose propaganda would spread, coordinated by Nazi Party leader Goebbels.

Liberals and conservatives debated Hitler's proposed policies, as the new Chancellor's voice spread racial hatreds over the country. More Nazi terror would soon break out: on February 28, the Reichstag (capitol building) was set on fire, and thus sealed the fate of the faltering Weimar democracy. Thousands of Nazi *Sturmabteilung* (storm troopers) marched from surrounding areas into Berlin and staged a massive torchlight parade, passing victoriously under the historic Brandenburg Gate into the center of the city. The Nazis were quick to blame the arson on the Communists and used the fire as a pretext to jail thousands of people — judges, opposition party leaders, and government officials. Most were to be brutally transported to concentration camps in Dachau, a Munich suburb, and Oranienberg, outside of Berlin.[2]

A confused and aging President von Hindenberg ordered emergency powers. "Until further notice" the basic civil rights of all German people – personal liberty, inviolability of the home, privacy of personal communication, freedom of the press, freedom of assembly and of association, and the constitutional guarantee of private property – were suspended by the "Decree of the Reich President for the Protection of the People and the State." Death penalties were imposed for a number of crimes that in the past had been punishable only by imprisonment.[3] Communists and Socialists were among the first persecuted: shortly thereafter, gypsies, liberals, and other "undesirables" were on the arrest lists. Over 800,000 Germans were jailed or sent to concentration camps.

Chancellor Hitler seized control. He dissolved the Parliament and installed a government of "national unity." By the time national elections were held on March 5, 1933, over half of the German electorate still refused to vote for the the National Socialist German Worker's Party.[4] But Hitler persisted; he had retained the Chancellory and Parliament by a slim majority, and prepared his Nazi Party for the swearing in ceremony of the new parliament in the Garrison Church in Potsdam.

Meitner and Hahn warily watched all of these events. Hahn was gone from Berlin during the months of turmoil following Hitler's election. He was working on a unique emanation method, on coprecipitation results, and on other radiochemical topics, which were all well advanced by 1932. He had the respect of colleagues abroad, and was honored by an invitation to deliver the George Fisher Baker Lectures at Cornell University in the spring of 1933. Thus Hahn had been planning for over a year to meet new American colleagues and spend the spring semester in Ithaca. With warm farewells to his wife, son, and Lise Meitner, Hahn, in high spirits, set sail for New York on February 14, 1933.[5] He did not foresee that during his absence it would become "legal" to dismiss thousands from their jobs in German universities without due process or academic review.

Thus, Lise Meitner had to assume leadership of the Institute for Chemistry during the months of civil unrest; the Nazi Party seized increasing control over life in Berlin. Meitner juggled administrative difficulties in between endless coffee, cigarettes and sleepless nights. She assumed that Hahn would keep up with the news through the American newspapers, and was cautious in writing to him on March 21, 1933 on official Institute stationary:

> ...I am glad you all in all feel well in America. I read with pleasure the beginning of your first course of lectures which had been tailored to American inclinations so very well. All respect!
>
> Here, of course, everything and everyone are under the impressions of political change. Today is the ceremonious opening of the new Reichstag in Potsdam. Already last week we received the order from the Kaiser Wilhelm Society to hoist the swastika banner next to the Black-White-Red one: the banner was paid for by the KW Society.
>
> A short while ago Frau Geheimrat Schiemann and Edith [Hahn] were here to listen to the radio-transmission of the Potsdam ceremony. It was harmonious and solemn throughout. Hindenberg spoke a few sentences and gave the lectern to Hitler, who spoke very moderately, tactfully and conciliatory. Hopefully, it will go on in this way. If the sensible leaders succeed, among whom especially Pappen is to be counted on, then one can finally hope for a trend towards positive developments. It is almost unavoidable that transitional periods cause all kinds of mistakes. Now everything depends on sensible moderation.
>
> It would have been very difficult for Haber to hoist the swastika banner. I was glad to hear that he had given Kühn the order to do it; this way it was much more dignified for him than if it appeared he had been forced.[6]

Haber was not the only one feeling isolated and contemptuous of the new regime. Now Goebbels' shrill voice filled the airwaves, urging "good Germans" to turn in anyone suspected of "communist leanings." Meitner did everything she could to avoid the news, but members of the Institute for Chemistry began arriving for work proudly wearing their brown shirts.[7] Soon intellectuals throughout the Reich, including scientists at Berlin–Dahlem, began to feel the wrath of the state directed against them as never before. Workers at the KWI turned up their radios to listen to the hate messages, and without Hahn as a spokesperson, Meitner's hands were tied.

Five days after taking office, Hitler lived up to expectations: he suspended the Weimar Constitution, and had the Nazis seize control of all state publications. On April 1, Hitler began his vendetta against "leftist-sympathizers," "non-Aryans," gypsies and other minorities, as well as those he identified as *Juden*. The official

"Boycott of the Jews" was accompanied by torture, arrests, and imprisonment of blacklisted "liberals." Lise Meitner spent the night at the home of Magda and Max von Laue. Their son Theodore later recalled that he would soon be sent to Princeton, NJ, to continue his education and that his father "rarely discussed politics with the family," although Lise Meitner was considered one of their close friends.[8] Von Laue seemed protected against the scandals of the "state sponsored boycott" because of his aristocratic Prussian family background. But he was concerned for Lise's fate: she wrote urgently to Hahn that "[Laue] is of the opinion that under the circumstances, we shall not be able to do without you for very long."[9]

Despite the political persecutions all around her, Lise Meitner continued to devote herself to her university responsibilities. Spring break was extended. Under the chairmanship of Schrödinger in the Physics Department, she was scheduled to deliver her yearly spring semester course at the university. Leo Szilard, thirty years younger than Meitner, was also expected to carry out his joint lectures in Professor Meitner's class along with Kurt Philipp (who began to openly pledge his allegiance to Hitler), and Drs. Kallman and London. Szilard, of Hungarian-Jewish descent, was active in and out of the physics department. He sensed future troubles on a global scale. Foregoing a promising academic future in Germany, he abruptly canceled his pending lecture with Meitner, parted from his fiancée, and left Berlin days before the bloody purges. He later wrote:

> I left Germany a few days after the Reichstag fire. How quickly things move you can see from this: I took the train from Berlin to Vienna on a certain date, close to the 1st of April, 1933. The train was empty. The same train on the next day was overcrowded, and was stopped at the frontier. People had to get out, and everybody was interrogated by the Nazis.
>
> While I was in Vienna . . . the first people [were] dismissed from German universities, just two or three; it was, however, quite clear what would happen.[10]

But to many, it was *not* clear what was "going to happen" since there had been so many forms of social unrest after the end of World War I. Meitner carried on, wondering how she would manage both her Institute and University tasks.

Meitner's early reactions to the Nazis were no doubt tempered by the fact that she was legally an Austrian citizen, and hence not subject to German laws or policies. Over the years, she had weathered many political storms in Berlin, so that it was not altogether unreasonable for her to believe that, given time, order would eventually prevail. It was hard for others, however, to believe that reason would triumph; so many Germans appeared eager to accept the new "faith" in Hitler's *Volk und Blut* (folk and blood). Growing numbers of people chose emigration over confrontation during that ominous spring of 1933.

It did not take long before hundreds of professors, scientific researchers, and university assistants throughout the country were fired because of the racial laws. In April, 1933, the civil ordinance *Gesetz zur Wiederherstellung des Berufbeamtentums* (Law for the Reestablishment of the Professional Civil Service) was put into effect. Two controversial sections, paragraph 3: dismissal of civil servants of non-Aryan descent and paragraph 4: dismissal for "political" reasons, provided the Nazi Party with a legal basis, however tenuous, for attacking the many academics who did not agree with their policies.[11] A great number of Hahn's and Meitner's close colleagues across Germany, including Hertha Sponer, George von Hevesy, and Albert Einstein, were openly dismissed from their university positions. Other academics, dropped from faculty positions for being Jewish, were threatened with acts of violence from both rabid students as well as police, and were compelled to leave university towns. Some, such as Max Born in Göttingen, remained in their homes, "emigrating" silently, and began to pack their cherished libraries. Few dared to protest against the Nazis; many that did were immediately arrested and degraded. Hahn's close friend in World War I (a fellow physicist whom Lise Meitner also loved and respected), James Franck, spoke out: he resigned from his professorship with a vocal public protest against the fate of his colleagues.[12] But many others just watched, waited, and remained fearfully silent.

Those scientists who had come to German universities during the Roaring Twenties to teach and conduct research during the "golden age" of physics and cultural renaissance in Berlin were amazed and outraged by brown-shirted student leadership. On May 10, 1933, students with lighted torches marched down Unter den Linden Avenue to the University of Berlin, and there set fire to a large pile of books. The works of such "undesirable" German writers as Thomas and Heinrich Mann, Stefan Zweig, Erich Maria Remarque, whose novel *All Quiet on the Western Front* had described the tragedy of dying for the "Fatherland," as well as works of "corrupt" thinkers such as Einstein, Kafka, Helen Keller, H.G. Wells, Marcel Proust, and others were cast joyfully into bonfires.[13] A number of assistant professors who had also declared their allegiance to the new Third Reich also took part in this shocking destruction.

Meitner's news of the nationwide university dismissals came as a shock to Hahn, but he was still optimistic in public. In a lengthy interview published in the *Toronto Star Weekly* on April 8, 1933, he denied American newspaper reports which, to him, seemed to "sensationalize German persecution of the Jews." In this same interview, he voiced his *hopes* for the "new Germany" he had read about in letters from some of his friends and colleagues.[14] This comment amazed and outraged some of his peers. Hahn hedged his public expressions for a reason; he knew that his trip was being scrutinized, and under his expression of optimism and support for the "new German society" lay his hope, expressed to American colleagues in private, that someday Hitler would be deposed. A few days after this interview, a more somber Otto Hahn learned that the *Reichskommissar* of the Prussian Ministry of Education had moved swiftly to act against Albert Einstein, who at this time was also on a lecture tour in America. Einstein's "crime" was that he had been vocally criticizing the arbitrary and racist government pogroms as well as persecution of intellectuals.[15] Hahn kept quiet during the rest of his visit in America, although the press continued to pursue him.

In her letters of April and May 1933 to Hahn in America, Lise Meitner alerted her long-time colleague that a fairly large "National Socialist cell" had developed in the Institute, and that everything now functioned "according to law." He also learned the sobering details concerning academic dismissals throughout Germany from members of the Kaiser Wilhelm Institutes administration. Meitner also sent an urgent letter to Max Planck, then vacationing in Sicily, about the Civil Service ordinances of April 1933, and the immediate gutting of entire academic departments.[16] Planck was informed that Fritz Haber, Jewish by background, German by loyalty, had been dismissed from his other academic posts on April 1, 1933, and intended to resign from his Directorship of the KWI for Physical Chemistry as well.

Younger members of the KWI began to feel the Nazi sting. Chemists still new in their careers wrote a strong letter to Haber, protesting their inability to "work with a government demanding prescribed racial antecedents for scientific work." On April 21, Haber received a call from the Prussian Ministry of Science, Art, and Education advising him that work at the Institute could not continue "with the present staff." Chemists A. Freundlich and Michael Polanyi were among those ordered to vacate their posts immediately. Haber, undaunted, immediately replied in writing that in this case, he too would resign, whereupon the officials retreated. "You are a famous man. We could not think of such a thing," came the Minister's apologetic response.[17]

But Nazi officials were not impressed by seniority, veteran status, or scientific reputation. On April 23, Friedrich Glum, acting Administrative Director supervising all of the Kaiser Wilhelm Institutes, called on Haber to say that the entire affair had been a "mistake." Only lower-level Jewish assistants had to be released; but

Freundlich and Michael Polanyi (the latter became famous for his work in physics as well as philosophy) resigned under protest. Hence, Haber saw no other way to protest the racist policies than to resign himself. He too urgently wrote to Max Planck, asking for advice and the date of his return from Italy.

Haber sent his official letter of resignation to the Society on April 30, requesting release from his duties on October 1. Although Kaiser Wilhelm Institute members were not officially exposed to the Civil Service laws, as were all state-salaried university professors, Haber wrote:

> According to the law of April 1, 1933, I am derived from Jewish parents and grand-parents.... I tender my resignation with the same pride with which I have served my country during my lifetime.... For more than forty years I have selected my collaborators on the basis of their intelligence and their character and not on the basis of their grandmothers, and I am not willing for the rest of my life to change this method which I have found so good.[18]

The Institute for Physical Chemistry had been under Haber's direction since its official opening by the Kaiser in 1913. A controversial figure in the international scientific community after the First World War because of his work and research on poison gases, Haber was nevertheless highly respected by German physicists and chemists, and loved by the many postdoctoral assistants who worked for him at the Institute for Physical Chemistry. His work on nitrates assured fertilizer production for generations, and his fame had spread for these discoveries. He was respected by all, including Max Planck. Planck was then 75 and needed more rest and time away from his duties, but he was still President of the Kaiser Wilhelm Society. At Meitner's insistence, he returned to Berlin at the end of April to attend to Haber's resignation. Meitner hoped that the Nazi Ministry of Education would respect Planck and reverse the dismissals of many younger colleagues. She heatedly discussed this with von Laue as well. Without Hahn at the helm of the Institute, her own office and lab had also become a focus of discussion with university members coming and going in their uniforms. Historian Fritz Krafft points out:

> No one but Lise Meitner, on the basis of long years of cordial friendship with her old teacher [Planck] and because of her courage and feeling for fairness, seems to have been capable of the "presumption" demanded by the times. She did *not* have a position that allowed her to take any action, and as a "Jewess" she was hardly able to satisfy her feeling of equity by action. But haste was needed, and her friends Max Planck and Otto Hahn returned too late to reverse what had happened.[19]

They all discussed Haber's resignation. Planck seemed locked in his loyalty to idealism. Meitner recalled: "Planck once said, truly desperately, 'But what should

I do? It is the law.' And when I said: 'But how can something so lawless be a law?' he seemed visibly relieved."[20] However, even the President of the Kaiser Wilhelm Society could not halt Hitler's decrees.

The sudden purge of the universities and civil service caught scientists at different stages of their careers. Lise's nephew Otto Robert was working with physicist Otto Stern at the University of Hamburg, and was scheduled to continue postgraduate research at Enrico Fermi's laboratory in Rome. He had received the coveted Rockefeller International Education Board fellowship for travel to and residence in Italy. Under the new German laws, however, Frisch could not leave the country because he would have no job upon return; ironically, a contracted permanent position was one of the requirements of the fellowship, and no Jewish scientist was eligible for hire in Germany after April 1933. Many years later Frisch recalled:

> (When) Hitler's racial laws came into force, the University of Hamburg, rather later than most others, finally had to sack its honorary [Ph.D.s working on postdoctoral research] staff members, myself included.[21]

Meitner, distressed that her nephew had to cancel his trip to Rome, nevertheless assured her sister by telephone that a good research position might be found for 29-year-old Otto Robert. Privately, she was frantic for his safety. James Franck and Niels Bohr also began lobbying for him with American and Dutch physicists.[22] But it was not until half a year later, in early autumn, that Lise Meitner was assured of Frisch's safety and funds for departure: he was offered a grant by the Academic Assistance Council of Great Britain, and a place with Blackett at Birkbeck College in London.[23]

Despite the political turmoil, Meitner decided that it was her duty to remain at the Institute for Chemistry and continue her leadership there. As an Austrian, she was assured, the Civil Service Laws could not touch her position, and besides, she had Planck's support, colleagues counseled her. Other Jewish or part-Jewish scientists, including Bohr in Copenhagen, pressed scientists around the world for assistance for dismissed German colleagues. Frisch had met Bohr in Hamburg before leaving for London. They spoke intimately of Lise Meitner, Otto Hahn, and the situation under the Third Reich. Frisch recalls:

> Hitler had come to power and Bohr was traveling about and talking to his colleagues in order to find out how many German physicists would be dismissed under the new racial laws, and how one could best organize help for them. I had just succeeded in measuring the recoil which a sodium atom suffers on sending out a light quantum. To me, it was a great experience to be suddenly confronted with Niels Bohr, an almost

legendary name for me, and to see him smile at me like a kindly father; he took me by my waistcoat button and said: "I hope you will come and work with us some time; we like people who can carry out thought experiments!"

You see, the recoil of an atom had been discussed a good deal in the early days, but few people thought that it could be measured. By the time I did that, it was no longer difficult; Otto Stern, under whom I worked in Hamburg, had created the necessary techniques. That night I wrote home to my mother (who was naturally worried about my future) and told her not to worry — the Good Lord himself had taken me by the waistcoat button and smiled at me! That was exactly how I felt.[24]

Few young German physicists were as fortunate as Frisch in finding research opportunities abroad. Most remained out of work, waiting and hoping – like Meitner – that the political mess would soon "blow over." But many, many other Germans *knew*, as the industrial machine geared up for armaments production and profits rose through military contracts and expenditures, that Hitler's internal policies were creating an economy destined for future expansionism and war.[25]

The English and Danes had to face the German situation earlier than most other countries. German scholars flooded them with letters, petitioning for aid and academic assistance. Grants were initiated, but not all scholars were "permitted" to leave Germany and accept offers elsewhere. In his essay on refugee physicists in America, Charles Weiner observes:

> The unprecedented situation vividly impressed scholars and others abroad on May 19, 1933, when the *Manchester Guardian* devoted a full page to a list of the professors who had been dismissed from their posts in German universities during the three-week period between April 13 and May 4. The list covered institutions from every part of Germany and read like a Who's Who of scholarship. Compiled for the most part from announcements in [the] German press, it included 196 entries and was by no means complete.[26]

By mid June, things became even worse. On-the-scene reports by visiting scientists from the US described the despair of those affected by the racial laws and alerted their colleagues in America and abroad to the growing crisis. Selig Hecht, an American biologist on leave from Columbia University in 1933, visited Germany in June. In a letter to a colleague, he wrote:

> It is really a terrifying situation when viewed close at hand. Not merely the obvious agony which the dismissed Jewish professors and others are up against, but the extraordinary callousness of the rest of the population in the face of it. Most people don't give a darn; a large proportion is rather glad it all happened. Those extremely few who are upset by it are disinclined to do or say anything publicly or privately. In Munich everyone says not to speak too loudly or one will land in Dachau.[27]

In another letter written on the same day, Hecht noted:

> With the exception of a few outstanding instances of men who have been untouched or who have been dismissed, they must absent themselves from their institutes, they cannot give lectures, and they cannot have any contact with their assistants and graduate students.... The situation of the *Privatdozenten* is much simpler...they are being rapidly dismissed since they have practically no tenure. The assistants are worst off. They have usually been given one or two month's notice and have been dismissed. People working on funds from the *Notgemeinschaft* are in the same situation. Many of the appropriations have already terminated and, like those which terminate July 31, are not renewable for Jewish applicants.[28]

Members of all the Kaiser Wilhelm Institutes found themselves in a strange predicament. Many Jewish staff members, including senior research scientists, had never sought university positions because of quotas imposed on Jewish applicants even before the Nazis came to power. Technically, employees of the industry and government-funded Institutes were not subject to the same civil service codes and law affecting university members. Yet an American observer commented on the situation he observed at the KWI for Physical and Electro-Chemistry in Berlin–Dahlem:

> ...one reason for [the] concentration of Jewish scientists at the Haber Institute lay in the fact that his able non-Jewish coworkers had been one by one picked off by this or that university...it was to meet just such a situation that the "Non-Aryan" clause in the new Civil Service Law was made. Consequently the regime asked Haber to exchange his "Non-Aryan" for "Aryan" research assistants; Haber himself was exempted because of his great services to German industry, especially during the war, when he succeeded in inventing a process which enabled Germany to do without nitrates from Chile (the hydrogen-fixation process) during the blockade. Briefly, the official view is that Haber was *not* dismissed because of state intervention but rather because of his own "obstinacy."[29]

Checkmate seemed inevitable. Lise Meitner viewed the situation differently: she felt particularly vulnerable when her own research assistants also came under racist scrutiny since, unlike Haber, she could not protest. She gladly left Berlin during a June break, and wrote to Otto Hahn in far-off America from Vienna:

> Many thanks for your card of May 24th. I traveled to Vienna for a few days and feel here how little one can *imagine* the atmosphere in Germany (even when one had personally experienced it) as soon as one is not in the country. This time is very difficult for Germany: the question of anti-Semitism is only one question — there are also serious, or even much more serious, questions and everyone who clings to Germany must be worried about how it all ought to be resolved.

You did not answer any of my recent letters: i.e., you never responded to what I had written so that I do not know whether you even received my letters. Now you will come back soon but you will surely need some time to be able to form a proper judgment. In such revolutions, the lives of individuals play less of an important role, but of course, it has also appeared who is "good and who is not" here. That my housemate belongs to the latter group did not surprise me. As for Planck, [von] Laue, Schlenk and many others, you can watch them with joy and admiration.

Until the holidays we were very busy in the Institute; you have probably seen that in the department letters. Franck was in Berlin once; it is surely very difficult not to be allowed to enter his Institute and unfortunately that is true for many others.[30]

Despite her courage, and the support of her senior colleague, Lise Meitner could not remain out of the clutches of the Nazi machine. That summer, she received a questionnaire from the Ministry of Education, pressing for details on her "racial heritage." She filled it out:[31]

(1) Name: Lise Meitner

(2) Birthplace and Date: Vienna, November 7, 1878

(3) Citizenship:

 (a) present: Austrian

 (b) at birth: Austrian

(4) Date of *Habilitation*: October 31, 1922

(5) Date nominated assistant, lecturer, or professor:

   Assistant to Privy Counsellor Planck, October 1, 1912

(6) Participation in the World War on the front for the German Reich or its allies: (a) From August 1, 1915 to November 1916, x-ray technician in Austrian hospitals in Lemberg and Trient

(7) Race of four grandparents: Non-Aryan

Signed,

Lise Meitner

Berlin, April 28, 1933

From that time, Meitner's presence as a "non-Aryan staff member" within the KWI became official record in the Prussian Ministry of Science, Education, and Popular Instruction headed by Nazi Party member Bernhard Rust. She had complied with the procedure of filling out the form so as to *not* make waves, still convinced that her Austrian citizenship would protect her.

Two years before Hitler's election, Einstein had drafted a long letter to Planck, which he never mailed; it was still in its original envelope when, in the spring of

1933, after Einstein's refusal to return to Germany, his papers were retrieved by diplomatic pouch through the French embassy. Einstein's foresight was clear. His letter read:

> You will surely recall that after the war I declared my willingness to accept German citizenship in addition to my Swiss citizenship. The events of recent days suggest that it is not advisable to maintain this situation. Therefore, I should be grateful if you saw to it that my German citizenship be revoked, and to advise me whether such a change will permit me to maintain my position in the Academy of Sciences (which I sincerely hope).
>
> Concern for the many people who are financially dependent on me, as well as a certain need for personal independence, compels me to take this step. I very much hope that you will understand and that you will not interpret this request as an act of ingratitude towards a country and an institution which have granted me enviable living and working conditions during the best years of my life. So far, I have always rejected offers from abroad, however tempting, which would have forced me to leave the scene of my work. I hope I shall be able to maintain this attitude also in the future.[32]

But during 1933 Einstein could not maintain this position. While on tour in the US, the news from Germany and the blockade of his return there dictated the necessity of extending his stay in America. Reluctant to lose all contact with his former life, he and his second wife, his cousin Elsa, decided to return to Europe in March. However, like countless refugees, they were undecided about where to settle. Finally, based on a personal friendship Einstein had developed with the King and Queen of Holland, he made the decision to emigrate to the Netherlands. In a long interview with Evelyn Seeley of the *New York World Telegram* on the eve of his departure from Pasadena, Einstein reflected:

> As long as I have any choice in the matter, I shall live only in a country where civil liberty, tolerance, and equality of all citizens before the law prevail. Civil liberty implies freedom to express one's political convictions in speech and in writing; tolerance implies respect for the convictions of others whatever they may be. These conditions do not exist in Germany at the present time.[33]

Einstein was to become *persona non grata* in Germany. His pacifist views and so-called "attacks" on the Prussian Academy of Arts and Sciences branded him an "enemy of the state." Professors who had lauded him as one of Germany's finest minds only a decade earlier now conducted a virulent anti-Semitic campaign against him. A local Berlin newspaper's headline read: "Good news from Einstein — he's not coming back." And a boastful editorial writer even bragged:

Relativity is in little demand by us now. On the contrary: the ideals of national honor and love of country which Herr Einstein wanted to abolish have become absolute values to us. So the outlook for Einstein here is very bad.[34]

During the spring of 1933 Einstein, like Meitner, looked to Planck as President of the Prussian Academy to somehow use his position of authority to counter such attacks against Academy members. But Planck was reluctant to act, and his stance to "uphold the ideals of Germany and her sciences" now cast his aloofness from politics in another light. In March of 1933, Einstein had written to him from Pasadena pleading that, as Chair of the Physical–Mathematical Section of the Prussian Academy, he [Planck] should consider a proposal by the astronomer George Ellery Hale of the Mount Wilson Observatory. Einstein affirmed in part:

> He said that he wants to make an attempt to cut out politics entirely from the work of international scientific cooperation. This is what he wants to do: First he wishes to set up a working committee of specialists from all countries, with the aim of transferring scientific methods from one discipline to another (for example, the transfer of physical methods into biological sciences). He naturally wants to give a share of this work to German research workers. But he does not want it to seem as if such invitations bypass, as it were, the German research workers. He wishes to attract from every country only those generally interested specialists who have no political axes to grind. He hopes that ready, well-disposed scientists will help him and also that the institutions of individual countries will approve of it, if one takes care to remove all political influences from the scientific work, and so to bring about again the state of affairs which was once taken as a matter of course.
>
> Herr Hale wants to know your view about this. He wants to know whether you would be prepared to submit the idea to the Academy in a friendly way. He thinks that I should write to you first since you, writing to me, can truly express your opinion. I will use your answer only in so far as you wish, and as it is needed to express your point of view.
>
> Please send your answer c/o Herr Ehrenfest, Leiden, Holland, as I am not yet sure where I shall pitch my tent.[35]

Planck's reply to Einstein, which was sent before the Civil Service Laws came into effect, arrived in Holland towards the end of April.[36] While Planck had, both in the Nobel process and university proceedings, considered himself a friend of Einstein, in a less than friendly manner stated that *resignation* from the Academy was the only way "to resolve the situation honorably" — and that this would "spare his friends immeasurable grief." No response was given to the suggestions from Pasadena: it was obvious that Planck was more concerned about Einstein's publicity than his continued international efforts to open physics to other disciplines. Soon a more formal, ominous announcement was issued by one of the permanent secretaries of the Academy, Dr. Ernst Neymann:

> The Prussian Academy of Sciences heard with indignation from the newspapers of Albert Einstein's participation in atrocity-mongering in France and America. An explanation is immediately demanded. In the meantime, Einstein has announced his withdrawal from the Academy, giving as his reason that he cannot continue to serve the Prussian state under its present government. Being a Swiss citizen, he also, it seems, intends to resign the Prussian nationality which he acquired in 1913 simply by becoming a full member of the Academy.
>
> The Prussian Academy of Sciences is particularly distressed by Einstein's activities as an agitator in foreign countries, as it and its members have always felt themselves bound by the closest ties to the Prussian state and, while abstaining strictly from all political partisanship, have always remained faithful to the national idea. It has, therefore, no reason to regret Einstein's withdrawal.[37]

Einstein's immediate rebuttal, and the many more that were to follow that year, ended on a note of protest and appeal:

> It would have been an easy matter for the Academy to get hold of a correct version of my words before issuing [that] sort of statement about me. The German press has reproduced a deliberately distorted version of my words, as indeed was only to be expected with the press muzzled as it is today. I am ready to stand by every word I have published. In return, I expect the Academy to communicate this statement of mine to its members and also to the German public before which I have been slandered, especially as it has itself had a hand in slandering me before that public.[38]

Indeed, the German public *was* something to be wary of that year. Some of Lise Meitner's colleagues did not wish to entangle themselves with the Nazis and simply avoided their Jewish colleagues; others, like Edith Hahn, later told friends that while persecuted, they should know that the Jews had "justice entirely on [their] side, while *we* bear the disgrace and inextinguishable, irreparable shame for all time."[39] However, few Germans admitted as such. There *were* those who were adamantly unwilling to see their names associated with the policies and practices of the new regime. Schrödinger, for example, prepared to leave Germany. He covertly communicated with colleagues in Great Britain and Ireland about his plans to relocate. Also, he could afford to: he was at the height of fame, having been awarded the Nobel Prize in Physics in 1933. Schrödinger first went to Oxford; three years later, he packed his books and journeyed back to Austria after he found scientific faculties and administrations in England deluged with academics fleeing Germany.[40] Austrians were proud to have a Nobel laureate back on their faculty. But by 1938,

Schrödinger would be forced to leave Austria, as would hundreds of thousands of other "undesirables" in Hitler's expanded Reich.

Upon Hahn's return in July 1933, he took a stand he felt would demonstrate his contempt of the regime: he withdrew from the faculty of the University as a sign of protest.[41] By then the colloquium had dwindled to a third of its size: Einstein's theories were not allowed to be mentioned, and few hard working scientists cared to travel to the university for lectures that sounded like diatribes. But no German scientists experienced the violent wrath of the Nazi Party as did those Jews who secretly or openly left Germany for good in the spring of that year. On April 2, Einstein's Berlin bank accounts were seized, his apartment door was padlocked, and shortly thereafter his beautiful summer house at Caputh was taken over. At his retreat in LeCoq, Belgium, a photo album was sent to him, containing pictures of "leading opponents of the Nazi Government." On the first page was his portrait.[42] The caption read:

> Discovered a much-contested theory of relativity. Was greatly honored by the Jewish press and the unsuspecting German people. Showed his gratitude by lying atrocity propaganda against Adolf Hitler abroad. [And then, in parentheses, were the words, *'Noch ungehangt'* — Not yet hanged.][43]

Colleagues and friends were stunned. Einstein would never return to Germany again, and spoke sharply against Hitler to the American press. The attacks grew worse. Self-declared "Aryan physicist" and German Nobel Prize Laureate Philipp Lenard, wrote in the *Völkischer Beobachter*:

> The most important example of the dangerous influence of Jewish circles on the study of nature has been provided by Herr Einstein with his mathematically botched-up theories consisting of some ancient knowledge and a few arbitrary additions. This theory now gradually falls to pieces, as is the fate of all ideas that are estranged from nature. Even scientists who have otherwise done solid work cannot escape the reproach that they have allowed the theory of relativity to get a foothold in Germany because they did not see, or did not want to see, how wrong it is, quite apart from the field of science itself, to regard this Jew as a good German.[44]

Proponents of "Aryan physics" had long claimed to have Hitler's backing, and the Führer never persecuted the physics community in a direct way. He had formulated his views on education and civil service in *Mein Kampf*: "as necessary as chemistry, physics, mathematics, and such subjects may be in this 'materialistic' era of technology, it is *dangerous* to devote more and more schooling to these disciplines. What Germany needs is education [*Bildung*] based not on the materialistic egotism fostered by science but on individual sacrifice to the community."[45] It was obvious, as the university was drained of its intellectuals, that "sacrifice" would fall on an entire generation of students who would be denied modern physics for over a decade.

With such rhetoric all around them, German scientists faced a bewildering morass of personal, political and special interest conflicts by challenging the dismissals of their colleagues. Some took a moral stand privately but did nothing publicly. The basic attitude was mirrored in Planck's "passive assumption" that service to the state through "dedication to the institutions of German science" was distinct from "service" to National Socialism. Others, like Lise Meitner, felt disgust for the Nazis, but felt that she could "hold center" amid the tempest.

Hitler's followers had not yet seized control of all of Germany's vital functions, so the pursuit of science continued to be the main priority for many despite the political climate. The dangers of this position would be revealed in the years to come. When asked by a colleague some time later why she did not return to Austria during the early 1930s, Meitner replied curtly that she had never held a post there . . . that only in Germany was she firmly established professionally and institutionally . . . that she had seen political turmoil after World War I . . . and that she would not let anti-Semitism drive her away from her scientific and administrative responsibilities. Others shared her sentiments.

Some counseled other measures for political protection. Heisenberg quietly discussed the Nobel Prize as potential immunity. In December 1932, Planck had nominated Lise Meitner and Otto Hahn for the 1933 Nobel Prize for their pioneering research in radioactivity and radioelements, months *before* the dangers stemming from Hitler's election arose.[46] Meitner was grateful for the nomination and waited anxiously throughout the year for the Swedish Academy's decision. She continued to turn to Planck for his support to remain in Germany and continue her research. Perhaps Planck felt, in his later years, that a Nobel Prize would be the best protection for Meitner against the Nazis. However, it was the team's pioneering work that most impressed Planck in 1932: he always made it clear that the Meitner–Hahn *team* deserved the Nobel Prize, and it was his duty to nominate the best Germany had to offer.

But on September 6, 1933, bad news arrived. Lise Meitner received an ominous notice from the Prussian Minister of Education that referred directly to the questionnaire she had so cautiously filled out that spring:

Immediately! For the reason set forth in paragraph 3 of the statute concerning the reinstatement of the professional civil service, I hereby revoke your authorization of professorship at the University of Berlin as of April 7, 1933.

Signed

The Prussian Minister of Education

Siegel[47]

Meitner was shocked. This blow was unexpected, and the fall semester was just weeks away. Nothing could be done, it seemed, to challenge the illegality of this act: after all, she was Austrian, she insisted to colleagues, and she had lectured at the university for over fifteen years! She telephoned Planck, but he did nothing. The question still remains: Why did Planck wait so long to petition university officials and other legal channels on Meitner's behalf to redress this dismissal? Did Planck quietly counsel her to remain at the Institute for Chemistry, a privately funded position, rather than face trouble at the university due to the issue of her Jewish heritage? Or was *confronting* the Nazi bureaucracy an act many Germans such as Max Planck wished to avoid?

By the time Hahn returned to Berlin, his Institute was in an uproar. His wife and son were upset, and Edith would soon suffer a nervous breakdown. For weeks, Hahn and Meitner were counselled by colleagues like von Laue to follow the rules and avoid direct confrontation with the Nazis: half their staff were now Party members. Hahn had resigned from the University, but Meitner would not let her university privileges be revoked without protest. She pressed Hahn to "officially" confront the bureaucrats in the Ministry of Education. As the situation intensified, and Hahn's official resignation from the University was accepted by Third Reich officials, he too petitioned his direct supervisor, President of the Kaiser Wilhelm Society, Max Planck, for support. As Hahn later recalled, "It is well known that, although we did not ourselves actually suffer from any dismissals [in the Institute], we were far from being enthusiastic supporters of the official policy."[48]

Finally, in a letter dated August 27, 1933 to the Prussian Minister for Science, Art and *Volksbildung* [Popular Education], Otto Hahn carefully addressed the withdrawal of Lise Meitner's lectureship privileges and long-term position at the University of Berlin:

> Highly esteemed Minister! In agreement with Geheimrat Planck, President of the Kaiser Wilhelm Society, who is now out of town, I would like to inform you of the following:
> It concerns the non-tenured full professor at the University of Berlin, Dr. Lise Meitner. Professor Meitner is not of pure Aryan descent. Nevertheless, I believe that the discharge

provisions of the April 7, 1933 law pertaining to (tenured) professional positions cannot be applied here for the following reasons: the provisions for carrying out this law of April 7, 1933 state that anyone who had fulfilled all requirements for obtaining his first regular position by August 1, 1914, can be considered equivalent to a regular (tenured) employee. Professor Meitner was a regular Assistant at the Institute for Theoretical Physics under Geheimrat Planck on August 1, 1914. This position was normally filled by an instructor/assistant professor because, in 1914, women could not yet receive the Habilitation. Besides her assistant's position at the university, Professor Meitner already had the regular position of a scientific member of the Kaiser Wilhelm Institute for Chemistry in Dahlem, and the Kaiser Wilhelm Society was already during this time closely connected with the Prussian Ministry of Culture. In any case, the position of Professor Meitner in Dahlem corresponded to that of a full professor at the university. During the war, Professor Meitner volunteered for front-line duty and held the position of X-ray technician in the Austrian army. She included a notarized copy of a document concerning this when she filled out the second questionnaire.

Professor Meitner enjoys a reputation as a leading radium researcher in this country and abroad, directly next to Madame Curie, the Nobel Prize winner in Paris. By chance, Professor Meitner is now on leave from the university for the following reasons:

She had announced a seminar on atomic theory with Drs. Szilard, Kallmann, and Philipp. Since Drs. Szilard and Kallmann were affected by the law pertaining to non-tenured lecturer positions, the seminar was canceled. On the basis of a ruling by the Ministry in July of this year, Professor Meitner and Dr. Philipp were placed on leave for the coming winter semester, although the latter is a pure Aryan and a veteran, and Miss Meitner too, according to the facts outlined above, should not fall under the provisions of the ruling.[49]

Hahn was indeed upset about the Ministry's harassment of his colleague; Planck also voiced his concerns in writing. In his letter to the Minister for Science a few days later, Planck was more direct than Hahn in getting to the heart of Lise Meitner's professional status in Germany. He wrote from his home office in Grunewald:

Dear Sir:

At the risk of telling you something you already know, I am taking the liberty of calling your attention to a matter which appears to me to be of grave seriousness in the interests both of the science of physics and of the Kaiser Wilhelm Society. I have been informed that the Ministry is currently discussing the idea of revoking the right of non-Aryan instructors to lecture. You have already graciously allowed me to submit to you a list of non-Aryan physicists whose continued functioning in their current roles appears of absolute necessity for scientific research. Of the names that appear on that list, beyond that of Max Born, I would like to emphasize especially only that of Miss Lise Meitner. I set aside the question of whether her authority does not already protect her — before the war she was employed as a regular assistant professor at the University Institute for Theoretical Physics in Berlin; even more important, she, along with O. Hahn, is the soul of the Kaiser Wilhelm Institute for Chemistry. That Institute owes those two its reputation and world renown; their continued association with the University is an unconditional prerequisite for the further success of its scientific work, not least in the interest of stimulating university instruction and the preparation of the next generation of scientists in the area of atomic physics through the employment of doctoral students. She is a primary authority in this field; her name has been respected in Germany and in other countries; for several years she has accepted distinguished invitations to lecture on science, among others in Copenhagen, Zurich, London, and Rome. Her departure would cause a sensation in far-reaching circles, no less than did that of (Dr.) Haber; in her case, the circumstance would contribute to the sensation that her resignation would, unlike his, not be a voluntary one. The foreign reaction to the situation may be judged from the fact that two Rockefeller fellows who had announced their intention of working this winter in Miss Meitner's Institute have withdrawn their applications "in consideration of the political situation." We hope most urgently, and are thus in agreement with the Chancellor [Hitler], that in good faith, we will return to the stable conditions which are, after all, absolutely necessary for the peace of mind which accompanies scholarly cultivation of knowledge. Those who are behind the incitement to disorder ought to be stopped.

I remain, with expressions of the greatest respect.

Very truly yours, Dr. M. Planck

cc: Professor Lise Meitner
Berlin–Dahlem, Thielallee 67
Return receipt requested.[50]

Petitions of this kind were to no avail. Arguments based on past service or a sense of honor did not sway the Nazi bureaucracy; Planck's petition went unheeded. Party membership and well-placed informants of the Reich had become the new "authorities," and pressed the Institute for Meitner's resignation. Meitner had delivered her last University lecture in the fall of 1932; her spring 1933 lectures in conjunction with Szilard were canceled and her fall class dropped.

Institutes throughout Berlin–Dahlem had to regroup in the wake of the purges. Planck asked Hahn to serve temporarily as the provisional head of Fritz Haber's Institute, while the Kaiser Wilhelm Society members sifted the ranks to find a "suitable" replacement for Haber. Hahn bore the extra work under strain; by autumn, he also tried to lobby for financial compensation for Jewish colleagues who had been dismissed from Haber's Institute. On October 16, 1933, he wrote to Planck:

Dear Geheimrat,

You ... charged me with the leadership of this Institute. Now that my responsibility for the Institute has come to an end, please permit me to refer once more to some points, the successful settlement of which seems important to me for political and human reasons.... They mainly concern a number of employees who were dismissed on the grounds of the law concerning Civil Servants [which pursuant to S-14 was accordingly applied also to employees], whose change to other livelihoods would be facilitated by severance pay....[51]

Hahn's plea and repeated attempts to seek "proper" administrative redress were not effective. As Institute Director, he continued to rely on established channels of recourse and the civil procedure of administrative petition. But soon he himself became falsely allied with the propaganda that spread quickly across Germany in the darkening years of the 1930s. He later recalled:

> There was at one time in the Third Reich a traveling anti-Semitic exhibition called "The Wandering Jew," which displayed among other things a list in which my name was included of dismissed Jewish professors. When, to their horror, our general administration heard of this they asked me, in great consternation, if it was true that I was "non-Aryan." The explanation was quite simple — in 1933 I had resigned my lectureship at the University of Berlin to avoid having to attend the many official party meetings. In doing this I risked nothing, since as a member of the Prussian Academy I had the right to lecture wherever I liked. As a matter of fact, I had known my name was on the "Wandering Jew" list, but I had not bothered to do anything about it.[52]

Hahn's spirit of public compromise rather than righteous indignation was demonstrated again and again, yet little could be done about smears to his own reputation. Others fared far worse. It is during this period that his acts and opinions provide a telling insight into the hesitancy on the part of many educated Germans towards becoming "overly involved" in the growing frenzy around them.

The employment gaps created by the exodus of Jews from Germany increased and allowed many to take advantage of the situation. Fritz Krafft states bluntly:

> Too many people saw an unexpected chance for furthering their own personal careers by filling the gaps left by dismissed Jewish staff, feared for their own fortunes, or were enthusiastic about the new national ideology with all its consequences and brutalities. And far more people thought that way than were ready to admit it later on.[53]

Shoring up the Institute for Chemistry against the rising Fascist tide proved a difficult job. The head of the Guest Department in the upper story of the Institute for Chemistry, Kurt Hess, was a National Socialist Party member. Hahn recalled that Hess, whose "hope of inheriting the entire Institute had become even greater after the events of 1933, was hardly willing to make concessions."[54] Hahn continued to watch out for his administrative attacks. The Institute staff was divided, Hahn also remembered: "The representatives of the party, [Kurt] Philipp and [Otto] Erbacher, cut a great figure politically at that time and the rest had to hold back."[55] Gottfried von Droste, a chemistry assistant to Hahn, had joined the S.A. in 1933 and usually appeared in uniform at staff functions. Bernhard Rust, Prussian Minister of Science, Art, and Education had announced in the press that "it is less important that a professor make discoveries than train assistants and students in the proper views of the world."[56] Interpreted bluntly, his formal message was that faculty politicize their teachings into ideology for the state. Meanwhile, Lise Meitner watched as one by one graduate assistants and doctoral candidates began leaving the Institute after the universities were purged, or shifted their time into political meetings, not after-hours research.

Meitner was now entirely cut off from university participation. Both her work and influence were confined to the first floor Physics Section of the Institute. Hahn recalled that although she had retained her position in the Institute, "where she was not disturbed at all," she could no longer play any role in the scientific community "outside," and "she could no longer go to the University physics colloquia of [von] Laue."[57] Despite feelings of depression, she maintained her daily research schedule at the Institute. One of her brightest postdoctoral assistants, Max Delbrück, also remained at the Institute, and later fondly recalled Meitner's habit of "working her assistants into shape."[58] They all were working extra hours late into the night. Her frustrations, of course, stemmed from the overall situation, and were not directed towards her students in general, for they too were undergoing tests of endurance in trying to pursue academic careers. Max Delbrück later recalled:

> While I was the assistant of Lise Meitner I also tried to become a lecturer at the University; it was not really a job because you didn't get paid for it — you got "permission to lecture." This procedure was made more complicated very quickly as the Nazis divided it into two steps. One, you were supposed to get an advanced degree, the Dr. *habilitation*; that means essentially presenting all the publications that you made, [demonstrating] that you are scientifically, scholarly, qualified. In addition, you were, however, supposed to also pass some political tests. To do so you had to go to a *Dozentenakademie*, an indoctrination camp, which was quite a fascinating thing. We had "free discussions," (lectures on the new politics and the new state) and after three weeks of "free" discussions they decided whether you were sufficiently politically mature to become a lecturer at the University. I went to two of these. The first one, Nazi Party members had run themselves, and that was a lovely place opposite Kiel.
>
> There we lived on a very nice estate with a large park; they housed us all, and we also had our daily discussions and other exercises and social events. There were about 30 of us, I would say, three of us to a room. And in a way it was a marvelous thing, because it was the first time in my life I got thrown together closely with people from other disciplines. I was together with an economist and a psychiatrist and we got to know lots of other people. I learned more about other sciences at this academy and at the next one

than anywhere else. But of course there was also the business of having these wonderful lectures by reliable party members, and everybody was terribly nervous because you really didn't know what was going on, and what you could say and couldn't say, and so forth. Anyhow I obviously was too incautious, and I was informed afterwards that I wasn't quite mature enough but that I could try again.

So I tried again. The next time it was in an equally beautifully place, Thüringen, which is in the central part of Germany [later East Germany], beautiful mountains there. There things ran much more smoothly; everybody knew by then what he could say and couldn't say and everything was much more relaxed. But still I must have shot my mouth off. It must have been apparent that I wasn't in great love with the new regime, so I don't know whether I was officially informed that I wasn't mature enough, or whether they just didn't answer my letters.... Anyhow, it was pretty clear that a University career was not likely to be open for me. I went to considerable lengths to prove that I was not Jewish (also part of the business), which involved supplying real authenticated copies of all the baptismal certificates of your four grandparents, and their Christian marriage certificates, maybe even to the great-grandparents. So when (an opportunity for) a Rockefeller Grant came around, it seemed like a good idea to see something of the world and see what was going to happen — because at that time it was anybody's guess how long the (political) mess was going to last. Some people said six months and some people said much longer. I was immensely lucky that I had this opportunity.... Many nasty things have been said about those who could have left and didn't leave, like Heisenberg, he's the most outstanding case; I don't agree at all with these derogatory comments. I don't think that it was anything to my credit that I left at all. I think it was a question which could be answered one way or the other, and there is great merit on both sides.[59]

Delbrück's experiences are especially revealing in the context of Lise Meitner's career. It was then twelve years since she had given her inaugural lecture on cosmic physics and was granted Lectureship status in the university. But with University teaching denied, she too had to face the possibility of leaving Germany. Her only avenue at this time was to continue her daily research at the Institute. Fritz Strassmann, an analytic chemist who had begun to serve as primary assistant to Hahn and Meitner, was also feeling the political pressure: when offered a position in an industrial firm, he was required to become an official Nazi Party member. He refused, and consequently lived on the scant quarter-time income of a first-level assistant at the Institute throughout the 1930s.[60] Strassmann and Meitner used to discuss potential research together over coffee while Nazi members were absent from the Institute's common room. Both felt that their on-going research work at the Institute was vital and agreed with Hahn that they *would* remain, despite their tremendous financial difficulties and carry out their radiochemical analyses under the auspices of the politically quiet Institute. Later, when Strassmann began to hide in his apartment a friend who feared for his life, the strain became even greater. He was to become an invaluable aide later on.

Reactions outside of Germany to Nazi actions were mixed. An American observer from Harvard, E.Y. Hartshorne, was shocked by what he saw, yet he noted:

> Whatever the personnel losses in the practical sciences, it is improbable that the subject-matter has undergone much change. The Nazis know as well as the industrialists that there is no substitute for scientific technology; they are merely less keenly aware than their business associates of the rarity of genuine inventive talent. There is probably an exaggerated impression abroad of the extent of Nazi influence in these fields, because of the currency of such terms as "German Science" as opposed to "Jewish Science."
>
> In a number of cases scientists dropped from State research or educational institutions have been taken over by the laboratories' business concerns. In other cases the displaced men have received unsolicited support from certain industrial leaders, who sought in this unostentatious manner to maintain the scientist's capacities for work and even to induce him not to accept offers from abroad.[61]

Hartshorne, however, had not taken into account that many businesses run by Jews were being boycotted or even destroyed, and that Germans were pressured not to hire "non-Aryans," even those professionally trained in the sciences.

Many Americans during these years deliberately attempted to avoid "the politics of the situation in Germany," and took a non-interventionist position. Yet most academic departments were deluged with petitions for assistance. Scholars abroad often emphasized the need to "uphold the ideal of the international nature of scholarship," but when the realities of hiring new colleagues or assisting emigrants arose, some turned a cold shoulder. Economic problems beset everyone during the Great Depression, and few knew where to turn for financial assistance during the 1930s.

Vague proclamations were issued by professional organizations but did little to counter Hitler's policies. In the spring of 1933, for example, the American Association of University Professors issued a public resolution that was transmitted to the Committee on Intellectual Cooperation of the League of Nations:

> The council has no wish to express any opinion on the political life or ideals of any nation, but science and scholarship long since have become international, and the conditions of intellectual life in every important country are a matter of legitimate concern to every other.[62]

This statement was interpreted as an expression of solidarity with "members of the profession who have been subjected to intolerant treatment in these difficult times." But abstract sympathy was all the League of Nations' Committee could offer; funds were scarce, and colleagues who had sent food packages and other goods to Europe at the end of World War I now had troubles of their own.

Many in Europe and Great Britain forecast a seriously dangerous situation and feared that the very freedom of science had been threatened or destroyed. In 1934, a group of 14 European scholars from various fields issued "a call to scientists" in which they condemned the uses to which science was being limited under the Third Reich:

> In that country [Germany] the exact sciences have been openly degraded to jobbing for war industries. During the education of young physicists and chemists, much time is devoted to lectures and practical exercises in "defensive science:" gas protection, air protection, study of explosives, war intelligence service, which have no relation to the scientific significance of this field. Moreover, only such investigations are favored which are likely to bring about a direct technical advance . . . .[63]

Charles Weiner notes that the initiation for such a call to scientists to protect "free international science" was induced by a statement from the Nazi Minister Rust, who maintained that "National Socialism is not unscientific, only hostile to theories." This twisted logic was disturbing, and elicited this response from the "Call to Scientists" signers:

> . . . all branches of physics which cannot be made to serve political and economic imperialism are therefore hampered and restricted. Studies which have contributed essentially to the broadening of our concept of the physical universe are thus thrust aside openly as vain and fruitless intellectualism.[64]

Signed by J.B.S. Haldane, Ernest Rutherford, Paul Langevin, Julian Huxley, and many others, the position paper denounced Nazi-style eugenics, the pseudoscience created "in the service of racial doctrine." Such practices were directed towards the "development of German character and will," and collided with the international scientific community as German bureaucrats tried to harness science to "serve" the State's racial programs.

But the international scientific community had not yet been entirely closed off to freethinking Germans. In August 1934, Meitner and Hahn, together with H. Wieland, O. Honigschmidt, Friedrich Paneth, and other select scientists from the Berlin circle, were invited to the Soviet Union to attend a large convention. Meitner, feeling left out from the exchange of ideas at the University for over a year, was happy to prepare to travel abroad. Her work had taken a dramatic turn when Wolfgang Pauli, proposing what he humorously labeled a "desperate remedy," had challenged the "continuous" primary beta spectrum results emerging worldwide, and postulated its cause: an entirely neutral particle that was dubbed the neutrino.[65] There was still much to discuss with colleagues abroad about such a new particle, since Pauli had not yet published his results. Russian colleagues agreed and looked forward to the arrival of the German contingent, and Meitner packed many notebooks with the latest findings to share with her physics colleagues.

When the paperwork and visas were complete and Kaiser Wilhelm Society approvals finally obtained, the group sailed by ship for the Soviet Union under the watchful eye of Nazi Party officials, who were highly suspicious of "Communists" in and out of the Reich. Before the International Mendeleev Conference gathered, the Berlin group toured Moscow and Leningrad for two weeks, visiting the Radium Institute founded upon the Curies' work and animatedly discussed Fermi and Bohr's work with curious Russians. They also found translators for their own publications, rebuilding bridges that had been burned when Hitler came to power. The conference began in mid-September. A somber Otto Hahn later recalled:

> For us "Western capitalists" the glimpse we had of Russia at that time was most instructive. There was in general still great scarcity of all ordinary goods, but one could see that the country was on the way up. What particularly disconcerted us was the exaggerated anti-religious propaganda in the churches: what one saw in place of sacred images were posters, partly derisive, partly obscene, directed against the priesthood.[66]

During the Congress and through the evening hours, in poorly lit lodgings and under terraced cafes, Hahn, Meitner, and their colleagues shared views on politics and research, and openly discussed Hitler. Hahn's family meant the world to him, and he informed his colleagues that they had decided to weather the political storms and remain in Germany. If the grandeur of "old Russia" was impressive, the fact that their Soviet colleagues lacked modern equipment, laboratories, and the political wherewithall to lobby industry was very surprising to the well-equipped German contingent.[67] There was no comparison with their Berlin state-of-the-art laboratories, they agreed privately over scant meals. The sympathetic contingent felt that they could perhaps aid their Russian colleagues through shipments of equipment and supplies once the Nazi era had passed. None foresaw, however, that there would soon be a complete blackout of news and mail to the Soviet Union, a foreshadowing of the dreadful years of war between the two nations.

Immediately upon their return to Berlin, new challenges awaited Hahn and Meitner. In Rome, Enrico Fermi had published results that were puzzling and surprising. "I remember our coworker Max Delbrück," Hahn later reflected, "expressing his amazement after receiving the exciting news about Fermi's work in Italy that we could even sleep a wink before trying to *repeat* the experiments."[68] The article's title even eluded to "potential" new elements: Fermi had hedged his bets, entitling it "Possible Production of Elements of Atomic Number Higher than 92."[69]

Hahn and Meitner quickly got back to work, assigning Strassmann several chemistry tasks. He had been on the Institute staff since 1929, and now took over the careful chemical analysis of specific by-products. The experiments Delbrück was referring to were based on a bombardment of the nucleus, a physics experiment Meitner had also initiated in the early 1930s, utilizing alpha particles to penetrate the nucleus of known radioelements. Strassmann recalled her work clearly, and was a solid supporter as they combed through the Rome findings. The "exciting news" was that Fermi's group, by bombarding uranium with a neutron source, were claiming (in *Nature*) to overcome the Coulomb barrier, the electrical force that we know today repels incoming particles. As Fermi stated:

> Until recently it was generally admitted that an atom resulting from artificial disintegration should normally correspond to a stable isotope. M. and Mme. Joliot first found evidence that it is not necessarily so; in some cases the product atom may be radioactive with a measurable mean life, and go over to a stable form only after emission of a positron.
>
> The number of elements that can be activated either by the impact of an alpha particle (Joliot) or a proton (Cockcroft, Gilbert, Walton) or a deuteron (Crane, Lauritsen, Henderson, Livingston, Lawrence) is necessarily limited by the fact that only light elements can be disintegrated, owing to the Coulomb repulsion.[70]

This fact would be key in thwarting Lise Meitner's search as well, and five years later, she would return to this puzzle before deducing the process of the nuclear fission of uranium. But Fermi's challenge in 1934 was this: Was the process of disintegration involved in creating the so-called "transuranic" elements? He ended his article with the following:

> The possibility of an atomic number 94 or 95 is not easy to distinguish from the former, as the chemical properties are probably rather similar. Valuable information on the processes involved could be gathered by an investigation of the possible emission of heavy particles. A careful search for such heavy particles has not yet been carried out, since they require for their observation that the active product should be in the form of a very thin layer. It seems therefore at present premature to form any definite hypothesis on the chain of disintegrations involved.[71]

The neutron itself, a nuclear particle with mass but no charge, had been difficult to obtain as a bombardment "bullet," but Fermi had produced a new neutron source using radon and beryllium. Meitner urged Hahn to collaborate with her once more.

Twelve years had passed since they had last published a joint article — now Fermi's artificial "radio-nuclides" demanded both their chemistry and physics expertise. Upon their return from Russia, it had been Delbrück who urged her to follow up on Fermi's "challenge." Hahn later recalled the fall of 1934:

> I do not know whether Delbrück's remark alone would have been enough to make us put aside all the work we were engaged in. But when my one-time coworker Aristide von Grosse, who was by then in the US, raised the objection that one or even both of the thirty- and ninety-minute substances that Fermi had declared to be "transuranic" elements were nothing of the sort, but just isotopes of protactinium (that is, of element ninety-one), we felt ourselves bound to find out which of the two was right, Fermi or von Grosse.[72]

Meitner and Hahn, codiscoverers of the element protactinium (element 91) in 1917, were of course familiar with its chemical properties, and set about the task of ascertaining whether or not Fermi's 13-minute element was a protactinium isotope. It was decided that Fritz Strassmann should be brought into the project as an assistant; late in 1934, Strassmann began the careful chemical analysis of the products emerging from the neutron irradiation of uranium and thorium.

Radiochemistry, beginning in the early 1930s, had been transformed into nuclear chemistry, changes largely brought about after Fermi's neutron bombardments produced some "transuranic" elements: that is, elements "beyond" uranium. In the search for these transuranics, the Hahn–Meitner team was one of the few in the world prepared to confront the chemical and physical difficulties involved and still stay abreast of the findings coming from the Joliot–Curie laboratory in Paris, Lawrence's lab in Berkeley, and the Fermi lab in Rome.

But the political realities presented major problems for Hahn and Meitner. Hahn, it is true, had resigned from his university lectureship, but he was able to continue to deliver lectures during the winter semester 1933–34, summer semester 1934, and late winter semester 1934–35 on themes relating to the team's ongoing radiochemical investigations.[73] Lise Meitner, however, was granted no access to these lectures. The only way she was able to keep informed of research in Berlin was through reports from Hahn or phone calls with international colleagues.

Hitler ordered mandatory military training at German universities and continued his rhetoric against intellectual reason. Biographer Allan Bullock writes:

> Hitler always showed a distrust of argument and criticism. Unable to argue coolly himself, since his early days in Vienna his one resort had been to shout his opponent down. The questioning of his assumptions or of his facts rattled him and threw him out of his stride, less because of any intellectual inferiority than because words, and even facts,

were to him not a means of rational communication and logical analysis, but devices for manipulating emotion. The introduction of intellectual processes of criticism and analysis marked the intrusion of hostile elements which disturbed the exercise of his power. Hence Hitler's hatred of the intellectual. To Hitler, the German masses would triumph over the intellectuals in a "historic racial purge": he stated, "instinct is supreme and from instinct comes faith.... While the healthy common folk instinctively close their ranks to form a community of the people, the intellectuals run this way and that, like hens in a poultry yard. With them it is impossible to make history; they cannot be used as elements supporting a community."[74]

Hitler had clearly stated his hatred of intellectual thought in speeches taken directly out of *Mein Kampf*, written in prison ten years before:

> False ideas and ignorance may be set aside by means of instruction, but emotional resistance never can. Nothing but an appeal to hidden forces will be effective here. And that appeal can scarcely be made by any writer. Only the *orator* can hope to make it.[75]

Clearly, propaganda was the heart of the Fascist movement. Growing numbers of students, staff, and fellow academicians cast their lot with the Nazi platform. Professors, even in science departments, were often unheeded, and oratory became common for physicists who, Meitner shrewdly noted in private, "couldn't keep up with the advances of quantum physics."[76] Nobel Laureate Johannes Stark gave *speeches*, not physics lectures, and allowed his prestige to be used by younger Party members to draw large crowds in the university's lecture halls. "Aryan physics" began to replace modern science as the curriculum shifted from science to rhetoric.

The time had long since passed for leaders to speak out. In early 1935, Max Planck decided that the time had come to voice his views again publicly. His appeal to reason, delivered in the Kaiser Wihelm Society's lecture hall, was deeply felt by at least one member of the audience who kept herself inconspicuous: Lise Meitner. Years later, she recalled Planck's idealism:

> He himself repeatedly emphasized the noble meanings of truthfulness and justice. So it was in a lecture held in 1935 "Physics in Struggle for a Worldview." He stated: The scientific irrefutability and consistency of physics comprises the direct challenge of truthfulness and honesty. Justness is inseparable from devotion to truth. In the same way that the laws of nature are logically consistent in the great as well as the small, so too does a society of people require the same privilege.... Woe to a community when its sense of the guarantee of justice begins to vacillate, when consideration of position and descent [i.e., ancestry] play a role in lawsuits.[77]

Meitner's loyalty to Planck never wavered. During the last years of her life, she reflected:

[Planck's] words, delivered in a lecture accessible to the general public, were definitely a warning regarding the National Socialistic methods of justice and must absolutely have been so perceived. Without a doubt it cost Planck a definite effort to overcome his inborn reserved nature, and to articulate these words in a public lecture — but his inner truthfulness compelled him to do so.[78]

Virtues such as these remained within the small community which continued to gather. The physics community and scientific circles in Berlin were Lise Meitner's "family," and Institute matters and affairs were *her* affairs. After extensive interviews with former staff members thirty years after the war, Fritz Krafft wrote:

> Members of the staff of the Institute never went to Hahn with their personal problems, but to Lise Meitner, and if he ever heard about these matters, it was for the most part Lise Meitner who had told him! She was the true life and soul of the Institute and her motherly care for the welfare of the members of her former staff is expressed in many of her letters to Otto Hahn.... Lise Meitner was the one who could induce Otto Hahn to grant the penniless Fritz Strassmann a support of DM50 per month from mid-1934, taken from a private fund he had at his disposal for special contingencies. As Strassmann never talked much about his personal affairs, his "boss" probably did not even realize his need.[79]

Many German women, like Meitner, who had been forced out of their professional positions or decision-making roles, continued to help friends, fellow colleagues, and others hiding "behind the scenes." By 1935, Meitner supervised postdoctoral assistants Arnold Flammersfeld and Kurt Sauerwein, as well as Gottfried von Droste (in his brown shirt SA uniform), among others. But she also prepared food packages and wrote numerous letters for those less fortunate than her German coworkers. Women under Hitler were more and more being regulated out of the work place and into their so-called "natural roles" as baby-makers for the Third Reich. This infuriated Meitner and others who had struggled so long to enter the professional realm, and now found themselves even more conspicuous as women in the shrinking work force.[80]

The right wing tightened their control over the other Institutes in Dahlem by appointing a Nazi sympathizer in Haber's place. Meitner and Hahn were among the many who were shocked by Haber's dismissal. A number of scientists attempted to petition the Ministry of Education on his behalf, but Nazi officials refused to recognize appeals for the "dirty Jew." The "official version" for Haber's dismissal, the American historian E.Y. Hartshorne noted, was that Haber was:

> a competent scientist . . . unfortunate in that he was Jewish and had stocked the Kaiser Wilhelm Institute of Physical and Electrochemistry, of which he was for twenty-two

years the Director, largely with Jewish scientists, thus blocking the path to advancement for many promising German ("Aryan") scientists.[81]

Meitner's urgent appeals to Planck resulted in his attempt to officially intervene with the government concerning the unjust dismissal policy. As always, his approach was a conservative one. He felt Haber "deserved well of Germany" (although he had not spoken this way concerning many others dismissed that spring). Some modern historians see this as passive collaboration; others do not.[82] Planck, as President of the Kaiser Wilhelm Society, made a key decision: he would arrange to meet directly with Chancellor Hitler concerning Haber's dismissal.

Planck detailed this ominous meeting years later in an article, "My Meeting with Hitler."[83] Looking back, Planck recalled the contents discussed at the meeting of 1935:

> Hitler answered me literally: I have nothing against the Jews. But the Jews are all Communists, and these are my enemies. My life is against them.

When Planck brought up the fact that many "old families," representing the best of German culture, were Jews, and that one must differentiate between these groups, Hitler worked himself into a rage. Planck recalled him screaming:

> That is not right. A Jew is a Jew. All Jews hang together. Where one Jew is, there are others of their species.[84]

Hitler ranted on, leaving Planck sitting silently, until the physicist could take his leave. The fate of Haber and of Jewish people throughout Germany seemed sealed. Planck's silence continued for the next 15 years; he would not attempt to intervene again for any member of the Kaiser Wilhelm Society and lived quietly with his wife in their stately home until his death, several years after the war.

Fritz Haber, who had been ill with heart problems when he was forced to resign from his Directorship of the Kaiser Wilhelm Institute for Physical Chemistry, was offered a position in England. However, obtaining a visa was not easy. Hence, he decided to only visit England in late 1933 and then go on to Palestine, where his sister was living. But in January 1934, Haber died suddenly of heart failure in Switzerland before he could reach "the Jewish homeland." Hartshorne noted in journalistic fashion:

> The fact that Haber chose rather to share the fate of his fellow scientists rather than to continue to lead a staff composed of men, almost 90 percent not of his own choosing, awakens little official interest. But the fact that he died only months after his dismissal, in Switzerland [which made the protest even more dramatic], has induced the Government

to adopt a petty revengeful attitude to this man which is so at variance with justice that it is little wonder that some of the officials in the Ministry suffer from uneasy consciences.[85]

Throughout 1934, a number of influential people favored organizing a memorial ceremony to mark Haber's death. Max Planck took personal charge of the preparations and the invitations for the ceremony, which was to take place in early 1935, featured a long list of speakers. Invitations were sent out between the 10th and 13th of January 1935 to the German Physical and Chemical Societies in Berlin, whose members were among the most eminent in the German scientific community. Planck and others wished to honor in a formal way the scientist whom they could not save. Many cautioned *against* such a public memorial ceremony, warning of its dangers. However, in quiet circles, most Berlin scientists agreed it was their best means of protest, and a fitting memorial to a well-respected colleague.

On January 17, a widely read Berlin newspaper published a notice announcing the ceremonial program. First the Rector of the prestigious Berlin Technical *Hochschule*, then the Nazi Minister of Education Bernhard Rust and other Party members, reacted angrily; they warned that the ceremony could not and *would not* take place as planned. Hahn later recalled:

> Professor Bonhöffer was forbidden to deliver the planned commemorative speech, and a few days before January 29, the Rector of Berlin University, who was also director of a Kaiser Wilhelm Institute in Dahlem, telephoned to tell me that, as Rector, he must, to his regret, prohibit my planned commemorative speech. I could reply to Professor Eugene Fischer that I had left the Philosophical Faculty in 1933 and therefore no longer belonged to the University (I had only been an extracurricular professor). Professor Fischer answered, obviously with a certain degree of relief, that, in that case, he had no more orders to give me.[86]

Lise Meitner also decided to attend the ceremony, arriving at the hall with Fritz Strassmann and a few other courageous Institute members. The entire affair had become a "bone of contention" for the Nazi Party. Max Delbrück later recalled:

> The Nazi government tried to prevent it, and forbade any state employee, that means any professor, to attend. The principal speaker at this memorial, one of four speakers but really the main speaker, was supposed to be my friend [Karl Friedrich] Bonhöffer, then a [professor] in Leipzig. So here he was; he had come to Berlin and received this strict order from the Minister not to attend this meeting [or he would lose his professorship]. So what was finally arranged was that Otto Hahn would read Bonhöffer's speech. And Bonhöffer and I walked around the place and tried to decide whether he should go in or not; finally he decided not to go in, but I went in and sat in the back row.... Actually, it was a very dignified and well-attended affair.[87]

Hahn noted further:

> The following grotesque situation existed on the 29th: Planck picked me up at my home, and we drove to Harnack House. . . . We did not know whether we would be hindered by force from entering the building. But nothing happened. Every seat in the beautiful hall was taken. The participants were a few relatives of Haber, but above all, Geheimrat Bosch and leading men from the I.G. Farben corporation, whom Bosch had invited by telegram; many wives of the professors who were not allowed to attend the ceremony; His Excellency [Minister of State] Schmitt-Ott; Geheimrat Willstätter, who had come from Munich; and other close friends of Haber.[88]

Despite a capacity audience, there was silence in the hall. Haber's relatives (the few still in Berlin) listened silently in the front row as Otto Hahn read the speech Bonhöffer had prepared.

Many of the participants there were women, wives of professors denied attendance: "They came as representatives of husbands who were deprived by a brutal prohibition of their opportunity to say a last farewell to an important man and scientist." Hahn later reflected: " . . . in early 1935, this obvious resistance to the regime was still possible."[89] The final instructions of the Ministry of Science, Education, and Popular Culture concerning the service were that no reports of the event and no publications of the speeches be distributed, but the program had already been printed, and was passed out to the solemn crowd. In the very back row sat Fritz Strassmann, Max Delbrück, and a subdued Lise Meitner.[90]

Meitner's stance that she was an Austrian citizen and therefore not "subject" to German laws became increasingly precarious. On September 15, 1935 the "Nuremberg Laws" were passed, depriving all Jewish people of their German citizenship. The laws were carried out with rage and terror by the Brown Shirts and S.S. troopers all over Berlin, against businesses, schoolchildren, and prominent Jewish men and women. Hitler was "liquidating the past," historian William Shirer has observed, "with all its frustrations and disappointments. Step by step, and rapidly, he was freeing Germany from the shackles of Versailles, confounding the victorious Allies and making Germany militarily strong again."[91] Jails filled, and so did the concentration camps.

Many still did not condemn the notorious acts of the Third Reich. When the journal *Nature* in a December 1934 editorial entitled "Nazi-Socialism and International Science" questioned "whether an organization of international standing and repute" such as the German Astronomical Society should "come under the control of an anti-Semitic clique," science editors were still hesitant to step directly into the political arena to call for intervention. The editorial concluded with the following compromising position:

> It should be added in fairness to the Germans that internal persecution of a minority is not limited to their country, nor even the use of international gatherings to spread nationalist political propaganda. Both these undesirable practices are to be found in more than one country: we have no sympathy for either. But when these tendencies spread and a nationalistic movement tries to secure the control of international scientific work, it is time to call a halt.[92]

The question of how to *stop* the Fascists and Nazis faced the world. The majority of the German people seemed to feel differently from those abroad. Jobs, not freedom from fascism, was the topic on their minds. By the autumn of 1936 the vast unemployment had diminished; Hitler was filling the concentration camps and putting people to work in a system of armament factories across the Third Reich. Freedom to protest, strike, organize, or collectivize was a thing of the past. William Shirer notes, "One heard workers who had been deprived of their trade-union rights joking over their full dinner pails that at least under Hitler there was no more freedom to starve."[93] But Meitner, still employed, walked past these workers every day for the next four years. Incomprehensible as this may be to us today, the "blinders" went on, and she went on with her work.

CHAPTER VII

The Transuranic Maze: 1934–1938

The fruitfulness of newly-discovered factual connections for the renewal of existent knowledge, and the application of such knowledge to the facts, do not derive from purely logical or methodological sources, but can rather be understood only in the context of real social processes.

Max Horkheimer
Zeitschrift für Sozialtheorie Forschung I, 2, 1937

CHADWICK'S DISCOVERY of the neutron (1932) was a major event for physicists. About two years later (1934), Irène Joliot-Curie and Frédéric Joliot discovered artificial radioactivity. Meitner, soon thereafter, wrote an article summarizing this history: "Fermi recognized how neutrons, through their absence of charges, were directly suited to penetrate heavy elements — that is, elements high on the Periodic Table, and how suitable they were to the release of nuclear reactions."[1] Fermi conducted a series of experiments with the neutron and the irradiation of uranium; his findings were published (mid-1934) in an article entitled "Possible Production of Elements of Atomic Number Higher than 92,"[2] and in other articles that appeared in *Nuovo Cimento* and *Nature*. Uranium behaved strangely when bombarded with a stream of slow neutrons. Lise Meitner later noted:

He and a host of young coworkers, some of whom were trained by him, irradiated all possible elements with neutrons and thereby obtained a series of new radioactive isotopes from heavy elements. Most interesting were the results of the irradiation of the then heaviest element, uranium. Fermi was of the opinion that this would lead to the higher elements with atomic numbers 93 and 94 — thus trans- (above) uranium.[3]

Fermi's published articles of his experiments fired the imagination of scientists around the world. There was competition to create transuranic elements as other laboratories — for example, the Joliot–Curie team in Paris and soon a small team in Berkeley rushed to try and repeat the Fermi experiments. Lise Meitner recalled:

> I found these experiments so fascinating that, upon their appearance in *Nuovo Cimento* and *Nature*, I immediately persuaded Otto Hahn to again resume our direct collaborative work – which had been interrupted for several years – in order to dedicate ourselves to these problems. So, in 1934 we again began, after more than twelve years, our cooperative work....[4]

Joining their collaboration was the young analytic chemist Fritz Strassmann who had earned his Ph.D. in physical chemistry from the Technical University in Hannover.[5] Hahn recalled:

> Thanks to the protactinium isotope uranium Z that I had earlier discovered, we soon ascertained that Fermi was right [elements above the atomic number of 92 could be artificially produced by bombarding them with neutrons]; but we also found that what happened when uranium was bombarded with neutrons was very complicated indeed.[6]

German chemist Ida Noddack, together with her husband Walter and another chemist Otto Berg, discovered the element rhenium in 1934. They published their results about the transuranics in the journal *Angewandte Chemie*. Ida Noddack declared that she was *not* convinced of the existence of transuranic elements "as long as the irradiation products of uranium had not been shown to be similar to the isotopes of all known elements." In her findings she speculated that "one can imagine that when heavy nuclei are bombarded with neutrons, these nuclei break apart into several large fragments, which are indeed isotopes of known elements but not neighbors of the irradiated elements."[7] This was the first insight into nuclear fission.

Noddack's speculation, however, defied all previous experience in physics. Her main point seemed to be that Fermi's tests had not *excluded* known elements such as polonium to be the actual "transuranics." The then-accepted theories regarding atomic structure limited the products of natural and artificial radioactivity to particles of mass no greater than that of an alpha particle. But, Lise Meitner retorted, Ida Noddack had no empirical proof to support her theory that nuclei might "break apart." So the Berlin team continued their radiochemical analysis of the products of the "transuranics" without taking Noddack's ideas seriously. Furthermore, the Noddacks were considered very "brown" politically — a reference to Hitler's brown-shirted party members who made themselves conspicuous at every official gathering. At a German scientific congress, the Noddacks presented their evidence in a slide show and lecture, but they seemed ill-prepared, and the audience was skeptical. Could nuclei "break apart?"[8] Hahn–Meitner–Strassmann ignored speculations about the uranium nucleus "bursting apart" and pressed on with their analysis of elements above uranium.

What was taking place during debates about the redefinition of the atomic nucleus and its "disintegration" was what historian of science Thomas Kuhn calls a "crisis in the established paradigm" or accepted body of information. In *The Structure of Scientific Revolutions* he states:

> When, for these reasons or others like them, an anomaly comes to seem more than just another puzzle of normal science, the transition to crisis and to extraordinary science begins. The anomaly itself now comes to be more generally recognized as such by the profession. More and more attention is devoted to it by more and more of the field's most eminent men. If it still continues to resist, as it usually does not, many of them may come to view its resolution as *the* subject matter of their discipline. For them, the field will no longer look quite the same as it had earlier. Part of its different appearance results simply from the new fixation point of scientific scrutiny. An even more important source of change is the divergent nature of the numerous partial solutions that concerted attention to the problem has made available. The early attacks upon the resistant problem will have followed the paradigm rules quite closely. But with continuing resistance, more and more of the attacks on it will have involved some minor or not so minor articulation of the paradigm, no two of them quite alike, each partially successful, but none sufficiently so as to be accepted by the group.[9]

The new series of "transuranic" elements now being created in three different laboratories left the physicists struggling to interpret the findings in light of ever-changing models of the nucleus. At first the leaders were Fermi's group, with Meitner and Hahn hot on their trail. But Fermi's enthusiasm for the transuranics diminished as fascist power intensified and many of the fine graduate students had to worry about their jobs or survival under Mussolini's reign of terror. Also, Rome's physicists knew that the Paris and Berlin teams were triple-checking their work. Emilio Segrè recalled:

> The first thing Hahn and Meitner did was to confirm all our results . . . and the more experiments they did, the more things became 'strange.' You see, they started to find things that looked like lanthanum and radium, and the whole thing became more and more complicated and messy. In Rome, we had stopped working on transuranics for these reasons. We discovered the slow neutrons in 1934, and that was a big discovery and of much greater interest to Fermi than the transuranics. Then the chemist De Agostino, who was working with us for years, left because he wanted to produce insecticides. I became a professor in Palermo where I had no sources and no way of working on transuranics. So more or less by default, we stopped the transuranic work.

143

I came to Berkeley in the summer of '36 and I urged people at UC Berkeley to study uranium because there was something there that one couldn't understand – and of course, Berkeley had a million times stronger source than we had in Rome, and the technical difficulties which were tremendous with a weak source wouldn't exist for a strong source. But nobody worked on it very much in California – I mean, some desultory attempts were made to do some work but they were not done well, not understood – left to students.[10]

Thomas Kuhn observed in his aforementioned classic that part of the "different appearance" of a scientific field entering crisis simply results from a new "fixation point" of scientific scrutiny. Although Meitner and Hahn had studied the various radioactive properties of elements for years, never before had the entire Periodic Table been extended. Elements above uranium simply had not "existed" before the Joliot–Curies' work, nor had elements been "transmuted" before Fermi's research. As Kuhn further notes, divergent *partial* solutions begin to arise during a scientific revolution when concerted attention is focused on an anomaly available. The transuranics posed this kind of problem to many now senior physicists like Meitner and Fermi.

Since 1931 Hahn and Meitner had been involved in an unpleasant controversy with Aristide von Grosse, a former graduate student, over priorities concerning the discovery of protactinium. *Die Naturwissenschaften* had published both sides of the debate.[11] Meitner found Grosse's new challenge, that Fermi's proposed element 93 behaved strikingly like protactinium, just what she needed to help her "recover" from the loss of her university lectureship. What mattered, however, was that she was an expert not only in analyzing such substances but in unraveling their complex physical nature as well.

By 1935 new models explaining the transformation of the nucleus were all the rage, and international rivalries over the transuranics became particularly acute. The curious new data on nuclear phenomena coming from the Joliot–Curies' laboratories provoked the competition. Based on their careful experimental and radio-chemical analyses, the Joliot–Curies also added new *physical* explanations of the transuranics. Meitner animatedly translated their findings from the French for Hahn and Strassmann. Despite her exclusion from "official" university affairs, she was still at the heart of the discussions, conferring often with Hahn, and spurring on younger colleagues to examine their research. The transuranics were so complex as to require precise radiochemical analysis as well as a physical understanding of the decay patterns created in these short-lived elements. During late nights at the KWI, working with her own neutron data, Meitner was still puzzled by the findings of the French team.

Carl Friedrich von Weizsäcker clearly recalled the heated discussions across Berlin concerning various physical models of the atomic nucleus in the late 1930s.[12] He had inherited his part-time physics assistantship to Lise Meitner from Max Delbrück, who had quietly left Germany late in 1934 after receiving a Rockefeller fellowship. Von Weizsäcker was only one among many bright young theoreticians who contributed fruitful ideas to Meitner and her staff. The Institute for Chemistry remained a hub of conversation and debate within a growing fascist city in which open forums for ideas, controversy, or originality were rapidly being snuffed out. While not openly critical of the regime, Hahn did his best to keep the doors open to all.

In 1933, Meitner had coauthored a small book, *The Structure of the Atomic Nucleus*, with Delbrück,[13] introducing and summarizing then-current physical models of the nucleus. The book described the known nuclear constituents and gave a clear analysis of current decay schemes of radioactive elements and their products. Meitner speculated on possible elements "above" uranium, but there were no new major hypotheses published. However, the debates that were to erupt by 1935 once the transuranics entered the game disturbed Meitner's prestigious authority in the field. Nevertheless, Meitner was pleased with her book and continued to insist that *physicists* would lead the way toward an understanding of the complex radiochemistry of the "trans"-uranic elements.

Fermi himself had doubts about his "discovery" of new elements. However, the Mussolini government was anxious to publicize his findings as an additional "feather in the cap of scientific prestige" for Fascist Italy. Segrè recalled that Fermi was particularly embarrassed by the attention the Fascists lavished on his results;[14] international scientists continued to insist that it was truth they were seeking, not national honor.

By 1935, scientific communication and cooperation were far from open between Germany and Italy. Personalities such as von Laue, Hahn, and Planck continued to work under the assumption that politics would return to "normal" in the long run, but Fermi knew that his government frowned on "sharing" research findings. Hahn also continued to publish his research in applied radiochemistry, studying the crystallization of iron oxides, which, he was quick to point out to Institute officials, were "important for heavy industry." Reich government officials scrutinized the Institute's budget each quarter. Hahn could have pursued this work exclusively for several years if it had not been for Meitner's strong urging that they resume their collaboration to examine the radioactive properties of uranium. Known to be anti-Nazi, Otto Hahn felt that if he and Meitner could discover more new radioelements, there would be, in the long run, practical applications in biomedicine or radiochemical work that might please their industry sponsors. Hahn (and others) could assuage political administrators and financial backers by pointing out that their own research interests actually paralleled that of the State and its "industrial emphasis." Meitner knew better.

The Institute's – and Hahn's – dependence on and collaboration with German industry is important to examine. After the inflation of the 1920s, Hahn's persistent cultivation of industrial backers from Berlin and elsewhere helped the Institute for Chemistry expand during the 1930s and at the same time maintain a certain distance from the rivalries of state politics and bureaucracy. But as German industry began to focus on the "interests of the State," Hahn's research was directly influenced by the needs and dictates of Institute benefactors. Hence, Lise Meitner's fascination with the transuranics and her active solicitation of her former partner's insight into the radiochemistry of Fermi's "trans-uranics" may have unintentionally taken Hahn out of the industrial spotlight during the mid-1930s — and, as the Nazis grew stronger, out of the Party and military scrutiny as well.

Many factors accounted for Meitner's decision to pursue research involving uranium. The first, of course, was the familiarity she and Hahn had with the nuclear field itself: recall their discovery of protactinium (1917). Hahn later remembered that "it was our work on the chemical properties of this substance [protactinium] that gave rise later to our keen interest in investigating the irradiation of uranium with neutrons."[15] A tremendous amount of collaborative work had developed between them in the nearly twenty years since that discovery, but their last joint publication was ten years prior, in 1925. Since then, Meitner had published an average of six papers a year on a wide-ranging spectrum of nuclear-related topics, and her international reputation had grown.

Hahn's publication record had also kept pace with his broad interests, so that, with Fermi's challenge of the transuranic controversy in the 1930s, it was natural for the team to resume working together. Strassmann contributed his analytical talents, and experiments were arranged after hours, when Institute duties permitted. The slow process of bombarding uranium began. They used a standard "indicator" technique in their first radiochemical analyses, a well-proven method characteristic of much of their earlier work. They irradiated uranium with neutrons and then tested the products chemically after dissolving them in an acidic solution. Their excitement came when the new elements, the transuranics announced by Fermi and the Joliot-Curies, were produced in tiny amounts, with an atomic number higher than 92.[16]

There was no method to chemically precipitate such a minuscule amount of radioactive substance in the 1930s, so it was customary for radiochemists to add a similar element as a "carrier." The radioactive atoms had to "follow" the carrier through the operation, and then had to be laboriously separated from the mother substance. To test for an element heavier than uranium, Hahn and Meitner had added a rhenium compound to their solution, and then introduced hydrogen sulfide. What was precipitated out was rhenium sulfide, proving that the new radioactive atoms were chemically similar to rhenium, and likely to be the "eka-rhenium" named by Fermi. A similar test with platinum, historian Spencer Weart pointed out, worked even better, "proving that Grosse was wrong: the radioactive elements, whatever they were, did not behave like protactinium."[17]

Meitner was disturbed by these findings. Although they seemed to prove that they were on the right track, she had found a long line of beta-ray disintegrations after the analysis of radiochemical results. Minute quantities of "transuranics" seemed to be transposing themselves through radioactive decay into other products, so she urged Hahn and Strassmann to continue their investigations. But whatever the puzzle, she decided it would be best to write up their findings and announce them.

Hence Hahn and Meitner published an article entitled "On the Artificial Transformation of Uranium by Neutrons" in 1935 in *Die Naturwissenschaften*.[18] As expected, the Italian and French teams were very surprised to read about their results. The puzzle on all their minds was intensified towards the spring of 1935: both the Fermi team and Hahn–Meitner were reporting substances with *three* different half lives! Certain of the accuracy of their results, both groups hesitated to publish how these products were related.[19] In Rome, Segrè recounted many years later, the political climate began to affect the Fermi team and they turned away from the transuranics despite the discovery of slow neutrons, which had provided new "bullets" to penetrate the nucleus. Many graduate assistants had become more interested in the results coming from Lawrence's cyclotron in Berkeley, California than those coming from Paris or Berlin concerning chemical separations. But Lise Meitner and Otto Hahn continued their collaborative radiochemical/physical research on the strange products emerging from uranium.

As Hitler's SS tightened its grip on Berlin, all Jews in Germany, including scientists, were in danger, particularly those who had supported the memorial ceremony for Fritz Haber. Nevertheless, Lise Meitner affirmed her wish to *remain* with her work and colleagues in Berlin. She dined often with the Hahns; Otto, Edith, and her godson Hanno (now a teenager), all discussed her political options rather nonchalantly. She was still comfortably situated in her apartment in Harnack House across the street from the Institute, and, she argued, that with Planck behind her no one would dare bother her there! It was neither for material reasons nor prestige that she insisted on staying in Germany, she reminded her friends, who did not have the heart to force her to make a decision to leave her home. Yet all signs – and eyewitness reports – pointed to the fact that Jewish and non-Jewish citizens were being arrested, detained, and even brutalized for no other reason than their backgrounds or political persuasion. Lise Meitner's dedication to the Institute, its members, and their work kept her in Berlin long after the spirit of German society had turned viciously against those who were once committed to its continuity. She returned to the laboratory day after day, confronting the transuranics with the same determination she showed her peers by remaining in Berlin. It was her focus and she would not let the research go.

Hahn was sensitive to her situation, and in the fall of 1936, refused an invitation to lecture in Erlangen to the German Chemical Society concerning "his" work on the transuranics. On October 13, 1936, he wrote to Rudolf Pummerer:

[In particular,] one hears only my name for investigations which Lise Meitner engaged in as fully as I. The reason for this is, of course, related to [political] conditions we can't do anything about. But I think it is not quite right when I take credit for overall intellectual property which is not mine exclusively.[20]

In 1936, Max Planck again nominated the Hahn–Meitner team for the Nobel Prize in Chemistry.[21] When Planck and von Laue pondered their yearly nominations to the Swedish Academy of Sciences, they took into account the protection offered by Nobel laureate status. Werner Heisenberg also suggested that Lise Meitner be nominated for the Nobel Prize in Physics for her work with the transuranics. The three – von Laue, Planck and Heisenberg – recognized that Meitner's Austrian citizenship might not protect her much longer even if she did receive the prize. Journalist and pacifist Carl von Ossietzky, nominated by many international laureates, had received the Nobel Prize for Peace for his outspoken views on links between German industry and state payoffs. Despite international outrage over this exposé, the Nazis sent the Nobel laureate to a concentration camp.[22]

Work continued on at the Institute. Throughout 1936, Hahn, Meitner, and Strassmann worked long evenings on the transuranics, publishing eight articles between 1935 and 1936. The three were concentrating more than ever on their Institute responsibilities in an increasingly hostile environment and did not worry about winning the Nobel Prize or international accolades. Lise continued to maintain her close friendship with the Planck couple, visiting them at their home for weekends of music and conversation as she had done when she was a young *Doktor*. Planck remained steadfast in his belief that Germany would emerge from its political turmoil, but few of his friends or colleagues agreed.

When Planck had taken over the Presidency of the Kaiser Wilhelm Society in 1930, many felt that this "changing of the guard" was an important move. From the founding-president Harnack, a theologian, to Planck, a scientist, the reputation of the Society had been enhanced by the prestige of its leader. Maintaining an independence of the affairs of the various Institutes, Planck wanted to represent the entire constituency of the Kaiser Wilhelm Society at home and abroad. "While Harnack's secret had been his superior gift of oratory," a historian of the Society writes, "Planck was effective by virtue of his quietness and modesty which raised him above the bustle of political circles, but often above that of the scientific ones too."[23] Planck was kept abreast of current work by physicists, but he slowly withdrew from even the university colloquium and its debates over controversies in physics. In 1937, delivering his last speech under the swastika banner, the 79-year-old Planck officially resigned the presidency of the Kaiser Wilhelm Society, still believing that the National Socialist Party would rise above its diatribes against "Jewish science."

His successor was a more pragmatic German, the well-respected industrial chemist, Carl Bosch. Planck's public position "above" both the bustle of scientific research and inner-Party machinations also sheds light upon his personal distance from the immediate political dilemmas of Lise Meitner. He had done all he could to aid her and other Jewish colleagues politically, he gently reminded her during their quiet coffee hours together; now Bosch was at the helm.

After his retirement, Planck removed himself from further public disagreements with National Socialist policies. In December 1936, from his office in the University's Physics Department, he wrote to von Laue concerning the 1937 Nobel Prize:

> I am very sympathetic towards the proposal to suggest Fräulein Meitner [once again] for a Nobel Prize. I have already outlined such a proposal last year, by suggesting a split credit between Hahn and Meitner for the 1936 Chemistry Prize. But I am agreeable in

every respect with the proposal which you will be discussing in this direction with Herr Heisenberg.[24]

Twelve months earlier the Nobel Committee had rejected the nomination of Hahn–Meitner. In 1937, Planck and Laureate A. Deissmann nominated Meitner and Hahn for a joint Nobel Prize in Chemistry, while Heisenberg and von Laue nominated Hahn–Meitner for the Prize in Physics.[25] Did they work behind the scenes together in an attempt to rescue Lise Meitner from inevitable persecution? Or were these scientists more interested in international recognition of colleagues known for nearly thirty years, who had been turned down for the Nobel Prize again and again?

The Nobel announcements came as fierce winter storms beset Berlin. Once again the Hahn–Meitner team was disappointed; the British chemist Hayworth and Swiss chemist Karrer shared the Nobel Prize in Chemistry in Stockholm, while Davisson and Thomson were awarded the Nobel Prize in Physics.[26] The KWI group worked on, while Heisenberg and von Laue turned back to their respective university departments.

In a detailed study of the scientific work leading to the discovery of nuclear fission, Spencer Weart wrote:

> Nuclear theory in the 1930s was an inchoate, agitated blob of ideas that inched forward by grasping here and there onto an experimental result or a plausible equation. There were not many solid points of contact with reality, not many experimental processes that theory could actually calculate. The first great success, coming even before the neutron was discovered, was George Gamow's theory of alpha-particle decay. He imagined the nucleus as nothing more than a potential well with particles inside a bag, through whose skin an alpha particle might sooner or later tunnel. This first, and for some time only, successful explanation of a nuclear process profoundly impressed physicists. Yet it tended to inhibit the way of thinking that could lead toward understanding that a nucleus can split in half. From the outset Gamow's theory made clear that nuclear fragments even slightly larger than an alpha particle would have a hard time slipping through the barrier. An alpha particle had a charge of only two units, and Gamow showed that a fragment with a charge of, say, six units could never be seen to tunnel out. This picture was firmly embedded in the minds of nuclear physicists, including in particular the Berlin group. It was one reason they assumed that a uranium nucleus could never find a way to transform in one step into a substantially lighter nucleus: how could a substantial fragment escape?[27]

Proceeding with their irradiation of uranium, Hahn and Meitner found more and more substances with confusing yet distinct half-lives. It was a tedious job to classify these substances according to chemical behavior. To catch the shortest elements' radioactive emissions, they had to perform chemical separations and radioactivity counts at breakneck speeds, while for those newly-created elements with long half-lives, stretching to dozens of hours, they or their assistants had to take turns day and night watching the geiger counters. Gradually, they learned to distinguish which radioactive substance was produced from which, but an overall physical explanation of the process eluded them and particularly vexed Meitner. To make their task even harder, nobody knew just what the chemistry of "transuranium" elements should look like. The best guess was that the periodic table could simply be extended downward. Hence, if uranium was regarded as lying beneath the similar metallic element, tungsten, then the first element beyond uranium, number 93, should behave chemically like the element next to tungsten, namely rhenium. So they called their transuranic discoveries "eka-rhenium," then eka-osmium, eka-iridium, etc. However, chemists soon pointed out the danger of encountering a transition group such as the rare earths (a 15-element series beginning with lanthanum), all having almost identical chemical properties. In fact, the periodic table of this time placed actinium (atomic number 89) directly below lanthanum (57). Such a group might begin at atomic number 90 (thorium) or at uranium, or some other nearby element, Hahn and Meitner began to postulate. But until the transuranic elements were confirmed and their chemistries worked out, there was no way of telling for sure where they would fit into the periodic table.[28]

To compound the confusion, the decay scheme for the new substances did not resemble the familiar chains of decay followed by the natural radioactive series. Irradiated uranium produced complex products. As Hahn wrote Rutherford in 1937:

> Although we have definitely established three disintegration processes for uranium, and believe we have determined the chemical properties of the artificial atomic types with considerable accuracy, we are not yet certain of the correct interpretation of these processes.[29]

The "interpretation" of such radiochemical results ultimately rested with Lise Meitner, who as a theoretical physicist and meticulous experimentalist struggled daily with the conclusions coming out of the team's radiochemical analyses. In one model she envisioned that an incoming neutron might "chip off" another neutron in the uranium nucleus, and both free neutrons would then fly off. But this explanation was not plausible when interpreting why slow neutrons as well as fast ones could initiate such a process. Prompted by Meitner, her nephew Otto Robert also published research in this field: an article, written at Bohr's Copenhagen Institute, entitled "Velocity of Slow Neutrons" was published by *Nature* and summarized his findings.[30]

In the summer of 1936, Meitner sought another approach, summarizing the evidence for isomerism in uranium in an article for *Zeitschrift für Angewandte Chemie* in which she pointed out similar problems in analyzing the bombardment of indium with neutrons.[31] Her published work seemed convincing to many, particularly to those in Switzerland where, in the summer of 1936, she lectured on the team's results at the ETH Institute in Zurich. But even in this collegial setting no new theoretical models emerged to completely explain the puzzling transuranics.

Other physicists were beginning to have doubts about the transuranic scheme. Leo Szilard had earlier proposed a theoretical explanation to account for such phenomena without the need for the complex chemically "new" radioactive substances.[32] Von Weizsäcker, who began to appear frequently in Hahn's office (after his work assisting Lise Meitner in the large first-floor physics section of the Institute), also produced a theoretical explanation for Meitner's observed isomerism. He would soon publish a book in which he proposed his new models of nuclear structure.[33]

A particularly detailed alpha-particle model was developed by Wilfrid Wefelmeier, who often discussed his ideas with von Weizsäcker. Wefelmeier imagined nuclei as alpha particles "stacked" geometrically, and proposed that the Hahn–Meitner–Strassmann "transuranics" with their tendency to emit beta particles could be understood if heavy nuclei were stacked in a nonspherical lump, like a "nuclear sausage!" This was fondly called the "Kernwurst" model and once again provided a visual model for a process unseen to any observer at that time.

By late 1936 Gregory Breit, Eugene Wigner, and Niels Bohr jointly proposed a more sophisticated "compound nucleus" model that surprised even Meitner.[34] In Rutherford's words, Bohr had proposed that "it is impossible to deal with the movements of an actual particle within a nucleus. He therefore considered that the space inside the potential barrier is a sort of 'mush.' "[35] The barrier Rutherford spoke of was the strong electric Coulomb barrier, which surrounded the nucleus and, most physicists then believed, prohibited any particle except the electrically neutral neutron from penetrating the nucleus. Weart noted:

> Meitner and her colleagues tended to think not in terms of the pure compound nucleus model, but of some sort of hybrid picture in which the nuclear particles neither behaved as individuals nor were homogenized into a collective mush. Many physicists noted that since alpha particles were especially stable, the neutrons and protons in a nucleus should tend to clump together to form alpha particles. This picture conveniently explained the emission of an alpha particle when a neutron struck a heavy nucleus: the neutron simply gave up its energy to a pre-existing component of the nucleus, which now had enough energy to escape.[36]

The theorists, Weart explained, were not "ingenious" enough to point out to the experimenters processes that might lie outside the accepted norms of current nuclear models, what Kuhn called "novelty theories."[37] Yet, true to their profession, they were constantly supposing new schemes in an attempt to fit the anomalous data into new models of the nucleus.

The Coulomb barrier Rutherford referred to in 1936 was an energy barrier that combined the repulsive electric force and the stronger, attractive nuclear force. This barrier, it was commonly felt, would keep positively charged particles from penetrating the nucleus, and also "would hold the nucleus together" against the electric repulsion between protons in the nucleus. In their theory of quantum mechanical tunneling, George Gamow, Ronald Gurney, and Edward Condon, described the *tunneling* of alpha particles through this "barrier." They called it the Coulomb barrier after its French discoverer, but this was confusing because the term "coulomb" refers to the electrical force, whereas the nuclear barrier described by Rutherford, Gamow, Gurney, and Condon entailed, as we understand it today, both the electrical and the strong nuclear force acting in opposite directions.

At the Radium Institute in Paris, Irène Curie and her coworkers Hans von Halban and Peter Preiswerk were pressing experimental boundaries in the creation of their own approach to the problem of the transuranics. What had originally begun as an international quest initiated by Fermi now became kind of a nationalistic rivalry between Paris and Berlin, sparked by experimental results neither group would accept, which of course led to priority conflicts expressed both publicly and privately.

In early May 1935 Hahn and Meitner had announced the discovery of a radioactive decay series long sought in natural substances, which was a product of their continued neutron bombardment of the element thorium. However, later that month, the Curie–Halban–Preiswerk team announced that *they* had discovered the missing series.[38] Hahn wrote to Rutherford that he and Lise Meitner were "a little angry with Mme. Irène Curie for not having cited us properly." Referring to the political situation in Berlin (where dissent of all types was being brutally silenced) as well as Curie's oversight, Hahn added, "We regret this very much, for a scientific echo has never been as necessary to us as just now."[39]

His anger was warranted. When the 1936 Nobel Prize nominations failed to bring international attention to their team, Hahn grew more and more worried for Lise Meitner's safety. Irène Curie, moving toward the center of Hahn's and Meitner's field of specialty – the radiochemistry of irradiated "transuranics" – rejected the Berlin team's elaborate decay schemes as well as theoretical models. If isomerism in the transuranics was real or accepted, Curie curtly retorted, one could then "make a great variety of suppositions about the possible transformations." Lise Meitner sent one of her finest former graduate students, Gottfried von Droste, to search for alpha particles that "should have been emitted" when uranium was transmuted to thorium, but he found none.[40] Worse still, Irène Curie and Paul Savitch soon reported *another* product coming out of irradiated uranium, a type of transuranic product that Hahn and Meitner had consistently denied existed![41]

What Curie and Savitch found was a short-lived, 3.5-hr product that resembled an isotope of thorium. The Berlin group could hardly believe that they had missed such an important isotope during their years of work with the transuranics, and they jokingly named the 3.5-hr substance "Kuriosum." Strassmann set to work reanalyzing the filtrate using thorium as a carrier, but he failed to find any radioactive thorium as a product of irradiated uranium.[42] For over two years, they had been neck and neck in the race to publish their findings; now work seemed to halt. In his history of this period, Fermi's keen assistant Edoardo Amaldi recalled:

> Two very important theoretical papers had appeared [one year earlier] dealing with this ensemble of phenomena. The first one was presented by Niels Bohr to the Danish Academy on January 26, 1936 and appeared in *Nature* on February 29 (entitled "Neutron Capture and Nuclear Constitution").[43] It contained the idea that by capture of the incident neutron, a compound nucleus was formed in an excited state, the mean life of which was long enough for explaining the thin lines observed in the epithermal region.[44]

Although Fascist policies had nearly emptied his laboratory, Fermi and Amaldi conducted experiments that confirmed the very low velocities reached by slow neutrons. This work fed the Columbia University group and two British teams information necessary to measure absorption coefficients of a variety of elements, and was in clear contradiction with current theories of absorption of neutrons by nuclei, which had predicted a capture cross-section inversely proportional to the velocity of the neutron.[45]

As time went on, Breit and Wigner published a fascinating paper describing the resonance observed in the capture and scattering of slow neutrons by nuclei with a single resonant level.[46] Hence, Fermi, Meitner, and Curie had to turn back to their laboratories in search of explanations concerning the response of uranium, an anomaly that could not be easily explained. Amaldi later recalled that he and Fermi reasoned as follows:

> Since we knew from all other heavy elements that under neutron irradiation they undergo (n, γ) reaction, we interpreted our results as due to a transuranic element $Z = 93$ produced

by neutron capture by U238 which we suggested to be a beta-emitter decaying in element 93, which we naively assumed to be an "EkaRe." We did not consider the isotope U235 because it was discovered only about one year later.[47]

To complicate matters, the new 3.5-hr substance the Paris team had identified added one more "transuranic" to the growing Periodic Chart.

On January 20, 1938, Hahn and Meitner wrote to Irène Curie, noting details of their own experiments and politely suggested that she had "committed a gross error." Courteously, they suggested that if she would make a "public retraction" of her findings, they would not publish their criticism.[48] It was too preposterous: the "kuriosium" short-lived element, produced when uranium was bombarded by slow neutrons had never emerged in their radiochemical analyses, and they were, until then, the team the international community had relied on for verifiable results.

This was the situation between the French and German laboratories in the beginning of 1938. Other scientists around the globe were also taking note of the puzzling data that was leading to controversies in scientific communities across Europe. In the mid-1930s, the work on uranium and thorium had merited only a few cross-references in reviews and textbooks. But when Irène Joliot–Curie and her team announced their 3.5-hr decay product in 1937, others began to enter the research field in pursuit of the "transuranics."[49] A team at the University of Michigan generated neutrons through their cyclotron, and began to bombard uranium samples, which also yielded the mysterious 3.5-hr product — and others!

The Vienna Radium Institute, like the Ann Arbor group, was built on Hahn and Meitner's findings with thorium, but could not find alpha particles emitted after neutron irradiation.[50] In Berkeley, Lawrence set Philip Abelson on a search for alpha particles, and Abelson later recalled that the "unorthodox" decay schemes for the transuranic elements, with their multiple beta decay chains (so unlike the radioactive decay patterns of heavy elements) aroused "puzzlement and skepticism" throughout the physics department.[51]

In 1938 Abelson began to search for the X-rays emitted by the supposed transuranic elements, since X-ray spectral lines had long been recognized as an unequivocal way to identify an element. Similarly, Norman Feather and Egon Bretscher prepared apparatus at Cambridge University to search for X-rays in order to pin down the emission patterns of these interesting radioactive elements. Interviewed long after, Bretscher's widow recalled that the young physics community, many of whom were emigrants from Germany, felt that there was indeed something "very odd" about the transuranium decay schemes.[52]

Only a few isomers were known among the ordinary elements, but as the transuranics demonstrated to Hahn, Meitner, and a host of physicists who tackled the problem in 1938, isomers appeared in great numbers when elements were created "beyond 92." At both Berkeley and Cambridge, experiments began to turn up results that were hard to explain. The published literature, however, still showed throughout the year only confirmation of the work by the Paris and Berlin teams, although questions were being raised. According to Strassmann, their "rival teams" seemed at odds with one another every time a new article emerged. He recalled that Lise Meitner literally threw up her hands after reading a new Curie and Savitch article reporting that their 3.5-hr substance resembled actinium.[53]

Radiochemical analyses grew increasingly difficult that year in Berlin, yet yielded new evidence for a whole set of beta-decaying products not found previously. But much of the new data still had no explanation in existing physical theory. Then a retraction came from Paris even more disturbing than the French group's original claim in early 1938. In this article, they reaffirmed the existence of their 3.5-hr substance, which they postulated might actually be lanthanum,[54] nearly halfway down the periodic table from uranium.

Stepping back from the scientific rivalries of 1938, and looking at the events through Lise Meitner's experience, certain themes emerge from this period. The first is the elaborate nature of the physical models and theories then being developed from experimental evidence. In the case of the transuranics, we may note that the physicists literally depended on laboratory results, painstakingly gained, in order to "explain" shifts in their developing theories. Lise Meitner was considered an expert in her field, and she felt she could wholeheartedly trust the results of Hahn and Strassmann; if they said an element or radioactive substance was there in the test tube, it had to be so. Her students and colleagues consequently built their models on this evidence, and Meitner readjusted her theories as new evidence came out of the lab.

When we examine other subtle motives involved in the pursuit of these particular research findings, we see that more was at stake during the 1930s than the creation of new physical models or the discovery of radioelements. The Hahn–Meitner tedious extraction of protactinium, like Marie Curie's radium extraction from tons of pitchblende, made both teams experts in the methodology and identification of the chemical as well as physical properties of the numerous tiny quantities and characteristics of radioelements encountered in their research. Their decision to pursue Fermi's challenge concerning the bombardment of uranium with neutrons, and Lise Meitner's solicitation of Hahn's collaboration once again, may have been based in part upon the decision to participate in an international scientific pursuit *away from* the political storms threatening their very existence. It was also highly appropriate: Hahn and Meitner were considered one of the few expert teams in their field throughout the world. The prestige gained both scientifically and politically for both women scientists – Curie and Meitner – through the discovery of new elements or even new atomic processes may have been a major factor in Meitner's choice of research direction. And once involved in a challenging scientific quest, it was hard to turn away. In an article "Scientific Internationalism and Weimar Physicists," Paul Forman states:

> The scientist's persona, his 'image of himself,' usually makes him most reluctant to admit that he has chosen between his science and his nation. He will often prefer to equate the interest of his nation with those of science (and, perhaps, humanity) most easily and most often when his nation is both a leading scientific power and a leading military power. At the same time, in order to maintain that persona intact, he must take care to avoid any direct repudiation of the ideology of scientific internationalism. And those who have the most to lose from a cancellation of international competition and recognition are likely to take the greatest care.[55]

This statement seems ironically suited to both Lise Meitner and Max Planck, although they manifested their "choices" between nation and science in different ways. No longer actively involved in university leadership or Kaiser Wilhelm Society affairs after 1937, Planck withdrew from his role of "preserving the role of German science" in the face of growing hostility in Berlin. He supported the free spirit of scientific debate, but his steadfast loyalty to the State, even one controlled by Hitler, led him into compromising positions again and again against the growing vocal tide of "Aryan physics."

A relatively small band of politically active German scientists pushed their racial views into the mainstream of scientific congresses. In contrast, while Meitner had the most to lose in the face of the Third Reich, she continued year in and year out to take painstaking care with her team's research results. Meanwhile, the Jewish editor of the prestigious journal *Die Naturwissenschaften*, Meitner's dear friend Arnold Berliner,[56] had been silenced, forced from his job and spheres of influence. Nobel Laureate Philipp Lenard increased his shrill attacks on those physical theories he considered "Jewish" and therefore false (such as relativity), creating a climate of tension that threatened to tear apart the fabric of science in Germany.

Moreover, reports on this situation began to appear in the *New York Times* and alienated the German community even more, including those *not* in favor of Hitler's brutalities. On March 9, 1936, Otto Tolischus wrote for the *Times*:

> The most interesting controversy is that of the physics professors, illustrating the extent of the confusion wrought, even in eminent minds, when science is combined with politics. The exponents of "German" physics in this controversy are Professor Philipp Lenard, discoverer of "Lenard's rays," Nobel Prize winner in 1905 and now head of the Philipp Lenard Institute of Physics at Heidelberg, and Professor Johannes Stark, discoverer of the "Stark effect," Nobel Prize winner in 1919 and president of the German Research Association. Their opponents are Professor Max Planck, Germany's most eminent physicist, creator of the quantum theory on which modern physics is based, Nobel Prize winner in 1918; Professor Max von Laue, Nobel Prize winner in 1914, and Professors Erwin Schrödinger and Werner Heisenberg, Nobel Prize winners in 1933.
>
> The controversy started when a student of physics, Willi Menzel, whose scientific attainments are still to be revealed to the world, but whose party orthodoxy apparently is unchallenged, published a violent attack in the *Beobachter* against Albert Einstein and against all theoretical Jewish physicists or products of the Jewish spirit. He modestly admitted there was no National Socialist physics, but maintained that there was a "German" physics, which he defined as "experimental research into reality in inorganic nature caused by the joy of observing its forms of reaction." "Jewish" physics, as he defined it, "aims to make physics a purely mathematical thought construction, propagated in a characteristically Jewish manner."
>
> The idea of using the purely scientific contrast between theoretical and experimental physics for a National Socialist campaign against Jews, however, is not original with Willi Menzel. It is based on the violent diatribes in the same direction uttered repeatedly by Professor Lenard, who maintains that science "is conditioned by race and blood," and by Professor Stark, who denies that theoretical science has any merit whatever and denounces "the Jewish propaganda that makes Einstein the biggest scientist of all times and seeks to impose Jewish views as a measure of all things."
>
> Mr. Menzel's attack was answered by Professor Heisenberg, who declined to follow his opponent into the field of political anti-Semitism, but confined himself purely to the defense of theoretical physics, citing in particular Professor Planck as an authority, demonstrating how, through [Planck's theoretical work,] new experiments had been stimulated, and above all, their results had been coordinated and explained.
>
> This answer, however, was followed by a statement from Professor Stark, commending Mr. Menzel, expanding the attacks on Dr. Einstein to all those who support the Einstein idea or methods, and concluding with a demand that their influence be excluded in deciding future university appointments.
>
> In this controversy, the weight of numbers and authority seems to be on the side of the theoretical physicists, but the "German" physicists are winning out because they have greater party orthodoxy on their side. [57]

Party orthodoxy or not, scientific breakthroughs require the consensus of the international community. In the midst of political turmoil at home, Hahn and Meitner continued to have theoretical differences with the Paris team that had no political origins. In May 1938, Otto Hahn met Irène Curie's husband, Frédéric Joliot, for the first time. The event was the Tenth International Congress of Chemistry in Rome, and, as Joliot–Curie later recalled, despite Hahn's friendly overtures, he did not hide his "skepticism" about the Paris Radium Institute's results surrounding the transuranics. Hahn told Joliot that he would repeat each of Irène's experiments and prove that she was in error.[58]

In the fall of 1938, when Hahn received Curie and Savitch's paper announcing that their 3.5-hr substance clearly resembled lanthanum, his reaction, a visitor to Berlin recalled, was that "it just could not be" and that "Curie and Savitch were very muddled up."[59] Strassmann and Hahn now undertook a serious search within the minute products of irradiated uranium, hunting for any elements resembling thorium, actinium, or radium. Their diligence was rewarded, for they found a number of radioactive substances formerly overlooked, and published their results in *Die Naturwissenschaften*.[60] But despite the various incentives to continue their difficult analyses – subtle factors of scientific pride and identity, national rivalry, and the quest for international honor and prestige – still more ominous events intruded on the teamwork of Meitner, Hahn, and Strassmann in 1938.

CHAPTER VIII

Escape from Nazi Germany: 1938

The image is one of survivor as creator: the one who has known disintegration, separation, and stasis now struggling to achieve a new formation of self and world. This process may involve exploration, experimentation and risk as the search for new forms lead in unanticipated directions.

Robert Jay Lifton
Living and Dying

MARCH 12, 1938: The German Army crossed the border into Austria, as the Third Reich's Air Force screamed through the sky. With the forced "annexation" or *Anschluss* of their country into the Third Reich, all Austrians, wherever they resided, officially became German "citizens," subject to the laws of 1933 concerning race and other exclusions. Lise Meitner could no longer hope to rely on her Austrian birth to protect her from Nazi racial laws. The difficult decisions and dramatic series of events in Meitner's life that would follow the *Anschluss* were not unique in these times: countless other non-Germans, under the jurisdiction of Hitler's army, were also faced with life or death choices — to remain, escape, collaborate, or die.

The Austrian state had been brought into existence only twenty years earlier at the end of WWI when the Austro-Hungarian Empire was dissolved. In the newly independent state's constitution, notes historian Jurgen Gehl, it had been "foreseen that Germany-Austria should be a constituent part of the German Reich."[1] A week prior to the Nazi occupation, a plebiscite had been held; according to "official" results, 98% of Austrian voters had "approved" the unification with Germany. But that vote was clearly skewed: many Austrians had refused to participate in the mock election and others were imprisoned long before the vote was counted. Hitler arrived at his schoolboy home of Linz on March 12 and announced the completion of his mission to restore his "dear homeland" to the German Reich.[2]

During her 25-year residence in Germany, Meitner retained her Austrian citizenship, which had afforded her some protection during the 1930s. She was not dismissed, jailed, or forced to emigrate, as nearly 15 to 20% of German scientists had been compelled to do since 1933,[3] including her own nephew, Otto Frisch, who was discharged from his research post at the University of Hamburg in 1933. He left his "first real University job" unwillingly, and after a short stay with Oliphant in England, found a position in Copenhagen. Yet even at Bohr's Physics Institute, Frisch felt the sting of the annexation of Austria: "The occupation . . . changed me 'technically' from an Austrian into a German. My aunt . . . who had acquired fame through many years of work, now also had to fear dismissal, and there was also a rumor that scientists might not 'be allowed' to leave Germany."[4]

The Third Reich's anti-Semitic agenda was seen in various articles on science. Yet, the international press reflected the "hands-off" policy of other nations. For example, the London-based scientific journal *Nature* printed the following editorial on April 30, 1938 in its "News & Views" section:

The 'Jewish Spirit' in Science

In July of last year, several correspondents sent us copies of an article entitled " 'Weisse Juden' in der Wissenschaft" ['White Jews' in Science] from the German periodical *Das Schwarze Korps* [The Black Corps] of July 15. The main theme was that it was not sufficient to exclude all Jews from sharing in the political, cultural and economic life of the nation, but to exterminate the Jewish spirit, which is stated to be the most clearly recognizable in the field of physics, and its most significant representative Professor Einstein. "There is one sphere in particular,' the article says, 'where we meet the spirit of the 'White Jews' in its most intensive form and where what is common between the outlook of the 'White Jews' and Jewish teaching and tradition can be directly proved: namely, in Science. To purge science from this Jewish spirit is our most urgent task. For science represents the key position from which intellectual Judaism can always regain a significant influence on all spheres of national life. Thus it is characteristic that in a time which brings fresh tasks to German medicine and which awaits decisive achievements in the fields of heredity, race-hygiene and public health, our medical journals should, in the space of six months, publish from a total of 2,138 articles, 1,085 from foreign authors, including 116 Russians of the USSR. These articles of foreign origin scarcely concern themselves with those problems which seem so urgent to us. Under cover of the term 'exchange of experience' there lurks that doctrine of the internationality of science which the Jewish spirit has always propagated, because it provides the basis for unlimited self-glorification."

Several men of science of international reputation were named in the article as followers of Judaism in German intellectual life, and it was remarked that 'They must be gotten rid of as much as the Jews themselves.' *Das Schwarze Korps* invited Prof. J. Stark to express his opinion on this suggested purge on account of 'the importance of the problem for the future of German scientific work.' After reading Prof. Stark's reply, we asked him if it represented his considered views upon the relative values of experimental and theoretical studies, and if so, whether he would care to make them known to other men of science through the columns of *Nature*. He willingly consented to contribute the article.... At the moment we make no comments on the views expressed by Prof. Stark; and we gladly give him the opportunity of making them known to the scientific world. We should, however, be surprised if the inclinations which these new principles imposed on the pursuit of scientific truth are generally accepted as the highest or the best means of promoting the advancement of natural knowledge.[5]

To the editors, and many scientists around the world who had not yet directly felt the impact of these "limitations," the threat implied in these "new principles" still seemed remote. The conclusion of Stark's article, however, shocked many readers:

It can be adduced from the history of physics that the founders of research in physics, and the great discoverers from Galileo and Newton to the physical pioneers of our own time, were almost exclusively Aryans, predominantly of the Nordic race. If we examine the originators, representatives and propagandists of modern dogmatic theories, we find amongst them a preponderance of men of Jewish descent. If we remember, in addition, that Jews played a decisive part in the foundations of theological dogmatism, and that the authors and propagandists of Marxian and communistic dogmas are for the most part Jews, we must establish and recognize the fact that the natural inclination to dogmatic thought appears with especial frequency in people of Jewish origin. In establishing these facts, of course, I do not maintain that there are no Aryan men of science who are actively engaged in the dogmatic spirit in the realm of science; nor do I maintain that Jews cannot produce valuable experimental work carried out in the pragmatic spirit. I wish solely to make a statement on the frequency of occurrence of the natural tendency to pragmatic or dogmatic ways of thinking. It must also be taken into consideration that, by training and practice, Aryans can become accustomed to the dogmatic and Jews to the pragmatic habits of thought. I acknowledge scientific achievement in new discoveries irrespective of the nationality of the discoverer, and I combat the harmful influence of the dogmatic spirit in physics when I encounter it in my scientific work, regardless of whether the culprit is a Jew or not. Moreover, I have been engaged in this fight not only since 1933, for as long ago as the year 1922, I denounced in the strongest terms the formalism and dogmatism invading German physics.

Stark and "Aryan physicists" throughout Germany continued to broadcast these views, which were championed from platforms both within and beyond German universities with increasing vehemence during the spring of 1938.[6]

National Socialist politics weighed heavily on the senior scientists in the Kaiser Wilhelm Institutes, as the formerly peaceful atmosphere in Dahlem gave way to political rallies and nationalistic youth meetings. Lise Meitner's Austrian passport was invalid, but she still continued her routines at the Institute, attending to her transuranic research. Yet fear for her safety was a daily reality. She avoided those staff members who appeared in Nazi uniforms. The head of the new Guest Department of the Institute for Chemistry was a Nazi party member named Kurt Hess, who was vocal in his harassment of those who spoke out and sympathized with Lise Meitner's precarious position. Hess had ambitions of "inheriting" the leadership of the entire institute, and Hahn was also cautious around him. Hess was often heard to remark, "The Jewess endangers this Institute."[7]

The highest officials within the Kaiser Wilhelm Society itself had disputed Meitner's status in the Institute for Chemistry after the *Anschluss*; a report was filed with the Ministry of Education and her case had come up before the controversial head of the Reich Research Council, Rudolf Mentzel.[8] Otto Hahn found himself in a difficult position, and Lise Meitner never fully forgave him for his actions. As he recalled the incident:

> Then I too rather lost my nerve, and in a conversation with Hörlein (our treasurer) I spoke of Lise Meitner and the new awkward situation since the annexation of Austria. Hörlein suggested that Lise might resign from her position because nothing more could be done; perhaps she could continue working unofficially. I do not exactly remember the concrete proposals. Unfortunately, I told Lise about that conversation. Lise was very unhappy and angry with me, because now I had also let her down....[9]

Hörlein's "suggestion" hurt Meitner deeply, as she recorded in her diary: "Hörlein demanded that I leave. Hahn says I must not come to the Institute any more." The next day, she wrote, after going to the Institute anyway to record some of their irradiation results, "He has in essence thrown me out."[10] Soon she was asked to vacate her apartment and was moved to the Hotel Adlon "for safety."

But Meitner was not without allies in prominent positions. Carl Bosch, successor to Max Planck as president of the Kaiser Wilhelm Society in 1937, had always been friendly and supportive. He resolutely objected to any suggestion of her resignation when Hahn brought the matter before him. Bosch firmly asserted that all Institute affairs were still under *his* jurisdiction, not the government's, and that Meitner *could* remain. Yet the inevitable faced them all — Lise Meitner would have to leave Germany to save her life. Hahn recalled that April and May were very tense months:

> The situation became more and more unbearable. Lise was very angry with me and could not get over my having spoken about her to Hörlein. On the other hand, I was continually being told that Lise ought to have given up her post long ago, because she ought to see that she endangered the institute, etc....[11]

The uncertainty of her employment was a serious concern for political as well as personal reasons; if she were no longer employed, any safe exit from Germany would be even more difficult. Sympathetic colleagues like Max von Laue worried that she would face the same dangers as thousands of others who had attempted to cross the border, many of whom had been detained, imprisoned, or worse. These warnings, coupled with the knowledge that her case had been brought to the attention of the Nazi Ministry of Education, made the strain her colleagues labored under even more complicated. When the Schrödingers, for example, came from Graz to celebrate Planck's 80th birthday, never once did the Austrian physicist bring up the subject of his own situation.

Outside of Germany, friends and colleagues had already communicated their concern for Meitner to their Berlin colleagues. Two days after the annexation of Austria, physicist Paul Scherrer had written to Meitner from Zurich, and formally invited her to lecture there, all expenses paid. James Franck wrote from his new position at the University of Chicago, pledging his financial support and even took the first steps for her future emigration to the US.[12] However, Meitner did not wish to go to any country where German was not spoken or understood. Niels Bohr sent her an invitation in April to come to Copenhagen. The tone of his letter was cordial but serious. In early May, she reached a decision: she would go to Bohr's Institute. But the next day she was refused a travel visa at the Danish consulate since her Austrian passport was no longer valid. Other options fell through, and feeling that she had no alternatives left, Meitner turned to the the Kaiser Wilhelm Society.

After a long discussion with Carl Bosch, it was agreed that he would assist her in applying for an official state exit permit to "travel" abroad. Both felt that the most important thing was to get her safely out of the country, after which she could take time to consider her relocation options. The need for extreme caution in these proceedings was only too obvious. On May 22, Bosch wrote to the Reich Minister of Education Frick, requesting permission for "...the famous scientist Lise Meitner to leave for a neutral foreign country: Sweden, Denmark, or Switzerland." His appeal continued:

Miss Meitner is non-Aryan; nonetheless, with the concurrence of all the official departments, she has been allowed to retain her position as she is engaged in important scientific discoveries, in particular, as a physicist of special capacities. In ongoing joint scientific research with Professor Hahn, who works in this area as a chemist, she has solved several formidable scientific problems. Miss Meitner is an Austrian national. With the repatriation of Austria, however, she has become a German national, and it is to be expected that the question of her emigration in the long or short term will become an acute question. As Miss Meitner has established a highly esteemed reputation in the scientific world, in the interests of the Institute, I find it very desirable if it would be made possible to come to a smooth resolution of this case. On her part, Miss Meitner is prepared to emigrate at any time and to take another scientific post abroad. Responsive inquiries [concerning her relocation] have also been made on her behalf.... It is only a question of obtaining for Miss Meitner, who has an Austrian passport, notice that she may return to Germany – otherwise, travel abroad for purposes of employment is impossible – or that Miss Meitner be issued a German passport.

Heil Hitler![13]

As Meitner waited for the Minister's response, she was cheered by a reunion with Bohr and his wife, Margrethe, who were visiting Berlin. Bohr took his mission seriously: there was no time to lose in helping physicists to emigrate. Bohr went with Meitner to see his old friend Peter Debye, who was serving as director of the Kaiser Wilhelm Institute for Physics. Somehow, all the activity brought back Meitner's confidence: she noted in her diary, "D[ebye] told B[ohr] there was time, there was no great hurry."[14] Bohr had previously alerted Dutch colleagues to Meitner's predicament, and soon after, she had received an invitation from Dirk Coster to teach in Groningen, the busy university town where she had lectured in 1923.

Lise's growing despondency became apparent, as doors slammed shut.[15] Coster and H.A. Kramers in Leiden began notifying other physicists throughout Holland that Meitner needed urgent help. Coster wrote to his colleague, Adriaan Fokker in Haarlem, on June 11:

Lise Meitner will probably be thrown out of Berlin–Dahlem shortly.... It would be splendid if she could work in Holland for a time.... Perhaps we can tap colleagues for regular contributions.... If Lise Meitner could work in Groningen, there would also be a grant of about f.500 per year out of Groningen [University] funds.... I would like to have her here, but would not make my personal commitment dependent upon that....[16]

Meanwhile, concerned friends persuaded Meitner to give up her longtime comfortable apartment across the street from the Institute, and move to the Hotel Adlon where Bosch was also staying. Reluctantly she packed her things and took a cab there. Soon after, Bosch received a reply from the Ministry of Education, from which it was clear that there would be no "smooth resolution" to the situation. He nervously telephoned to her room and read the following letter to her. Meitner jotted down the text in shorthand on Hotel Adlon stationery, the ink running dry at the end of the scribbled note, a sign of her own despondency:

> By order of the Reichminister Dr. Fr[ick], may I most respectfully inform you in answer to your letter of the 20th of last month that there are political objections to the issuing of an *Auslandpass* [i.e., a passport for traveling abroad] to Prof. M[eitner]. It is considered undesirable that renowned Jews [underlined!] should leave Germany for abroad to act there against the interests of Germany according to their inner persuasion as representatives of the German sciences or even with their reputation and experience. The Kaiser Wilhelm Society will certainly find a way for Prof. M. to stay in Germany after her retirement and under certain circumstances.... The Reichsführer-S.S. and the Chief of German Police in the Reich Ministry of the Interior [Heinrich Himmler] in particular have advocated that opinion.[17]

Alone, she reread the hastily scrawled letter. There was no denying that her life would have to change dramatically. Some colleagues had suggested she consider moving to Cambridge or Liverpool where her nephew had been made welcome. But, she had formerly protested, her English was minimal; besides, when the situation in Berlin was resolved, she wanted to be as close as possible to Dahlem and her lab. The hope, however faint, that things would soon "blow over" persisted — still. Her Dutch friends continued their efforts on her behalf, communicating secretly with colleagues in Berlin; Debye smuggled letters to them through Hahn.

With the help of Adriaan Fokker and W.J. de Haas, her "departure" was secretly arranged. A Dutch immigration officer would let Meitner enter Holland despite her invalid passport and lack of an entry visa. In Berlin–Dahlem, Bosch contemplated appealing directly to the Reichsführer of the SS, Heinrich Himmler, but was advised against the idea by other staunch anti-Nazis, including Paul Rosbaud, a friend, editor of *Die Naturwissenschaften*, and physics consultant to the publisher Springer–Verlag. A forged passport was also considered. Rosbaud telegraphed his Swiss colleague Paul Scherrer about Meitner's case. Sensitive to the risks involved, Scherrer immediately telegraphed a veiled message directly to Meitner: ARE YOU COMING FOR A PHYSICS WEEKEND 29 JUNE TO 1 JULY?[18] The day before, Debye had written guardedly to Bohr in Copenhagen:

> When we last spoke [June 6]...I assumed everything was quite all right, but in the meantime it has become clear to me that circumstances have substantially changed.... Even a very modest offer would be considered and followed up if only it provides the possibility to work and to live.[19]

Debye then stressed that "even the most dispassionate observer of the situation would *not* come to a different conclusion." The "concerned party," meanwhile, noted in her diary on June 15: "Went for information. Hear that technical and academic [people] will not be permitted to get out. [Max von] Laue hears the same from the legal faculty."[20] As tense days passed, she continued anxiously to await news from friends abroad, languishing in the humid Berlin summer heat.

Bohr was himself involved in relocating many political refugees, but took a personal interest in the fate of his long-time friend Lise Meitner. From Copenhagen, he wrote to Fokker on June 21 that Meitner should leave Germany as soon as possible, and enclosed Debye's letter. [21] He urged Fokker to assess "...exactly how the matter stands in Holland, and regretted that he could not bring Meitner to Copenhagen, as it would be "quite impossible to obtain the necessary permission of the authorities on account of the great number of foreigners already working in this institute." He continued on a positive note, "...a person with her unique qualifications should hardly [have] difficulties [in finding] some...solution for the long run, if only she can get out of her present most precarious situation."[22]

Fokker, meanwhile, wrote to Coster, confiding that one week was not enough time to fill Bohr in with their plans for Meitner's "relocation" since they themselves did not have funding, housing, or the promise of a job for her. He replied to Bohr's June 21st letter cautiously, and questioned "whether it is *certain* that Lise Meitner will be dismissed, and whether she is living badly or in fear." He further said that he thought about asking Hahn, but was concerned about Nazi censorship of Hahn's mail. Then he hedged: "I think if the Nazis let her keep her position, then we should not try to get her here."[23]

Fokker had heard that biochemist Otto Warburg was being allowed to continue his Berlin work "undisturbed," and wanted Bohr's assurance that their attempts to aid in Meitner's escape were necessary. He had also received a rather cool letter from Professor D. Cohen in Amsterdam, which, while sympathetic, outlined some of the difficulties involved in getting Meitner into the Netherlands. Dr. Cohen thought that the Dutch Ministry of Justice as well as the official Ministry of Education should be contacted. He wrote:

> The letter...should...convince them that her presence in the Netherlands is of great importance. I fear that even then there will be difficulty, in that the regulations in fact apply to Germany and not to Austria. But if you and your friends push hard enough it seems to me you can show that this is a very special [case].... I hope from my heart that you succeed.[24]

167

Discouraged, yet well aware that every step they took might save Meitner's life, Fokker summoned Coster to Haarlem. Together they had only been able to raise a fifth of the funding they had sought for Lise Meitner's escape and maintenance. Coster, however, clearly saw that time, not money, was the most important issue. Four days later, they filed a formal request to the Dutch Ministry of Justice to allow Meitner to enter their country. Meanwhile, the Ministry of Education advised them that their request would be viewed more favorably if it came directly from a university with a *position* to offer Meitner. The irony of the situation, as Coster wrote to Fokker upon his return to Groningen, was that foreigners were not allowed to work for pay, but could only be granted lecturer status, which required faculty approval. This appeared to be near impossible during the summer months. Hence, they took the final brave step in requesting help for Lise Meitner: they formally wrote to their own government asking for assistance. Their carefully worded petition to the Minister of Justice was supported by representatives of the prestigious Royal Academy of Sciences in Amsterdam as well:

> Whereas it would be of great value and also esteem to the development of physics in the Netherlands if a scientist of the quality of Miss Meitner could work in this country. . .
> and whereas money has been donated thus far by private parties . . . she would not deprive any Dutch scientist of employment opportunities. . . ."[25]

During the Depression, jobs and financial support were the key fears on most minds in the Netherlands, as well as the potential of a Nazi invasion.

While physicists around the world had responded with sympathy, they had little financial assistance to offer as they struggled to feed their own families. Coster and Fokker turned to the industrial community for support as they tensely awaited a reply from their Justice Department. They wrote to a wealthy executive, head of the X-ray division of the firm NV Philips of Eindhoven for aid, reminding him that Meitner had lectured years before in the physics laboratory of Philips Labs. Fokker was blunt. Lise Meitner was likely to be fired, or worse, and if she left Germany, she would lose *all* of her pension rights.

Soon both small pledges as well as disappointing rejections came in. The reply came from influential Philips Petroleum director P.F.S. Otten that "internal budgetary considerations as well as considerations of a political sort" made it impossible for the corporation to help.[26] Some Dutch physicists worried that if Meitner were to leave Germany now, she might lose her entire pension, and worried that it would be difficult to support her for the indefinite future. Despite these reservations, many others did contribute to her fund.

On June 27, 1938, Coster wrote to Fokker that he had made a decision: he would not wait for the Dutch government action any longer. He would personally travel to Berlin. Only there would he privately inform Meitner that enough pledges had been collected to "support" her for a year, and, if possible, he would bring her back with him to the Netherlands. Fokker, understanding his colleague's determination, still expressed reservations: "Don't panic! Don't let your presence in Berlin... make L.M. leave too hastily. You must let her calmly make the decision herself." There were concrete reasons for hesitation, he concluded, as "it is always possible that the two of us will *not* get the money!"[27] Coster, however, had made up his mind. He wired Debye that he was coming to Berlin to interview an "assistant" to whom he could "offer a one-year appointment,"[28] and began to make urgent travel plans. Europe was on the verge of war.

In Scandinavia there were also many actions taken that summer to assist Lise Meitner. Bohr, upon returning from Berlin to Copenhagen in early June, immediately contacted his Scandinavian colleagues. The first communiqué went to Manne Siegbahn in Sweden, who was supervising construction on the new Research Institute for Physics, adjacent to the University of Stockholm. Siegbahn had fond recollections of Meitner from their work together in Sweden, where she had first lectured in 1921. The new Institute for Physics promised to be a leading physics center: Siegbahn was also supervising the design of a cyclotron, to be built on a model similar to Lawrence's in Berkeley.[29]

Physicists in Stockholm were proud of the Swedish Academy of Sciences' financial support for the new Research Institute. Not only did the Academy supervise and award the Nobel Prize process, but its physics committee had lobbied the organization for years for cyclotron funding. While his focus was on the first stages of organizational development, Siegbahn was petitioned again and again by Bohr concerning Lise Meitner. At last he agreed that, in the face of present dangers, he could offer Lise Meitner safe haven and work in Stockholm.

Such news could not be communicated through the censored mail. Hence, Siegbahn sent a personal messenger, Dr. Rasmussen, to Berlin–Dahlem, offering a one-year salaried position (the best offer that could be secured) to Meitner. On June 27, the same day that Coster wired Debye of his intentions (unknown to Meitner), the distinguished member of the Swedish Academy of Sciences, Dr. Rasmussen, arrived in Berlin-Dahlem. After a private conversation with this emissary from Stockholm, Meitner gladly accepted the offer from Siegbahn.

Lise was unaware of all the trouble her friends and colleagues in Holland had just gone through on her behalf. Debye had even arranged a clandestine meeting to tell her the good news: that funds had been secured for her after her arrival in Holland. But when Debye began to openly discuss the Dutch offer, she politely refused. Loyal to her friend Bohr and grateful for his efforts, she insisted that her Scandinavian colleagues had her best interests in mind: the continuation of her research at Siegbahn's new Institute in Stockholm. Unaware of the urgency with which Coster had pursued the Dutch option, she reasoned that there was no immediate danger to her life or work.[30]

Debye was confused as he said goodbye to Lise Meitner. Taken aback by Bohr's offer, he immediately wrote to Coster, referring to Meitner in masculine terms to protect her identity:

> Dr. Rasmussen has already been here on Monday [June 27] ... seeking an 'assistant' [for] Siegbahn's new laboratory .... I regret, actually, that I must write that in the end, Stockholm won. I would have preferred that it be Groningen, but I let myself be persuaded by the assistant himself [Meitner], who thinks he will be able to accomplish more in Stockholm.... Of course, I still let him know this morning what was in your letter.... I believed its contents would have a good effect on his spirits.... What a pleasure it is for me to see what a couple of splendid Dutch fellows like you and Fokker can do![31]

Of course, the Dutch physicists were depressed by this news. Fokker began notifying those who had contributed to the rescue fund that their help would not be needed after all. But on June 30 another twist occurred in an unnerving message from Bohr: it seemed that her potential position and visa to enter Sweden were not yet "in order." Were they to interpret that Meitner had been misled, or that Siegbahn had *not* secured funds, as they had in the Netherlands, for even minor financial support of a woman without a home or belongings?[32]

Plans sommersaulted. While the struggle to arrange her secret exit continued, Meitner still went daily to the Institute. Nervous, exhausted, but unaware of the flurry behind the scenes, she sought solace in the details of her research. During the first days of July, as summer holidays were beginning for many Europeans, she spent long hours in the laboratories with Hahn and Fritz Strassmann, discussing their transuranic results and publications. But soon, outside events would disrupt any semblance of normal routine.

On July 4, Bosch notified Hahn that the National Socialist policy prohibiting scientists from leaving Germany would be strictly enforced, and warned him that while they all loved and respected Lise, she was in danger, even in the Hotel Adlon.

Debye immediately met with Meitner; they agreed she must leave the country as soon as possible. No visa or entry permits had arrived from Sweden as yet; so once again, Debye in Berlin turned to Coster in the Netherlands, referring to Lise Meitner in his letter in veiled terms as a male assistant:

> The assistant we talked about, who had made what seemed like a firm decision, sought me out once again.... He is now completely convinced (this happened in the last few days) that he would rather go to Groningen, indeed that this is the only avenue open to him ... perhaps I may now have the pleasure of showing you my laboratory. If you come to Berlin, may I ask you to be sure to stay with us, and (providing of course that the circumstances are still favorable) if you were to come rather soon, as if you received an SOS, that would give my wife and me even greater pleasure.[33]

It took several days for this appeal to reach Coster, who immediately recognized the urgency of the situation. He lost no time sending an answering telegram: SAT 9 JULY / I AM COMING TO LOOK OVER THE ASSISTANT; IF HE SUITS ME I WILL TAKE HIM BACK WITH ME.[34] He then frantically telephoned Fokker, who immediately contacted the Hague, where the Dutch Ministry of Justice had received their appeal for Meitner's entry twelve days before. But the day was Saturday — the Ministry was closed. Fokker then turned to the office of the Dutch Border Guards, and fortunately reached the Director, who promised him an answer on Monday. Coster spent the next day, Sunday, packing and anxiously awaiting a reply from Fokker and the authorities.

On Monday morning, July 11, Fokker's telephone rang. The Ministry had granted Meitner provisional admission; written confirmation was being forwarded. Coster was notified immediately, and departed directly from his office in Groningen for the train station.

In Berlin–Dahlem, the weekend of waiting strained Debye to the breaking point. He telegraphed Fokker at noon that same Monday: WITHOUT ANSWER FROM COSTER / CLARIFICATION URGENTLY REQUESTED.[35] Fokker wired back immediately that Dirk Coster would travel to Berlin and arrive that evening.

Coster quietly boarded the train to Germany from a Dutch train station. On board, Nazi soldiers were everywhere, and passports were checked and rechecked when they reached the border. The landscape seemed strange, with swastika flags waving from the medieval buildings in every little town. Covertly, he glanced at the documents he had brought for Meitner. At last he arrived alone in Berlin, late that Monday night. Debye had received Fokker's reassuring telegram, and met Coster at the crowded station in Berlin. The Debye family welcomed him into their home, and after a comforting evening he retired, exhausted.

Lise Meitner, unaware of the turn of events, had been waiting anxiously for confirmation from Stockholm. That hot and humid Tuesday, she arrived early at the Institute for Chemistry, and was startled when Debye suddenly intimated there had been a change of plans. When Hahn called her into his office, she knew something had gone amiss with the Swedish plan. She was stunned when Hahn informed her that she would be leaving for the Netherlands, that financial arrangements had been made for a modest stipend, and government approvals had been obtained. He then broke the news: she would have the afternoon to pack and leave immediately the next day. Hahn had met with Coster and Debye privately, and they had agreed that she was to stay at his home that night before her departure. She composed herself; the long-term research partners agreed that they would work a "normal" day at the Institute, then leave for her hotel together to pack her things.

A decade later she sadly recalled the events of her last night in Germany in the Institute. She did not want to make any of the staff suspicious, so she stayed until 8:00 at night, proofreading an article to be published by a young colleague. Then she packed two small suitcases. In her diary that night she recorded that Otto Hahn was very nervous. By 10:30 pm, Rosbaud came. She spent a sleepless night at the Hahn home.

The next morning, Wednesday July 13, Hahn and Rosbaud went to Lise's room at the Adlon, and packed one more small suitcase for her. The hotel staff were not informed of their guest checking out; no one was told of her departure that day. The plan was to give the impression that she had gone on vacation, a typical summer holiday. At the moment of their farewells, Hahn gave her a ring that had been his mother's, "for urgent emergencies."[36] Rosbaud was to drive her to the station. Coster would not meet them anywhere near the Institute, but would discreetly board the train and "accompany" her from a distance.

Panic overwhelmed Meitner on the way to the station. Rosbaud recalled that, as he drove her to the station, she begged him to turn back.[37] He steadied his nerves and hers, and they arrived at the crowded terminal. Rosbaud was relieved to see Coster waiting; the three exchanged casual greetings. Amid the many Nazi and SS-uniformed passengers, Coster boarded the train. Rosbaud guided Meitner to the door and embraced her. Alone on the train, she contemplated her uncertain future. Not until they neared the Dutch border many long hours later would she discreetly meet up with Coster in the same compartment. Hahn recounted:

> With fear and trembling we wondered whether she would get through or not. We had agreed upon a code word by which we were to be informed by telegram on the success

or failure of the journey. Lise Meitner was especially exposed to the danger of repeated controls by the SS in the trains going abroad. Time and again, people who tried to go abroad had been arrested on the trains and brought back.[38]

After a seven hour trip, the German train reached the Dutch border. Anxious and frightened, Lise sat watching as uniformed Nazi officers checked documents; her own lack of passport or entry visa made her feel faint. But days before, Coster had travelled to the border town of Nieuweschans: he had met personally with immigration officers there, shown them Lise Meitner's official entrance permit from the Hague, and asked them to use "friendly persuasion" with the German border guards to let her through.[39]

Now the crucial moment arrived. The Dutch guards looked down at her over the shoulders of the German officers, and then passed her by. Thus she was permitted to cross the border without incident. Several hours later, the train arrived in the university town of Groningen. Triumphant but exhausted, Meitner and Coster unobtrusively met one another in the small station; he took her bags and led her to his waiting car. Meitner was completely drained — in all of her eventful life, she had never experienced tension so great.

The next day Coster wired Hahn as planned that the "baby" had arrived. On July 15 Hahn replied with a cheerful telegram, I WANT TO CONGRATULATE YOU / WHAT WILL BE THE NAME OF YOUR DARLING DAUGHTER?[40] Then he quickly relayed the clandestine message to their network of friends throughout Berlin: Lise Meitner had arrived safely in the Netherlands.

The shock of the escape itself diminished very slowly for Meitner. She ate the nice meals prepared for her by Mrs. Coster, and in her small guest room sorted through the few belongings she had been able to bring. Surrounded by kind friends, it nevertheless felt strange to hear only the unfamiliar Dutch dialect. Still disoriented, she was unprepared to receive visitors; but Fokker and others did come to welcome her and discuss plans for the coming weeks. Fokker, elated by the success of their efforts, noted the heavy toll the escape had taken on Lise Meitner. It was apparent that she had been "inwardly torn apart," and that something had to be done to enable her to focus once again on her physics.

After seven day of rest, Lise gathered her courage and wrote to Scherrer in Switzerland, who had assisted in her emigration. She thanked him heartily, and reflected on her future options. At age 60, she was exhausted. Slowly she began to explore her new surroundings, with Coster's encouragement, going out for her daily walks. Groningen was so completely unlike the Berlin of the last five years:

no marching soldiers, no Nazi diatribes over the radio, no swastikas, or tension — just a peaceful, quiet university town, where people her age still rode bicycles to the marketplace, and scientific study was a way of life.

Coster had received a flood of congratulations from the international scientific community. Wolfgang Pauli, wired from Switzerland, YOU HAVE MADE YOURSELF AS FAMOUS FOR THE ABDUCTION OF LISE MEITNER AS FOR [the discovery of] HAFNIUM![41] Coster too had needed a full week of rest after the strain. During the remainder of July, while Meitner stayed in Holland, she and Coster frequently communicated with Bohr in Copenhagen about potential work opportunities in the Scandinavian physics community. Bohr held out hope that the Swedish Research Institute for Physics might still be a possibility when construction was nearer completion. He reported on its progress and Siegbahn's ongoing struggles to obtain cyclotron parts.

Meitner was optimistic about a possible move to Sweden. Although Holland had a fine reputation for physics, she was excited about the new Institute in Stockholm; her all-important transuranic research could continue there. Bohr concurred — after all, the financial backing from the Swedish Academy *was* in place for a spacious Physics Institute to be built, and construction was underway. Instead of boarding with friends as she did in Groningen, she would be provided with a room (and possibly a salary). Plans kept Lise focused away from the pain of her forced emigration.

At last a tentative offer came. Lise Meitner was to travel to northern Scandinavia, as had been originally planned. After two more hospitable weeks with the Costers, she packed her few things and prepared for the journey to Sweden by way of Denmark. Bohr himself assured Coster over the telephone that he would personally meet their precious cargo, and that she would stay in Copenhagen with his family until she was ready to work again. Deeply grateful to her Dutch friends, Lise waved goodbye from the ship, making its pleasant summer journey around the coast of Europe to the Danish capital.

Bohr's son Erik later recalled the arrival of the small, dark, somewhat withdrawn Lise Meitner at the family home that summer. The gracious mansion, the "House of Honor," had been given to Bohr by the Carlsberg Brewing Company many years earlier, upon receipt of his Nobel Prize, for use during his lifetime. A special guest room had been prepared for Lise, complete with fresh flowers and a stunning view. The house was an active one, filled with children and grandchildren, and Lise's subdued manner gradually warmed under their influence. She was much cheered by the congenial atmosphere, and enjoyed Bohr's tales of family and physics around the dinner table. By late August, she began to feel that there was indeed life after emigration, and that she would soon be able to carry on her work.

In Berlin, Hahn and her other friends had not given up pursuing the few remaining channels to secure her retirement allowance. Lise had previously hired a lawyer, who had discovered that her contract had been renewed by the Kaiser Wilhelm Society in 1929, but backdated to April 1, 1928 in order to extend her status to that of a "life employee" of the Institute for Chemistry, entitled to the pension of a professor. Meitner's legal status was discussed in a meeting between Rudolf Mentzel, the Referee for Sciences and Technology in the Nazi Ministry of Education (and, like Johannes Stark, an "Aryan physicist") and the director of general administration of the Kaiser Wilhelm Society, Ernst Telschow. Telschow had not been made aware that Meitner had emigrated; even Max Planck had not been notified. Carl Bosch, unsure of his colleague's political motives, reassured Telschow and others at the Institute that Lise Meitner was off on the typical August holiday that so many Germans enjoyed. But a letter from the Nazi Ministry of Education to the Kaiser Wilhelm Society was cause for continued concern:

> Professor Lise Meitner, previously an Austrian national, is working as a guest at the Kaiser Wilhelm Institute for Chemistry. Since she has become a German national through union with Austria, it has to be established what percentage of Jewish blood she has. According to previous assessments, Mrs. [*sic*] Meitner is supposed to have 25% Jewish blood. Referring to the talk which my referee Prof. Dr. Mentzel had with Dr. Telschow, I ask for prompt comment on that matter.[42]

Telschow then wrote to Otto Hahn in his official capacity as director of the Institute for Chemistry:

> I enclose a transcript of a decree of the Reich Ministry of Education for your information. You may discuss the matter with Prof. Meitner and inform me about what is necessary for the reply. I suppose that, now that the Ministry has taken up the affair, it would be expedient if Frau Meitner were herself to apply for leave until the question of her retirement or staying at the Institute [is] cleared up. Such an application would certainly be of use for the future discussion of the Ministry.[43]

Lise Meitner, on her way from the Netherlands to Denmark at the time, was unaware of this correspondence concerning rights to her pension based on racial heritage.

At Bohr's Institute, Lise Meitner soon relaxed and began to refocus on physics. The facilities for nuclear research were excellent; and as always, Bohr was surrounded by enthusiastic students and young active physicists. But her nephew Otto Frisch later speculated, "It was probably her wish *not* to compete with those younger people that led to her decision not to remain in Copenhagen."[44] The Stockholm offer had tempted her — a new lab, a cyclotron, possibly a small home. She was 60 years old and exhausted. But she was still particularly absorbed in the latest publications of the Joliot–Curies, whose transuranic research had led to their Nobel Prize in 1935. She devoured the French scientific journals as they arrived at the Bohr Institute. Still questioning the existence of their "short-lived" transuranic substance, she was anxious to resume her own experiments as a challenge to Paris, despite the lack of Hahn's perspective.

In late August, the day came for her departure to Stockholm. The finances were in place for her relocation, and eagerly awaited research opportunities lay before her. Bidding farewell to her Danish hosts, colleagues and friends, as well as her nephew, she boarded the ferry for Sweden. Upon her arrival, she travelled by train up the coast to the small town of Kungälv, where she was met by her longtime friend Eva von Bahr-Bergius. A former colleague from Berlin, herself a pioneering woman in science, Eva urged Lise to resolve her affairs in Berlin officially by retiring from the KWI and requesting her pension. Meitner agreed that this was the most honorable step. Still unaware of the ongoing correspondence between the Nazi Ministry of Education and the Institute, she wrote to Otto Hahn on August 24:

> Dear Otto,
>
> What I have to tell you today will bring about an extremely decisive turning point in both our lives. Yesterday I asked Geheimrat Bosch to agree to my retirement. We need not speak about feelings; both of us certainly know about such things. I am sure no day will pass when I shall not think with gratitude and longing of our friendship, our joint work, and the Institute. But I don't belong there any longer, and when I reflect on the last months, it seems to me that my retirement will also comply with the wishes of the staff. It's no use talking about it too much: the facts are facts, you cannot pass them over. My Swedish friends offered to let me stay with them, and I have come to the decision that this is the only course left open to me. Actually I would like to express my thanks personally to all the colleagues and employees of the Institute for the years of excellent cooperation. But perhaps you think it would be better for you alone to tell them. They all know me well enough to know that I don't like talking about things that are irrevocable. Think it over calmly and let me know which form you consider suitable. However, please let's not talk about the fact of my leaving. Of course, my decision involves a number of practical necessities. I also wrote a very short letter to Dr. Telschow and asked him to discuss the regulation of my affairs, such as retiring allowance, etc. with Dr. Leist [her lawyer]. I'll ask Annchen to carry out the dissolution of my flat, with the help of a removal company, of course; I know very little about the legal formalities but I hope for the help of Dr. Leist. I have not yet made any detailed plans, I just thought that I would put my things in storage here. And then I must await

to see what the future brings. I also need inner calm to think over everything rationally. In my inner self I have not yet quite realized that what I have just written is real, but it *is* real. Let me hear from you soon. With very kind regards,

Yours forever,

Lise[45]

On August 29, the day after the Institute's vacation period ended, Otto Hahn reported Lise Meitner's decision to the Institute for Chemistry administration and staff in Berlin-Dahlem. Many were saddened at the news, while others rejoiced that the Jewess was gone and they could continue their scientific work in peace.

The Discovery and Interpretation of Fission: 1938

LISE MEITNER GREW INCREASINGLY DESPONDENT as she prepared to move from the western part of Sweden to Stockholm, taking only the few belongings she had carefully packed at the home of her friend Eva von Bahr-Bergius. Hahn had written to her from Berlin on September 12:

> Telschow [Director of the Kaiser Wilhelm Society] asked me today how the [Nazi] Ministry [of Education] knew about your 'departure' — the Ministry had actually asked him about it. [1]

Obviously, Meitner made her departure before "legal" channels were formally notified, thus sparing her from receiving the Ministry's direct orders not to leave Germany — or an even worse fate. Former allies were quizzed about her status now that Nazi officials were making inquiries. On October 1, Otto Hahn wrote again:

> ...Having tried again and again to telephone the Adlon Hotel [temporary residence of Carl Bosch, president of the Kaiser Wilhelm Society], I contacted Geheimrat [Bosch] the day before yesterday, and yesterday spent the afternoon with him: at first, just the two of us, and then with Dr. Leist [Meitner's legal advisor]. Bosch confirmed what Telschow had already said. The Kaiser Wilhelm Society will try everything possible for you and Otto [Heidelberg, another victim of the Civil Service Laws within the Institute for Chemistry]. However, Bosch does not know whether the applications will go through. As for you, perhaps a certain difficulty could arise in so far as in your

case a personal consultation with "dem obersten Himmel" [Heinrich Himmler] has been considered. If they should find out now that you are not in Germany, they would perhaps be cross ....[2]

Safely out of the range of the Nazi government, Lise was beginning to understand that she might have to *remain* in Scandinavia for some time: the war had changed everything.

In the months that followed Meitner's immigration to Sweden, she lived in a small hotel room in downtown Stockholm and slowly adjusted to the Arctic winter environment. She was aided by her friends at the University, Oskar Klein and his wife, as well as the local rabbi and Protestant minister. But she fretted a great deal because she was out of touch with her former work. Her long letters to Hahn continued.

Manne Siegbahn welcomed her into the Nobel-funded Physics Institute, but he was so preoccupied after the groundbreaking that Meitner often took the subway from Stockholm to the Institute only to find empty labs, busy colleagues conversing in Swedish, and no assistants in sight. However, in the small room given to her, she tried to carefully assemble experimental apparatus to continue research on the transuranics. The constant noise and lack of basic equipment frustrated her further. The Institute was still very much a "work in progress," as younger Swedish physicists eager to work on the cyclotron later recalled. Meitner wrote to Hahn:

> September 25, 1938. ... Perhaps you cannot fully appreciate how unhappy it makes me to realize that you always think that I am unfair and embittered, and that you also say so to other people. If you think it over, it cannot be difficult to understand what it means to me that I have none of my scientific equipment. For me that is much harder than everything else. But I am really not embittered — it is just that I see no real purpose in my life at the moment and I am very lonely ....

> October 15, 1938. ... There is little to tell [from here]. Experimental work is out of the question; it will take a very long time until even the simplest apparatus will be there, and then perhaps also a collaborator. Of course, Bohr's visit was very pleasant and probably useful, too, because he also spoke to Manne S[iegbahn]. But S[iegbahn] is not at all interested in nuclear physics and I rather doubt whether he likes to have an independent person beside him. B[ohr] suggested that I should have patience, what else should he do? And I can't do anything but live my life just as it is ....

> November 6, 1938. ... On Monday [12th], Bohr will deliver a lecture here. I hope I can speak to him a little on his own. Siegbahn's Institute is incredibly empty. An excellent building where preparations are being made for a cyclotron and many other large X-ray and spectroscopic apparatus; but experimental work can hardly be *thought* of.

November 26, 1938. . . . Your suggestions for the apparatus are hardly practical. It would be so much better if one could come to Siegbahn so to speak with a gift instead of a demand, and if it were only a drawing of a piece of apparatus. However, that doesn't seem possible . . . .[3]

In all her negotiations with Cambridge, England and Berkeley, laboratories were available to her for nuclear research. She had believed that such was going to be the case in Stockholm as well. But the construction of a new Physics Institute was slow. On top of this there was her language isolation, as few Swedes wanted to converse in German and Meitner spoke little Swedish — although she set about learning it immediately. On November 28, she turned sixty years old. After a small party she returned to her hotel room through a blizzard by herself. The small vestiges of a collegial mood had evaporated, although in the face of it all she kept her inner optimism about the future. The following letter to Hahn reveals her plight that winter:

December 5, 1938. There is little or nothing to tell about myself. . . . I often feel like a wound-up puppet that does certain things, gives a friendly smile, and has no real life in itself. By that you can judge how valuable my activity is. Nevertheless, I am very grateful for it after all, because it compels me to collect my thoughts, which is not always easy . . . .[4]

Fortunately, there was one chance that fall for Lise to interact with Otto Hahn. Invited by Bohr, Hahn went to Copenhagen where he and Lise had a tearful reunion. Before Denmark, however, Hahn travelled to Vienna to interview Josef Mattauch, a potential successor to Meitner's position; he also had dinner with two of Lise's sisters and their husbands in his hotel. On November 10, Lise Meitner hurriedly packed and took the eight-hour train ride across Sweden to Copenhagen and to Bohr's beautiful home where she met Hahn. There was a lovely reception there given by Margrethe Bohr. The next day the three colleagues got down to business, challenging Hahn about his chemical results and the puzzling transuranics. Hahn later recalled:

I gave a talk in Copenhagen in which I described our previous results and our most recent investigations of the artificial radium isotopes. Bohr was skeptical and asked me whether it was not highly improbable that two alpha particles were separated from uranium (with the charge of 92) if slow neutrons were used — in this way, radium with four fewer positive charges should be produced. I had to reply that there was no other explanation, since our artificial radium could be separated only with weighable quantities of barium as a carrier substance. So apart from the radium, only barium was present, and it was out of the question that it was anything but radium. Bohr suggested that these new radium isotopes of ours might perhaps in the end turn out to be strange transuranic elements. [5]

The guest book at the Copenhagen Physics Institute indicates that Lise Meitner and Otto Robert Frisch were present for these discussions, and obviously took part in debating *interpretations* of Hahn and Strassmann's chemical precipitations.[6] But Hahn's visit was brief. Administrative "duties" awaited him in Berlin and he was hard pressed to get back and attend to his wife who was seriously ill. A mental breakdown seemed imminent. Before Hahn caught the train back to Berlin, Meitner implored him to keep her informed, and Bohr urged him to triple check that their supposed radium was not some sort of transuranic element. Remember: Hahn was puzzled and had gone to the physicists – and his long term colleague – for the interpretation of this work.

While Hahn returned to his secure job and home, Meitner had little to return to, and became even further depressed during the nine-hour train ride back to Stockholm. She had begun her work in Siegbahn's Institute at the low salary level of an assistant; this did not change for years, and it was the best the Swedes could offer her.[7] Tragically, soon more emigration would beset the Meitner family. Lise began to make secret plans for her eldest sister Auguste (Otto Robert's mother) to join her. Auguste's husband, Justinian (Jutz) Frisch, a well-known lawyer in Vienna, had been recently stripped of his right to practice law after Hitler annexed Austria. Further, the day after Hahn had dined with them all, Jutz had been arrested and sent to Dachau. Hahn and Meitner were shocked. All their mail had been circumspect before this, but luckily, Hahn was able to hand a letter to a Danish colleague of Bohr's, who mailed the following to Meitner from Copenhagen on November 19:

> I can hardly tell you how depressed I also am by Jutz's situation. And perhaps mostly because of YOUR fear and worry about him . . . .The old von Weizsäcker is on vacation [State Secretary Baron Ernst von Weizsäcker, father of Meitner's physics assistant Carl Friedrich von Weizsäcker]. Anyway, it is unlikely that he would have intervened or would have been able to intervene. We are all of the opinion that with these mass arrests, it would be impossible to have the cruelty and harshness that was the case earlier, when undesirables came to the c[oncentration] camps. But naturally, this is small comfort . . . . [8]

Hahn's attitude about concentration camp victims – "human undesirables" as he referred to some of them – did little to soothe Lise Meitner. Hitler had unleashed the Nazi Party's fury on November 9, 1938 throughout his expanded Reich. Jewish lives were shattered like so many panes of glass in an organized pogrom known as

Kristallnacht. Three days later, the "Atonement of the Jews" completely eliminated the rights of all Jewish people in the Third Reich from engaging in any business or cultural life. By law, those who had Jewish grandparents, even if they had never practiced the religion, were unable to sell merchandise or engage in commerce. On November 15, Jewish children were forbidden by law to attend any German school, and all Jewish names were to have "Sarah" or "Israel" inserted for identification. Soon the Jews were required to wear the badge of a yellow star on each coat or sweater. All of Meitner's documents were duly changed to "Lise Sarah Meitner" by her lawyer, Dr. Leist.[9]

In dark and snowy Scandinavia where the sun did not rise until noon and cast a dim glow until it set at 3 P.M., Lise decided not to spend the holiday season alone. Her friend and colleague Eva Von Bahr-Bergius had invited her to the Swedish coast, and hence Lise decided to spend time in the quaint coastal village of Kungälv. The Christmas season had always been a mixed blessing and this forthcoming holiday seemed particularly empty without friends and family. But Otto Robert had also been invited to come from Copenhagen. Since his father's internment, even Bohr could not console him. Otto Robert would travel by boat from Copenhagen to join his aunt. He did want to talk about his own physics work with her. The two planned their trip by telephone.

Before departing from Stockholm, Meitner received a troubled letter from Hahn. He and Fritz Strassmann had begun experiments before the holiday and were facing some startling results regarding the troublesome transuranics, results which might have invalidated several years of earlier work. In a series of experiments with radium, actinium, and thorium, a "fractionalization" had occurred which called into question the whole transuranic schema. Hahn wrote to Meitner describing the new results and expressed his doubts about his own findings. It is essential to note the *tone* of his letter to Lise Meitner. He was frustrated and puzzled by the results and relied heavily on Meitner's physical interpretation of new results. The famous letter is dated December 19, "Monday evening, in the lab" and formed the backdrop for the first act in the most momentous series of events in both scientists' careers. It read:

Dear Lise,

...I must mention your affairs at the outset, since you do not know 'if someone is taking care of your affairs.' Before, we agreed that I would speak with a greater sense of urgency to someone at the Finance Office. I did that on Saturday. We were told that we should make new petitions, etc. When I had a small attack of my 'dizziness', things went better. I stressed the fact that the Kaiser Wilhelm Society is also urgently interested in the settlement of this matter. There is, however, no way around a short new petition, and this will be sent to the Finance Office tomorrow morning with a letter of interest by the Kaiser Wilhelm Society ... You can believe me when I tell you that this approach, which drags things out, is of great concern to me as well, especially when you really believe that we are just letting things lie.

... My name in the exhibition "The Eternal Jew" is now suddenly also bothering the administration. Therefore, at the suggestion of Cranach, I made a sworn statement today of assurance concerning my citizenship in the Reich, and will submit my papers in a few days. I have also had difficulties in the Mattauch matter up until now, because an aspirant to your apartment who doesn't belong to the Institute has been "presented" ... I hope this attempt can be turned aside.

... Meanwhile, we are working on the uranium substances – myself, whenever I get the chance, and Strassmann, untiringly, supported by Lieber and Bohne. It is now practically eleven o'clock at night. Strassmann will be coming back at 11:45 so that I can get home at long last. The thing is: there is something so ODD about the 'radium isotopes' that for the moment we don't want to tell anyone but you. The half-lives of the three isotopes are pretty accurately determined; they can be separated from all the elements except barium: all processes are correct. Except for one — unless there are some very weird accidental circumstances involved, fractionalization doesn't work. Our Ra [radium] isotopes behave like Ba [barium]. We can find no definite enrichments with $BaBr_2$ or chromate, etc. Now, last week on the first floor [Meitner's former Physics Dept.], I fractionated ThX, and it all went exactly right. Then on Saturday, Strassmann and I separated one of our Ra isotopes with MsTh1 as an indicator. The mesothorium was enriched according to plan, but not our Ra. This again might be the result of a very curious accident. But we are more and more coming to the awful conclusion that our Ra isotopes behave not like Ra, but like Ba. As I've said, other transuranic elements – U, Th, Ac, Pa, Pb, Bi, Po – don't fit the picture. Strassmann and I agree that for the time, nobody should know but you. Perhaps you can suggest some fantastic explanation. We ourselves realize that it can't really burst into Ba. What we want to check now is whether the Ac isotopes produced out of the Ra behave not like Ac, but like La. All rather tricky! But we must clear this thing up.

... We [Hahn and Strassmann] intend to write something about the so-called Ra-isotopes for *Die Naturwissenschaften* before the Institute closes, because we have achieved some very nice [decay] curves.

Now I have to go back to the counters. I hope I can write to you again in two days. But if the next letter should arrive after the 24th, then I would like to wish you today a reasonably pleasant Christmas. Write soon.

Very cordial greetings,

Yours,

Otto

And at the bottom of the December 19 letter was the postscript:

Very cordial greetings and best wishes also from me.

Yours,

Fritz Strassmann[10]

The essential experiment Hahn was so anxious about had actually been carried out with Strassmann on December 16, 1938. The two, prompted by Meitner, had irradiated uranium with slow neutrons and then performed a painstaking fractional crystallization yielding three different fractions, each containing the same amount of the carrier element barium. The puzzle was: why didn't more appear in the third result than in the first? Historian of science Spencer Weart reflects:

> In particular, they looked for the 86-minute decay that characterized the substance they called radium III. By the evening of Dec. 17th, they had found this activity to have about the same strength in each of the three fractions: it followed the barium. This could simply have indicated that the fractionation was NOT working properly, but they had undertaken a control, much like the one Curie and Savitch had used when they showed that their 3.5-hour substance differed from actinium. Watching the radioactive curves develop, Hahn and Strassmann could see by noon on Dec. 18 that the fractional crystallization WAS working properly. Their supposed 'radium' was behaving like no other radium, but like *barium*.[11]

What Hahn wrote to Lise Meitner was "we are coming more and more to the awful [*schrecklich*] conclusion that our Ra isotopes behave not like Ra, but like Ba." His laboratory notebooks, now housed in the Deutsche Museum in Munich, prove that he was speculating about his results, changing his notations back and forth.[12] Even in his late night letter to Meitner, he writes "radium isotopes" in quotations and then without, as if in his mind, he was uncertain as to whether these results were correct. And indeed this was a puzzle to Meitner as well.

Over twenty years later Hahn wrote that his 'over-cautiousness' stemmed primarily from the fact that, as chemists, we hesitated to announce a revolutionary discovery in physics."[13] But we must note the troubling professional hesitations Hahn was also having: if their findings challenged years of work, their carefully constructed "trans-uranic" edifice would come tumbling down. As Hahn later revealed:

> The separation of this active group [three artificial beta-active radium isotopes with different half-lives which, in turn, transformed into artificial beta-active actinium isotopes] was performed by means of a barium precipitate — not, however, in the form of barium sulphate, which, with its large surface, strongly adsorbs other elements, but as suggested by Strassmann, in the form of barium chloride, which crystallizes very well

from concentrated hydrochloric acid and which precipitates uncontaminated by other substances.[14]

Note that, despite Hahn's skepticism, Hahn had taken Strassmann's suggestion and analyzed the same puzzling 3.5 hr. element found by the Joliot–Curies, which had indeed been lanthanum, the daughter product of barium. As early as 1936, Strassmann later documented, he had performed independent experiments on uranium and even at that time found what appeared to be barium. Yet when he related this to Meitner, she was highly skeptical, and he threw his lab notes away.[15]

Hence, in December 1938, Hahn and Strassmann concluded that the Paris team's 3.5 hr. activity was probably a mixture of several radioactivities and that anything "carried out" with barium could only be isotopes of radium. They began to develop a scheme, reminiscent of the scheme for thorium, with three isomeric radium-actinium chains, each arising from two successive alpha-emissions, the day after Hahn wrote Meitner. And, urgently, he pressed Meitner to *interpret* these results. Hahn later recalled his blind spot that hectic week before Christmas:

> Neither of us voiced the crucial question whether it was not the *barium* instead of the radium that was to be regarded as the 'product' of the experiment. This just shows how wildly impossible it seemed to regard barium as the product of the reaction, and it shows how cautious Strassmann and I were when we too began by regarding our results with deep skepticism. [16]

Lise Meitner took their findings as a challenge. She replied immediately in her customary brisk style, as she packed for her Christmas journey across Sweden:

December 21, 1938

Dear Otto!

Cordial thanks for your nice letter and all your trouble. Believe me, I understand that you have all kinds of matters filling your head now. Did the conversation with Bosch end all right? How was the Institute Christmas celebration? And Edith's birthday? Did she manage well, or were you nervous? The affair with the exhibition [Hahn's inclusion in 'The Eternal Jew' mentioned in his prior letter] is a very stupid thing to happen. Do you have any idea who played that trick?

Your radium results are certainly very odd. A process in which slow neutrons are used and the product seems to be barium! Are you quite sure the radium isotopes came before the actinium isotopes? Because I seem to remember you once wrote that you had observed the increase of actinium from Ra. Is that so? And what about the thorium isotopes from it? What ought to be produced out of lanthanum is Cer. As things stand, I find it very difficult to assume such a degree of bursting [*weitgehenden Zerplatzens*], but we've had so many surprises in nuclear physics that one can't very well just say, 'it is impossible'. Incidentally, are higher transuranic elements, such as Eka Au or even higher ones, absolutely out of the question? After all, Pa behaves in a very similar way to Zr. So why shouldn't, for instance, an Eka Au or an Eka Hg behave in a very similar way to Ba? Or is that possible?

Has Droste continued to measure his so-called Pa preparations, and what did he find?[17] And do you have any idea how the experiments of Flammersfeld are going? He wanted to write about them but never did, and maybe he lost interest.

My so-called work – when I think about it – I can only say 'Oh God!' I have finally collected a single filament electrometer, a few usable counters, and an amplifier — a very bad example for simple [particle] counting, no help at all. My head is so full of other things that almost everything else is of no consequence to me. But somehow, a person must, after all, live her own life.

This is not really a proper Christmas letter. But I wish you with all my heart a few good and happy days of Christmas. Please write me about Edith's situation in detail. Friday I will probably go to Kungälv for a week; if you write in the meantime, please use that address. The weather is not as nice here as where you are. We have so much snow and sometimes real blizzards. We have to have the lights on practically the whole day, and it is freezing most of the time — at least it is for me, at any rate.

Many cordial greetings to all of you, and the best. Much love for the New Year.

Yours,

Lise [18]

Cautious but open-minded, Lise Meitner was prepared to interpret Hahn and Strassmann's results. On the journey from Stockholm to Kungälv on December 23, she was already thinking about nuclear structure and the questions barium presented. Meitner was clearly receptive to new models of the very fundamental structure of the atomic nucleus. "One can't very well just say, 'it is impossible.' " She wanted to urge Hahn to retest his work. The death of the transuranic scheme – and four years of research – was almost too much to bear.

By the time she and Otto Robert were reunited, it was the beginning of the Christmas holidays at Eva's festively decorated home. In Berlin, Hahn wrote up an article and without showing it to Strassmann, delivered it to Paul Rosbaud, editor of *Die Naturwissenschaften*, who confirmed having received it on December 22.[19] He originally had intended the article to be an account of their work with radium isotopes, but crossed this out and substituted "On the Alkaline Earth Metals" in the title, since their results so resembled this section of the revised Periodic Table. His article, soon to become world famous, was simply titled "Concerning the Determination and Relationships of the Alkaline-Earth Isotopes Derived from the Neutron-Irradiation of Uranium." In its concluding paragraphs, he revealed his hesitations to draw conclusions without a physicist's interpretation of their results:

> We agree with the findings of Curie and Savitch on their 3.5-hr activity (which was, however, not a single species) that the product resulting from the beta decay of our radioactive alkaline earth metal is not actinium. We want to make a more careful experimental test of the statement made by Curie and Savitch that they increased the concentration in lanthanum (which would argue *against* an identity with lanthanum) since in the mixture with which they were working, there may have been a false identification of enrichment. It has not been shown yet if the end product of the "sample" which was designated as "thorium" in our isomeric series will turn out to be certain.[20]

Otto Hahn was preoccupied during the holidays. Nazi sympathizers were interested in taking over the Institute. There was even competition for Lise Meitner's apartment. He had not told anyone (except for Strassmann) about his amazing results. Lise Meitner was still his closest confidant; his wife would soon be committed to a sanitorium after another nervous breakdown. He and his teenage son Hanno spent a quiet Christmas. Yet the chemical results he had written up would not let him rest. On December 27, Hahn telephoned the editor of *Die Naturwissenschaften*, and requested the following crucial paragraph be added to the article:

> Concerning the 'transuranic group,' these elements are chemically related but not identical to their low homologues rhenium, osmium, iridium, and platinum. Experiments have not been made yet to see if they might be chemically identical to the still lower homologues masurium, ruthenium, rhodium, and palladium. After all, one could not even consider this as a possibility before. The sum of the atomic mass numbers of Ba + Ma, e.g., 138 + 101, equals 239![21]

This additional paragraph would soon change the world.

Only after this speculation did Hahn write to Lise Meitner. She was indeed still an active partner in their complex, interdisciplinary research, and he asked for her insights as he had done so often in the past. His final article, which would be shortly distributed worldwide (in January 1939), bore the brunt of revising and toppling the transuranics. However, the key insight – nuclear fission – remained a mystery to Hahn, and was *not* at the heart of his publication. The key paragraph in his several page article stated:

> From these experiments, we must, as chemists, rename the elements in the above scheme, and instead of radium, actinium and thorium, write barium, lanthanum, and cerium. As 'nuclear chemists' who are somewhat related to physicists, we cannot yet decide to take this big step, which contradicts all previous experience of nuclear physics. It is still possible that we could have been misled by an unusual series of accidents.

It is intended to carry out further tracer experiments with the new radioactive decay products. In particular, a combined fractionation will be attempted, using the radium isotope resulting from fast neutron irradiation of thorium together with our alkaline earth metals resulting from uranium. At places where strong neutron sources are available, this project could actually be carried out much more easily.[22]

On December 21, after writing the first draft of his article, Hahn had written a late night letter to Lise Meitner in Sweden, and then closed up the Institute for the holidays. He mailed this letter to the von Bahr-Bergius' in Kungälv, and several days later, when Otto Robert came out of his hotel room to greet his aunt Lise, she was puzzling over it. Christmas mail was slow, and the crucial article Hahn referred to, which he had already delivered to *Die Naturwissenschaften*, was not enclosed. Hahn's stress was apparent in his Dec. 21 letter to Meitner:

> . . . How wonderful and exciting it would have been if we could have worked together as we used to. You might have been rather shocked by the enormous number of experiments, because we never had the time, or at least thought we never had the time, to measure everything to the last point. Recently the number of experiments has always been limited only by the number of lead boats and the three counters that are all we have. Yesterday we began writing up our Ba-Ra [barium-radium] evidence, today the lab's closed, and at 8:00 this morning we switched off the counters. Tomorrow Fraulein Bohme will be in to type out the part of the work that Fraulein Müller couldn't do, which we shall finish writing up in the morning. We must get the paper to *Naturwissenschaften* on Friday.[23]

Otto Frisch could not understand his aunt's anxiety on the morning he greeted her in Kungälv. He had come from Copenhagen and wanted to talk over his father's desperate situation, as well as other physics experiments he had been working on. He recalled: "I wanted to discuss a new experiment with her that I was planning, but she wouldn't listen: I had to read that letter [from Hahn]. Its content indeed was so incredible that I was at first inclined to be skeptical."[24]

Lise Meitner was intrigued. She and Frisch excitedly discussed the experiments and their possible ramifications, ruminating over interpretations of the chemical results Hahn described. Meitner was so troubled by the implications of Hahn's letter that she asked Otto Robert to walk with her in the woods. He wanted to cross country ski, so the duo set off into the Swedish countryside. Lise Meitner walked briskly beside her nephew on thin wooden skis. Frisch later recalled their heated discussion:

> The suggestion that they [Hahn and Strassmann] might after all have made a mistake was waved aside by Lise Meitner; Hahn was too good a chemist for that, she assured me.

But HOW could barium be formed from uranium? No larger fragments than protons or helium nuclei (alpha particles) had ever been chipped away from nuclei, and the thought that a large number of them could be chipped off at once could be dismissed; not enough energy was available to do that. Nor was it possible that the uranium nuclei could have been cleaved right across. Indeed, a nucleus was not like a brittle solid that could be cleaved or broken; Bohr had stressed that a nucleus was more like a liquid drop.[25]

A new possibility was dawning: perhaps the nucleus had literally "burst" in two! Yet Meitner remained skeptical. What could this mean in terms of nuclear structure, when barium, half way down the periodic table, appeared from uranium?

Creative processes often are accompanied by visual models. However, the heuristic model is often NOT an accurate picture of reality. This is why models change over time in the sciences, and are challenged again and again by mathematicans, physicists and often, new evidence. [26] In 1937, Niels Bohr had developed a "liquid droplet model" of the nucleus and even coauthored a paper with F. Kalckar on it.[27] As early as 1934, George Gamow, a bright Russian student of Bohr's – and friend of Otto Robert Frisch – had also suggested such a model and published his thoughts about it.[28] Credit is due to both these seminal physicists for their insights. But at the moment in the Swedish woods when Meitner and Frisch were confronted with Hahn and Strassmann's evidence, they took the first leap and interpreted a new process which this "liquid drop" nuclei might go through. Frisch recalled that they speculated:

> Perhaps a drop could divide itself into two SMALLER drops in a more gradual manner, first becoming elongated, then constricted, and finally, being torn rather than broken in two. We knew that there were strong forces which would resist such a process, just as the surface tension of an ordinary liquid drop resists its division into two smaller ones.[29]

The key, Meitner was quick to point out, was that the Couloumb barrier, the electrical charge which "holds" nuclei together, was known to diminish the effects of nuclear surface tension, perhaps with dramatic results if it was overcome. At this point, they stopped their discussion. Frisch unstrapped his skis, and they sat down on a snow covered log. Always the physicist, Meitner pulled out a scrap of paper and pencil from her winter coat pocket, and began to calculate. Frisch remembered clearly:

> The charge of a uranium nucleus, we found, was indeed large enough to overcome the effect of the surface tension almost completely; so the uranium nucleus might indeed

be a very wobbly, unstable drop, ready to divide itself at the slightest provocation, such as at the impact of a neutron.

But there was another problem. After separation, the two drops would be driven apart by their mutual electric repulsion and would acquire high speed and hence a very large energy — about 200 million volts in all. Where could that energy come from?[30]

It was Meitner who had the key insight: fortunately, Frisch recalled, she remembered the so-called "packing fraction" formula, and visualized an elongation of the nucleus, so that the TWO nuclei formed would, according to her calculations, be lighter than the original by approximatley one fifth the mass of a proton. She then applied Einstein's $E = mc^2$. One fifth of a proton mass was equivalent to – and here, she was astounded by her own figures – 200 MILLION electron volts. Frisch recalled their triumph: "Here was the source for all that energy; it all fit!"[31]

*   *   *   *   *

In the final December days of 1938, overlapping letters flew between the two partners, posing questions and exposing hesitations, doubts, and excitement. The exciting realizations acquired in Kungälv needed to settle in Lise's mind. She had agreed to telephone Otto Robert in Copenhagen periodically to suggest experimental tests to confirm their new ideas. But this would have to wait until after the New Year.

Hahn's initial letter announcing that his article would soon arrive had certainly startled Meitner and affected her entire vacation in western Sweden. She had racked her brain wondering what the article had stated, and what new conclusions Hahn had drawn that were worthy of publishing without her contributions. Christmas was over; the town's small post office was again open. She quickly mailed her response to Berlin, both praising the December radiochemical experiments of Hahn and Strassmann and reporting in detail the insights she and Otto Robert had arrived at during their extraordinary holiday together. Dated December 29, mailed without a signature, she scribbled from Kungälv:

Dearest Otto,

Many thanks for your lovely letter of the 28th. Your radium-barium results are very exciting. Otto Robert and I have already racked our brains; it's a pity the manuscript wasn't forwarded to me, but I have reclaimed it and hope to have it tomorrow. Then we can better reflect on it. The proof appears to indicate to me what *affects* the transuranics, which are the uranium isotopes of U23′ and U40″, and this necessarily points to the transuranics. Or don't you think that these reactions decisively tip the balance in favor of the existence of the uranium isotope?

In the letter, we should observe Meitner's caution, her reluctance to name the substance a transuranic, but still. . . . She shifts her questioning; his "proof" indicates what *affects* the transuranics, and not the result. Then she shifts her inquiry altogether, continuing:

> Further, how is it with the so-called actinium? Do you sever yourself from lanthanum or not?

Here is her challenge to Hahn: will he remain committed to their previously published affirmations, or come out in support of his barium results? Then she continues on, fretting about personal matters:

> May I confide a few private requests, and if they can't be fulfilled, then it is all right. Could I receive my index card file? Further, the "Manual" of Kohlrausch (page 9) and finally, a few reprints of your first *Naturwiss.* notice; I don't even have those, you have taken them with you again. And what is happening with my volumes of Institute work?
>   I would also gladly know more of the Mattauch affair. Perhaps you could also relate the circumstances of my affairs in general to Fräulein R. Once again, all the best in the New Year and many loving wishes to Edith. Did Hanno get the book I sent him?
>   Sincerest regards from O.R. [Otto Robert] and me. [without signature]
>   Heartfelt New Year's greetings and wishes.
>   12.30.38
>   Today the manuscript [Hahn's Dec. 21 carbon copy of the article forwarded to *Die Naturwissenschaften*] came — unfortunately missing the very important page 11 and all the diagrams. It is all very bewildering.
>   All the best for Edith, and very heartfelt greetings to Hanno.[32]

Perhaps Meitner felt that Hahn's mail had been censored, or perhaps she was just frustrated to be so far removed, even from this, the most exciting insight she had reached in many years. Because scientific advances are often not discovered in a straightforward fashion, and because dead ends, frustrations, and anomalous data that make no sense all play a part in the shifting of conceptual frameworks,[33] this time period is crucial to analyze. The drama had just begun. Meitner must have paced up and down her hotel room. She finally called her nephew and promised that when several matters were settled, she would visit Bohr's Institute; in the meantime, she would work out the physical calculations of their new idea while Otto Robert would put together a series of experiments.

Lise Meitner did *not* have the full Hahn article during Christmas while she and Frisch were working through the Hahn–Strassmann discovery. Indeed she was quite amazed when she returned to Stockholm and, on January 2 (after she had picked up her mail), found Hahn's bold assertions had already been submitted to *Die Naturwissenschaften*. As she stated to an American journalist many years later:

> On reading the letter, I was thoroughly excited and amazed and also, to tell the truth, uneasy. I knew the extraordinary chemical knowledge and ability of Hahn and Strassmann too well to doubt for one second the correctness of their unexpected results. These results, I realized, had opened up an entirely new scientific path, and I also realized how far we had gone astray in our earlier work![34]

Back in Berlin, we know that Hahn had not rested easy over the Christmas holidays, as he thought about his new results. He later wrote that during the entire second half of 1938, between his wife's illness and the psychological stress of Lise's departure, he was under "severe nervous strain."[35] On December 28, 1938, a week after submitting his soon-to-be-historic paper for publication, he again wrote to Lise, thinking perhaps that she was still quietly resting with her friends and that her mail would be forwarded:

> Dear Colleague,
>
> I will quickly write to you about my new Ba [barium] fantasy. Perhaps Otto Robert is with you in Kungälv, and you can talk this over a bit. The manuscript of our work has been sent to you. I still have some insignificant alterations to undertake; all the footnotes from now on must be changed since the transuranics cannot perhaps be Re, Os, Ir, Rt. And now comes my new fantasy: we have not proved that the transuranics are not "Ma, Ru, Rh, Pd." At least I for my part know so little about those substances that I cannot exclude them chemically. Strassmann is away. What is the possibility that uranium 239 could split into one Ba and one Ma ? One Ba 138 and one Ma 101 gives 239.
>
> ...Eventually, you can calculate these and publish. If some are thus surrounded in this way, the transuraniums "Ausonium" and "Hesperium" will die. I do not know whether I am very sorry.... I have not talked over this newest affair with anyone until now.[36] [The letter ends here.]

This letter is essential in tracing the process of discovery and resistance that Hahn passed through as he looked back over his December data and puzzled over its physical interpretation.[37] One week after the fractionalizations were completed, he became extremely worried that the transuranic scheme he and his partner had labored over for nearly four years would "die" in the face of his new data. Meitner was also worried about this: after all, four years of labor in constructing an edifice as elegant as the transuranics, accepted worldwide, would crumble if they postulated the "bursting" or splitting of the atomic nucleus. Hahn's writing clearly reveals his major doubts over the physical interpretation of their barium results. As nuclear chemists, he and Strassmann lacked the theoretical physics background that Meitner had always provided, and were correctly hesitant to announce any radically new models of nuclear structure. During 1939, however, Hahn's article would be read around the world.

Lise Meitner also set about composing a major paper that would reveal a scientist's struggle with the issue of intellectual uncertainty which often accompanies scientific "anomalies" that do not go away. Carefully constructed theories and "proven" empirical results (such as the elaborate transuranics) that conflict with new evidence play an enormous role in the intellectual opposition to changes in theories or shifts in paradigms — those universally recognized frameworks that for a time provide model problems and solutions to a community of practitioners.[38] Hahn was uneasy about Lise's reaction to whether the entire transuranic scheme would have to be altered, and was certainly unwilling to introduce a new paradigm, let alone challenge their previous work. Only a week after his famous experiments, as he states honestly in his letter to Lise, it was his "new fantasy" that he and Strassmann had *not* proved their transuranic scheme to be incorrect after all! However, the presence of barium, halfway down the periodic table, could not be denied by any of the team members, and the Berlin team turned to Lise Meitner for an explanation.

Hahn had sent off his article to *Die Naturwissenschaften* before facing his doubts, but they overtook him almost immediately. In the text of the *draft* of his December 22 article, a revealing document with strike-overs and words crossed out (submitted as a final draft, but clearly indicating changes from the article Hahn had mailed off to Meitner the evening before), he writes: "as 'nuclear chemists' who are somewhat related to physicists, we cannot yet decide to take this large step, which contradicts all previous experience of nuclear physics." It appears that earlier, Hahn had not realized how dramatically his radiochemical findings might affect the entire field of nuclear physics. This was the work left to Lise Meitner. And as she and Otto Frisch set down their ideas into article form for publication, carefully choosing their words, arguing over details as they packed their bags on New Year's Eve, they too began to realize how momentous this discovery was. But unlike Hahn and Strassmann, they would not have their results published until over a month later — with near disastrous implications concerning the priority credit for their idea.

Meitner had come to realize in Kungälv that, indeed, a "splitting" of the uranium nucleus had occurred, and she had even calculated the energy released in the process. As the New Year arrived at midnight in Sweden, Lise and her nephew had composed a friendly letter together, writing to Hahn:

January 1, 1939 12:30

Dear Otto!

I am starting this year with a letter to you. May it be a good year for everything that has to do with us.

We have read and considered your work thoroughly. Perhaps it is indeed energetically possible that such a heavy nucleus may have burst [*zerplatz*]. However, I believe that your hypothesis that Ba and Ma will be produced is impossible for many reasons. I would like to ask you a few questions first.

[She continues with two technical questions concerning other short-lived radioactive products that Hahn and Strassmann had found.]

...You will understand that the question of the correctness of the transuranics has a very personal side for me. If the whole work of the last three years was incorrect, it cannot be demonstrated from only one side [of the team]. I was also responsible for that work, and must find a way to be included in a retraction. A joint retraction is impossible [to implement], yet it should be issued simultaneously, a statement from you and one from me (possibly to be published in *Nature*), a written statement, of course, arrived at through mutual agreement. You are in a more pivotal position to realize this, while I have only three years of work to repudiate in case the transuranics disappear. It's not a good recommendation for my new beginning [in Sweden]. But that cannot be changed.

I am not at all convinced that the transuranics have completely collapsed. And I am, of course, very curious about what Strassmann thinks, and I hope that you'll write soon about what you are thinking, and how [the transuranic schema] can be reaffirmed.

I would be thankful, too, if once again you could look up the beta-ray tables of Fournier. They must be there, and are very important.

...I'll return [to Stockholm] late this afternoon. Please pay attention to the enclosed attachments.

Many heartfelt greetings,

Yours,

Lise[39]

Then scribbled on the bottom of the letter was a postscript from her nephew: "All good wishes for the New Year from me as well! If your new findings are true and correct, this is of greatest interest, and I am very curious to hear about upcoming test results. There was a beautiful snowfall here, and I went skiing too. Very heartfelt greetings to you, your wife, and Hanno, Yours, O.R." Frisch obviously did not doubt Hahn's findings.

Morning came all too soon on New Year's day. Lise said her good-byes to her friends and Otto Robert and boarded the train for the long journey to Stockholm. Back in her hotel room she immediately began to reconsider Hahn's experiments. The "midnight sun" set at 2:30 P.M. in the Nordic afternoon, and by the time she arrived at her hotel, it was quite dark. Hence, it was only the next day that she was able to pick up her mail which included Hahn's article. After skimming the new publications at Siegbahn's Institute, she hastily wrote the following as an addendum to their New Year's Eve letter and mailed it off to Berlin. She stated:

> Unfortunately, just now the Year Book of the Swedish Physics Society is publishing a lecture I gave about uranium and thorium. First thing tomorrow I'll see if adding a footnote is possible (the Ra-Ac test results I've mentioned). Would you agree if I pointed out the Ra-Ba test results in a footnote, and carefully say something like 'it is possible now that the transuranic series will have to be restructured?' Please answer immediately.
>
> Of course, the experimental test results remain basically true; only the classification of the nuclear surface tension would change. But I won't write about this, it is only between you and me as a minor consolation.
>
> I think that maybe only the U-23'-body leads to an eka-rhenium (that maybe emits alpha-rays). If this proves to be true, and I have certain theories about it, I could take this thought as a starting point for my repudiation (about such, we should be in perfect agreement before any publication), hence, my repudiation won't be completely negative.[40]

In this important letter, we see Lise's mind at work. She is concerned, even alarmed, that their carefully developed transuranic edifice will have to be retracted, "repudiated," and is already considering how to do so. Of course, she is rightfully concerned about an article already published and asks Hahn his opinion about stating "it is possible now that the transuranic series will have to be restructured." Notice her careful, even hesitant wording. And as she wrestled with the theoretical developments Hahn's radium-barium results posed, Meitner was also concerned that their work was not yet done, that there were more experiments to carry out. And she was right.

During the quiet days after her return to Stockholm, Lise Meitner carefully worked through her calculations concerning the splitting of the uranium nucleus. She attempted to fit Hahn and Strassmann's radiochemical findings into a new picture of the periodic series of the elements. Her mathematical confirmations of Hahn's results were indeed "very exciting!" Yet she continued to challenge her colleague in Berlin to take a stand concerning the conclusions of the Joliot–Curies' earlier publications. When, in her letter dated December 29, she asked Hahn pointedly, "Do you sever yourself from the lanthanum conclusion or not?" she was asking for a theoretical commitment, for Hahn to take a stand concerning their brand new theoretical model. If lanthanum was indeed found when uranium was bombarded with neutrons, then Hahn's radiochemical data could not "decisively tip the balance in favor of the existence of the *uranium* isotope" — the puzzle Lise was still trying to understand. Yet the possibility, or reality, of barium could only point to a radical division of the nucleus; and as they had boldly written to Hahn just a week before from Kungälv " . . . we have had so many surprises in physics," that she would not even discount "a possible 'bursting' of the uranium nucleus!"[41]

Meitner was still not willing to immediately discard years of work on the transuranic schema. When Hahn's *Die Naturwissenschaften* article was retrieved from the post office after Christmas, missing a page as well as diagrams (which Hahn had added to the publication after mailing Lise's copy), she may have had further grounds to be doubtful. Was Hahn unsure of his evidence, and so had deliberately omitted the critical pages in his packet to her, or had he only been careless in his hurry to mail his carbon of the article to her before submitting it to the editor for publication on December 22? Whatever the reasons, Lise was left to her own analytic devices, and hence was convinced that she and her nephew must also publish their own ideas and demonstrate them experimentally as soon as possible.

In her New Year's letter to Berlin, Meitner comes to grips with the major implications of Hahn and Strassmann's data — that such a "heavy nucleus" may have indeed "split." She is also concerned with herrolein the previous laborious years of developing the complex, but incorrect, "transuranic" schema, and the ethical responsibility they had to retract these findings, if necessary, "each for themselves."

Contemplating such a public retraction during this time was indeed a personal embarrassment for her, especially since she had literally lost all other frames of reference, prestige, and membership in the international scientific community. Thus, through Meitner's correspondence with Hahn during these critical weeks, we can see that she struggled as much with the "death" of the transuranic scheme as with the birth of her new concept of nuclear splitting. Yet Meitner correctly realized that their prior published results and speculations must be completely re-interpreted in the face of Hahn's radiochemical results. The New Year's message she wrote late at night conveyed both triumph and a sense of responsibility for the "explanation schemes," now proven incorrect, which she had helped to create, set against the background of the struggle by physicists around the world to make sense of the radioactive products emerging from the bombardment of elements with neutrons. And all this was happening during an ominous time when real bombs were being made by governments preparing for war.

CHAPTER X

The News of Fission Spreads: 1939

A new idea, a new conception, is born in the privacy of one man or woman's dreams. But for that conception to become part of the body of scientific theory, it must be acknowledged by the society of which that individual is a member. . . . Scientific knowledge as a whole grows out of the interaction – sometimes complex, always subtle – between individual creativity and communal validation. But sometimes that interaction miscarries, and an estrangement occurs between the individual and the community.

Evelyn Fox Keller
A Feeling for the Organism

EVENTS WOULD SOON UNFOLD much more quickly than either Otto Hahn or Lise Meitner would have imagined.

As Meitner pondered her New Year's challenges, Niels Bohr and his 19-year-old son Erik were tossing and turning onboard ship crossing the Atlantic. They had booked passage with a young professor from the University of Liege, Léon Rosenfeld, a 33-year-old physicist Bohr had been most impressed with when they first met and collaborated at the Copenhagen Institute for Physics.[1]

Just prior to making the trip, Frisch had arrived at Bohr's house on January 3. In their discussions Bohr had been reminded of a model of the nucleus he had proposed many years before, called the "liquid droplet model." The same day Frisch hurriedly wrote to Meitner in Stockholm:

3 January

Only today was I able to speak with Bohr about the bursting uranium. The conversation lasted only five minutes, since Bohr immediately and in every respect agreed with

197

us. He was only astonished that he had not thought earlier of this possibility, which follows so directly from the present conceptions of nuclear structure. He was also completely in agreement with our view that this disintegration of a heavy nucleus into two larger pieces is an almost classical process, which does not occur at all below a certain energy, but already occurs very easily a little above it. (One indeed has to require this in order to understand the great stability of natural uranium as compared to the very great instability of the not-so-very-much-more-energetic compound nucleus.) Bohr still wants to consider this quantitatively this evening and to talk with me again about it tomorrow.[2]

As they had scheduled, on January 5 Frisch called his aunt in Stockholm. By telephone, he and Lise agreed to follow Bohr's prompting and began immediately to draft what would emerge as a carefully worked out 1000-word article for *Nature*, the quickest form of international scientific communication they could agree upon to share their insights.

When Rosenfeld, Bohr, and his son set sail from Goeteborg, Sweden on January 7, 1939, Frisch's news about the "splitting" of the atomic nucleus had so fired up Bohr that he had ordered a blackboard installed in his cabin on board ship! Bohr had been looking forward to spending leisurely weeks with Einstein in Princeton; he had looked forward to quiet weeks of discussing questions of mutual interest in physics as well as sharing insights and information about Nazi Germany and emigrant friends and colleagues. But the new "insight" of Meitner and Frisch had so dramatically challenged the "compound nucleus" model, which puzzled the international community for so long, that he could not rest until he understood the deeper implications of the process. Onboard ship, he could concentrate only on Frisch and Meitner's news, and he and Rosenfeld stayed up late nights debating the possible "splitting" of uranium. Bohr worried that Lise Meitner's credit for such a remarkable interpretation would be overshadowed if she did not publish quickly.

In Copenhagen, Frisch was working hard both to refine his joint article with Meitner on the new "nuclear model," and to discuss their new ideas with other friends and emigré colleagues. In just one week after his departure from Kungälv, he mailed the draft of their paper with a cover letter to his aunt, reporting to her about the week's tumultuous events:

I wrote up a first draft on Friday [January 6] and on Bohr's request, rode out to Carlsberg [Bohr's mansion] in the evening, where Bohr once again discussed the matter thoroughly with me. He let me recalculate my estimate of the surface tension, and he was in complete agreement with it; he had already hurriedly considered the electrical term, but had not realized it would be so large. Concerning the [formation of U239 by] resonance, he did not want to express himself directly, but seemed to see no difficulty with it. Later, I again considered this point a bit as it arises in the conclusion of the note [article to *Nature*]; in any case, Bohr did not take a position on this. Bohr only made several recommendations during the evening for a clearer formulation on several points; otherwise he was in agreement with everything. I then [Saturday, January 7] started to type up the draft and was able to take only two pages to Bohr at the train station (10:29 am) where he put them in his pocket; he no longer had any time to read them.[3]

It was these "notes" – actually, the text of their historic paper – that Bohr would have several days to work through on his way to America. Lise promptly edited Otto Robert's draft and returned it to Copenhagen, entitling the short article "A New Type of Nuclear Reaction." But unfortunately, when it arrived in Denmark, Otto Robert took his time in submitting the article to the editors of *Nature* in London. In the same letter of January 8 to Meitner, he had confided that one of his close friends, George Placzek, had been "skeptical" about the results emerging from Berlin. Was fear of the international community's rebuttals giving Frisch second thoughts? Hahn and Strassmann had been very skeptical of their own results too, it seemed. Hence, he wrote to his aunt:

> Since Hahn and Strassmann's article appeared here yesterday [published in *Die Natur-wissenschaften*], I discussed the entire matter thoroughly with Placzek, who at the moment is very skeptical, but he of course always is. Early today he [again] flew back to Paris, and then will travel to America, to Bethe in Ithaca, where he has a position.[4]

Of course, Frisch was interested in reporting his friend's professional job offers from outside the sphere of fascism to his aunt; physics positions were few and far between in Europe that year, and such news seemed as important as their new article. But perhaps the discussion with the skeptical Placzek dampened Frisch's enthusiasm in publishing until after he could confirm their interpretation experimentally. And recall, Bohr had left for America with only notes in his pocket; Meitner and Frisch had not yet confirmed their insights.

Here is the crucial point. Frisch and his aunt had discussed various experiments together during their last days in Kungälv, and she had openly complained about the lack of any decent laboratory equipment for her to work with. To dispel any doubts that such a "nuclear splitting" process *was* occurring, Otto Robert Frisch spent the first two weeks of January assembling apparatus with which to *view* the ionization "pulses" that fragments of uranium would give off, if indeed, as Hahn had speculated and he and his aunt had surmised, the uranium nucleus was "bursting."

Meanwhile, the joint Meitner-Frisch paper "Disintegration by Neutrons: A New Type of Nuclear Reaction" sat on his desk, awaiting submission to *Nature*.

On Friday, January 13, Frisch first observed the anticipated pulses, and he confirmed this observation over the following three days: yes, uranium did emit *energy* when "split." He then wrote another brief "note" to *Nature*, entitling it "Physical Evidence for the Division of Heavy Nuclei under Neutron Bombardment."[5] It was about five hundred words, and stated in part: "By means of a uranium-lined ionization chamber connected with a linear amplifier, I have succeeded in demonstrating the occurrence of such bursts.... About 15 particles a minute were recorded when 300 mg of radium, mixed with beryllium, was placed one centimeter from the uranium lining."

During late nights in Bohr's Institute lab, Frisch had removed the neutron source and then the uranium lining. No "pulses" of radioactivity were recorded. But when he surrounded the source with paraffin wax, the pulses were enhanced by a factor of two! Thorium gave him similar results — something had transformed the nucleus into a neutron emitter when neutrons were fired at it, much like firing a single shot into a balloon and having a spray of shot come back! Frisch animatedly discussed his second paper with Lise by long-distance telephone; and soon, by January 16, two weeks since his arrival in Copenhagen, his individually-authored paper, complete with graphs on the experimental results derived in Copenhagen, was ready to mail to *Nature*. Hence, Frisch dated both his own paper and the longer article coauthored with Lise Meitner on the process of nuclear fission, "January 16, 1939," and modestly sent them off to the *Nature* editor without a cover letter. It did not occur to him to call attention to the fact that these articles were of "unusual" interest, although he had worked long hours alone confirming fission, and Meitner had spent every day for two weeks poring over details of their results. It had been a hectic week, just seven days after Bohr's departure to America and only three-and-a-half weeks since Hahn's and Strassmann's results were mailed from Berlin. Few then knew that Frisch's and Meitner's interpretation of nuclear fission would soon be received with dramatic impact around the world. Meitner and Frisch themselves truly underestimated the amazing worldwide response to their findings.

Nature's editorial staff confirmed receiving the articles from Frisch on January 17, 1939. Despite the fact that he had stayed up all night finalizing his work, on January 17 he wrote dutifully to his aunt.[6] Otto Robert was rightfully concerned about what Lise Meitner would think about mailing their results as quickly as possible to the journal, particularly since the Hahn-Strassmann article was now circulating internationally. But scientists do not control editorial decisions about publishing their work at any specific time. Regrettably, *Nature* did not find room in its pages until three long weeks after Frisch submitted their historic work. And a great deal of international controversy would arise because of this time lag; not until February 11 did their articles appear in *Nature*.[7] By then, Bohr had landed in America.

In their articles, Meitner and Frisch pointed out that Hahn and Strassmann had been forced to conclude that an isotope of barium had been formed as a consequence of the bombardment of uranium with neutrons. In their jointly authored article entitled "Disintegration of Uranium by Neutrons: A New Type of Nuclear Reaction," they openly admitted that Hahn's and Strassmann's results "at first sight ... seem very hard to understand." They then present their carefully worded physical explanation of the fission process:

> The formation of elements much below uranium has been considered before, but was always rejected for physical reasons, so long as the chemical evidence was not entirely clear cut. The emission, within a short time, of a large number of charged particles may be regarded as excluded by the small penetrability of the 'Coulomb barrier,' indicated by Gamow's theory of alpha decay.[8]
>
> On the basis, however, of present ideas about the behavior of heavy nuclei,[9] an entirely different and essentially classical picture of these new disintegration processes suggests itself. On account of their close packing and strong energy exchange, the particles in a heavy nucleus would be expected to move in a collective way which has some resemblance to the movement of a liquid drop. If the movement is made sufficiently violent by adding energy, such a drop may divide itself into two smaller drops. In discussing the energies involved in the deformation of nuclei, the concept of surface tension of nuclear matter has been used[10] and its value has been estimated from simple considerations regarding nuclear forces. It must be remembered, however, that the surface tension of a charged droplet is diminished by its charge, and a rough estimate shows that the surface tension of nuclei, decreasing with increasing nuclear charge, may become zero for atomic numbers of the order 100.

Now came their momentous announcement — and we can hear Meitner's carefully chosen syntax describing her world-altering insight:

> It seems therefore possible that the uranium nucleus has only a small stability of form, and may, after neutron capture, divide itself into two nuclei of roughly equal size, (the precise ratio of sizes depending on finer structural features and perhaps partly on chance). These two nuclei will repel each other and should gain a total kinetic energy of c. 200 MeV., as calculated from nuclear radius and charge. This amount of energy may actually be expected to be available from the difference in packing fraction between uranium and the elements in the middle of the periodic system. The whole 'fission' process can thus be described in an essentially classical way, without having

to consider quantum-mechanical 'tunnel effects,' which would actually be extremely small, on account of the large masses involved.

When Lise and Otto Robert had first discussed their physical interpretations together in Kungälv, they also realized that their "speculations" should be demonstrated experimentally. She would do so as well if she could get the equipment in her Stockholm lab to function. Frisch's night-owl partner in the compound nucleus work, Placzek, had prodded him in the right direction, to "show the existence of those fast-moving fragments you are talking about."

"Oddly enough, that thought hadn't occurred to me," Frisch later recalled, "until *after* Bohr's departure."[11] But it had occurred to Lise Meitner, and within several weeks, she would publish a separate article entitled "New Products of the Fission of the Thorium Nucleus," demonstrated in her laboratory at Siegbahn's Research Institute for Physics, explaining that indeed, fission did occur with unstable nuclei.[12]

*   *   *   *   *

In the meantime, on board the ship that was bringing him to America, Bohr could not rest until he was completely satisfied that he really understood the mechanism of the fission process. Léon Rosenfeld later recalled:

> As we were boarding the ship, Bohr told me he had just been handed a note by Frisch, containing his and Lise Meitner's conclusions; we should 'try to understand it.'[13] We had bad weather through the whole crossing, and Bohr was rather miserable, on the verge of seasickness all the time. Nevertheless, we persevered for nine days, and before the American coast was in sight, Bohr had a full grasp of the new process and its main implications. Bohr examined the problem in his meticulous manner from every possible angle, and eventually found what Frisch later called "a very simple argument showing why the fission of the nucleus, as soon as the forces tending to prevent it are sufficiently weak, will have a fair chance to occur, even in competition with the other possible and more familiar types of disintegration."[14]

Ironically, their ship docked at the Swedish-American Line's W. 57th Street pier in New York City the same day (Monday, January 16, 1939) that Frisch mailed the two historic articles to *Nature* in London.

Back in Stockholm, Lise Meitner was also thinking about the physical processes involved in the fission of the uranium nucleus. She had written to Hahn on January 14 after receiving a series of letters from him:[15]

> Dear Otto!
> Many thanks for your dear letter. My card to you has probably arrived in between. Unfortunately, you have not written how Edith is. I hope, because you have not written about her, that it signifies that all goes well....

...I am naturally very anxious about the progress of your investigations. One must actually expect that barium and radium are associated with thorium. As there are only 2 units less of nuclear charge and a pair of neutrons lacking, it will therefore behave much like uranium: it will also burst asunder. Our *Nature* notice is finally being sent off tomorrow and then presumably published in approximately three weeks.

It is certainly nice that Siegbahn is getting the counter. Once again, he didn't tell me a word about it, I finally asked him. It is all very depressing, but I don't see any possibility of changing things. What with living for months out of half-unpacked suitcases in a single room – and when I need to write, having to spread paper all over suitcases and chairs and the bed – that my outward life is uncomfortable is of no great significance.

I have totally retracted my [transuranic] paper as of 14 days ago. I also wrote you of that. Since, in fact, I no longer believe in the earlier interpretation of our experiments, and I of course don't want it published, even in the odd annual publication. The new explanation is also more beautiful and much more clearly comprehensible; it really is a wonderful thing.

Has Delbrück actually written you? I haven't heard anything from him for months. I received a lovely and distinctly contented letter from Franck [in Chicago] a few days ago. He has much better opportunities in his choice of work, and Ingrid at least appears to feel much better.

Very heartfelt greetings to you all,

Your,

Lise[16]

Two days after Lise mailed this letter to Otto Hahn, Niels Bohr's ship entered the harbor in New York City. The New York dock, with its hustle and bustle, was crowded as usual. Twenty-seven-year old John A. Wheeler elbowed his way through the crowd that was waiting at the chilly pier. He had been looking forward to seeing Bohr ever since the year he had spent in Copenhagen as a National Research Council Fellow in 1935. He was excited about Bohr's planned visit to Princeton, where the two could continue the discussions they had begun about problems surrounding the compound nucleus model. And there were others awaiting Bohr's arrival as well. "I was there too at the pier in New York when Bohr arrived," Enrico Fermi's wife Laura later recalled. "There were quite a few [emigrant physicists].... All the questions revolved around the political situation in Europe."[17]

Laura Fermi was of Jewish descent, and the Fermi family had only narrowly escaped the Fascist Italian government's noose — in part thanks to Niels Bohr. Prior to the official word from the Nobel Prize committee in Physics, Bohr had informed Fermi confidentially in October that his [Fermi's] research on the transuranics would be awarded the Nobel Prize, which would help him leave Italy. Bohr cautioned him

not to tell others about this news. Two weeks later, on November 10, when confirmation came via the traditional telephone call from Stockholm,[18] Fermi and his
family were prepared for their secret departure. Fermi had been awarded the 1938
Nobel Prize in Physics for his 1934 production of new radioactive elements by the
bombardment of neutrons (the first "transuranics"). He had informed Mussolini's
officials that his family would travel to Stockholm with him for the Nobel Prize ceremonies, and they quickly prepared their last boxes of belongings for their departure.
As soon as they were finished with the solemn Nobel ceremony in Sweden, they
were secretly transported to an ocean vessel, eventually arriving safely in America.

Enrico Fermi had been offered a professorship at Columbia University for the
academic year 1939. He and his family were still settling in, adjusting to life in a
huge American metropolis, when the news of Bohr's trip reached them. Extremely
grateful to Bohr for his well-timed warning about the opportunity to leave Italy,
Enrico and Laura were of course on hand to welcome him to America that January.

With such a large crowd at the New York pier, Bohr had no opportunity to talk to
Fermi about Frisch's news of nuclear fission, but Fermi would soon learn about the
breakthrough. Léon Rosenfeld left the ship and Bohr in New York City, his head
filled by the days of discussion concerning the discovery that the nucleus could
indeed be "split." He was happy to see young physics colleagues he had met in
Copenhagen, and one of these men, Wheeler, embraced him and led him away from
Bohr's entourage towards Grand Central Station.

Rosenfeld traveled with Wheeler from New York to Princeton by train, and
during the ride animatedly repeated the gist of his "interpretations" with Bohr.[19]
Apparently, during his discussions on board ship, Niels Bohr had not mentioned his
own promise to keep the fission breakthrough to himself until Frisch and Meitner
could develop their article![20] Rosenfeld, Frisch later mentions, was in fact under
the impression that an article had just appeared or was forthcoming in January.

As a graduate student, Wheeler was then in charge of the Princeton physics
students' Monday evening "Journal Club," and it was the custom, he later recalled,
"to get three things reported there. And here was something hot, as I had learned
from Rosenfeld on the train." Excited by the "hot" news, Wheeler ushered in his
guest as a surprise "key speaker" at that week's Club meeting, and he and Rosenfeld
proudly recounted the story of the trans-Atlantic journey and of the "splitting" of
uranium. Of course, this news was received with considerable excitement. Soon this
news was told to their professors; but not everyone was pleased with this "word-of-
mouth" dissemination. As Otto Frisch later recalled:

Bohr was annoyed when he heard about this premature disclosure, and knowing our leisurely ways at the Institute (in Copenhagen), he tried to convey to us a sense of the urgency of the matter by sending us telegram after telegram asking for further information and suggesting further experiments. Some of these we managed to perform, but we had no idea of the reasons which could prompt Bohr to such an unusual impatience.[21]

In fact, Bohr acted only on a suspicion of what could possibly happen, but he at first did not know that a fantastic race was already going on in a number of American laboratories, where the news had spread from Princeton, to pursue the same easy experiments I had already done to detect the fission fragments. A meeting of the American Physical Society would soon take place in Washington, and they all wanted to have sensational findings to report. Bohr had business to transact during his first days in New York, and after this was completed, he learned of the news of fission traveling through the "grapevine" of the East Coast scientific community. He rode the train with his son Erik from Manhattan to New Jersey, then settled into his residence at Princeton with an uneasy conscience.[22] What was to have been a leisurely trip to Princeton University where he had planned to discuss with Einstein the intricacies of quantum mechanics had suddenly turned into an intense focus upon the continuation of Bohr's own work on the liquid droplet model of the nucleus.[23]

Bohr began staying up late at night once more, working through physics calculations with his younger colleague, John Wheeler, who was always ready to debate the subtleties of theory. By January 20, Bohr had still not received word from Frisch at the Institute in Copenhagen about how his experiments to confirm fission had come out. Hence, he decided to send Frisch his own understanding of the process based on calculations that he and Rosenfeld had made aboard ship during their voyage. This draft article, along with a cover letter dated January 20, was mailed to Copenhagen from Princeton via registered mail. Bohr asked Frisch to have his secretary forward the article to *Nature* "if, as I hope, Hahn's article has already been published, and you and your aunt's note has already been submitted to *Nature*." He told Frisch that he was looking forward to learning "the latest news in this connection and how the experiments are proceeding at the Institute, which I, despite the distance, follow in my thoughts." He added this "p.s." before mailing the article and letter: "I have just seen Hahn and Strassmann's article in *Die Naturwissenschaften*, which naturally has caused much discussion here at the Institute."[24]

Finally, two days later, though Bohr's letter and a second one after it still had not yet arrived at the Institute in Copenhagen,[25] Frisch found time to write to his mentor and colleague. In that letter, dated January 22, he noted that he was "currently planning various new experiments on these 'fission' processes." This was the first time that anyone had used this terminology: in a late night conversation, Frisch had asked his biochemist friend, Dr. Arnold, for the name of the process describing how a living cell divides, and his colleague scoffed and replied: "Every biologist knows that: it's called fission." The term stuck, and Frisch began to use this term in his writings from then on.[26] Frisch was young — he was excited about his experiments but had no idea that others in America would be racing to be the first to confirm the process of "nuclear fission" after Rosenfeld's premature announcement.

Meanwhile, Bohr's tension increased as he continued to hear no news from his home Institute. Would Meitner and Frisch receive due credit, he worried, if their findings had not yet been submitted for publication? The Fifth Annual Conference on Theoretical Physics would be convened at the Carnegie Institute in America's capital, generously funded by George Washington University and the Carnegie Foundation.[27] Thirty-six names were on the invitation list by January 6, and the theme of the conference, "Magnetic, Electric and Mechanical Properties of Matter at Very Low Temperatures," had attracted a small number of European physicists as well. Many others also decided during mid-January to attend, and by January 26, with a number of local Washington physicists as hosts, Bohr arrived to a full conference with his exciting news.

Well aware that the news he had intended to be "kept under wraps" was now common knowledge among many East Coast scientists, Bohr was distraught even moments before it was his turn to speak. I.I. Rabi, as Wheeler later recalled, had actually been present at Princeton when Rosenfeld announced the news of fission, and had carried the story back to Columbia University in New York weeks before.[28] Fermi later recalled that he learned about Hahn and Strassmann's results from Willis Lamb, who had also been on a brief visit to Princeton.[29] The day before the Washington, DC conference, Bohr had actually traveled back to New York; but, instead of finding Fermi at Columbia University, he met Herbert L. Anderson, with whom he discussed the pressing issues surrounding fission and the importance of priority credits for Frisch and Meitner.[30] He then left for Washington, convinced that if he did not announce the news, others would surely publish their experimental findings that would both confirm and undercut the priority of the startling findings of Hahn and Strassmann. Fermi came into his office soon after Bohr had left, and Anderson thought he would surprise him with the story of the Hahn–Strassmann discovery — but Fermi had already learned of it from Lamb.

Hence, on January 25, Anderson and his colleagues in the Columbia Physics Department repeated experiments similar to those that Frisch had also conceived, and detected fission fragments flying from the uranium nucleus when it was bombarded by neutrons.[31] Fermi and his colleagues stood transfixed — here was the energy only speculated about for decades, and the literal "transmutation" of elements they all had been working on since the early 1930s. Fermi packed his bags for the trip to Washington that night, convinced that history was in the making.

On January 26, Bohr and Fermi conferred in their Washington hotel. They took the floor of the Fifth Conference on Theoretical Physics immediately after the official opening remarks, before a single report on the conference's topic had been delivered. The group was quiet, respectful of the two famous Nobel laureates from Europe, and Bohr then publicly announced the news: Hahn and Strassmann in Berlin–Dahlem had achieved radiochemical confirmation that barium was produced when uranium was bombarded by neutrons, and Frisch and Meitner had interpreted these results as indicating that a process of "nuclear splitting" had occurred, releasing a tremendous amount of energy!

The announcement sent the room into instant turmoil. Many physicists bolted from the meeting, rushing back to their laboratories to measure the ionization energy of the fission products. Others ran to the phone to relate the news of fission to colleagues at home and abroad. The few bored reporters attending the meeting besieged the scientists present, begging them to "translate" what had just electrified the crowd.[32] Bohr witnessed the spectacle and gazed at the confounded conference organizers: scientific discovery was much more pressing than their carefully arranged programs, and the senior scientist took full responsibility for the charged announcement.

The scene at Carnegie Institute's laboratories, as described by Rosenfeld, was typical that week, with physicist Merle Tuve "closely monitoring an apparatus which registered fission fragments," while simultaneously relating to a newspaper reporter on the phone that they were "witnessing" uranium burst into two fragments: "Now, there's another one!"[33] Reports soon appeared in the Washington daily newspapers about "Bohr's exciting news," and Otto Robert Frisch was even reported to be Niels Bohr's son-in-law (especially remarkable since Bohr had no daughters)! "I can see how it happened," Frisch later reflected humorously. "A journalist asks, 'And how do you know of this news, Dr. Bohr?' and Bohr replies, 'My son wrote to me.' The journalist mutters: 'His son? But his name is Frisch; must be his son-in-law.' "[34] But it actually was one of Bohr's sons, his eldest, Hans, also a physicist in the Copenhagen Institute, who had reported in a letter to his brother Erik, staying with his father in America, that Frisch was indeed conducting *numerous* experiments through the pressing suggestions of his aunt, Lise Meitner. Erik related the news to his father, who took it as a sign that matters were proceeding well at home through his prompting. And Otto Robert Frisch, unaware of the stir his work had created, was content to slowly continue his experiments and proudly discuss his results with Institute friends and colleagues for several weeks in leisurely conversations after he had sent his first January 16 packet off to *Nature*.

The pace was not so slow in America. R.D. Fowler and R.W. Dodson of Johns Hopkins University were informed about the Hahn-Strassmann findings by their colleagues attending the Theoretical Physics conference. By Saturday morning January 28, they too had confirmed the results experimentally and began to write up their results.[35] The Carnegie Institute's own Department of Terrestrial Magnetism (of which Tuve was director) had swung into action as soon as Bohr and Fermi announced the news of nuclear fission, and they too confirmed the discovery on January 28.[36] A more dramatic cross-country series of events led Luis Alvarez and G.K. Green to experimentally observe the predicted ionization pulses from uranium on January 31. Alvarez was reading the *San Francisco Chronicle* while having his hair cut in the student union at UC Berkeley when he read the announcement of the "discovery" of fission. Rushing out of the barber shop and across campus to the Radiation Lab, he broke the news to graduate student Philip Abelson and then telegraphed George Gamow in Washington, whom he had known since the early days in Copenhagen, for further news.[37] By February 1, several other laboratories had also confirmed the splitting of the uranium nucleus experimentally.

Bohr returned to Princeton crestfallen. Despite his efforts, he feared that this exciting event might be blotted by a miscarriage of credit — that the priority credit, so deserved by his dear friend Lise Meitner and his younger colleague Otto Frisch, both of whom had suffered so many injustices under Hitler, might be usurped by others eager for the scientific limelight. At last, on February 2, Frisch's letter of January 22 (with carbons of the two soon-to-be-famous articles) arrived for Bohr. The next day, on February 3, Bohr, in his characteristic, precise, kind manner, wrote to the Frisch:

> I need not say how extremely delighted I am by your most important discovery, on which I congratulate you most heartily.... The experiments of Hahn, together with your aunt's and your explanation, have indeed raised quite a sensation not only among physicists, but in the daily press in America. Indeed, as you may have gathered from my telegrams and perhaps even, as I feared, from the Scandinavian press, there has been a rush in a number of American laboratories to compete in exploring the new field. On the last day of the conference in Washington, DC where Rosenfeld and I were present, the first results of detection of high energy splitters were already reported from various

sides. Unaware as I was, to my great regret, of your own discovery, and not in possession even of the final text of your and your aunt's note to *Nature*, I could only stress (which I did most energetically) to all concerned that no public account of any such results could legitimately appear without mentioning your and your aunt's original interpretation of Hahn's results. When Hahn's paper appeared, information about this could of course, for your own sake, not be withheld and was, in fact, the direct source of inspiration for all the different investigators in this country. When I came back to Princeton I learned from an incidental remark in a letter from [my son] Hans the first news of the success of your experiments. I at once telephoned this information to Washington and New York, and succeeded in obtaining a fair statement in a science service circular of January 30, of which I have sent a copy to my wife, but I could not prevent various misstatements in newspapers. This is of course regrettable, but without any importance for the judgement of the scientific world, which here even more than in Denmark is accustomed to such happenings.[38]

He thanked Frisch for providing his and Meitner's notes before the rushed departure to the US. Bohr added that, based on these notes, he was making a "few corrections" to his own article on fission that he was developing conjointly with Wheeler for *Nature*. He enclosed "a new copy" for Frisch to mail to London for publication, reassuring his younger colleague that they would work together to assure priority credit. This was a trusting gesture on Bohr's part. He ended his letter, "Quite apart from the question of how much or little news the note contains, I think that its appearance at the earliest possible opportunity will contribute essentially to clear up the confusion with regard to the history of the discovery and its theoretical significance."

However, several weeks would pass before these articles would appear in *Nature*. Meitner and Frisch's "Disintegration of Uranium by Neutrons: A New Type of Nuclear Reaction" was published February 11.[39] Frisch's "Physical Evidence for the Division of Heavy Nuclei under Neutron Bombardment" appeared February 18,[40] and Bohr's note, dated 20 January, revised 3 February, was published February 25.[41] But much excitement over fission would be generated by other "discoverers" that winter before they read their February issue of *Nature*.

By early February, even the general American public had been informed of the scientific breakthrough in a *Time* magazine article, which read in part: "Last week the Hahn report reached the United States, and physicists sprang to their laboratories to see whether they could confirm it. Early this week, the laboratories of Columbia, Johns Hopkins, and the Carnegie Institution announced confirmation."[42] Nowhere in the international popular press was a woman physicist mentioned.

CHAPTER XI

Chain Reaction and the Dawn
of the Nuclear Age: 1939

If we look back at the progress accomplished by science at an ever-increasing rate, we are justified in thinking that research workers, breaking or building up atoms at will, will be able to produce nuclear chain reactions. If such reactions can be propagated through the material, one can imagine the enormous amount of useful energy which will be liberated.

> Enrico Fermi
> *Nobel Prize ceremony lecture, 1938*

My future is cut off: shall the past also be taken from me? . . . I have done nothing wrong; why should I suddenly be treated like a *persona non grata*, or worse, someone who is buried alive? Everything is hard enough as it is.

> Lise Meitner to Otto Hahn
> Stockholm, September 6, 1938

BETWEEN DECEMBER 1938 and February 1939, the news of nuclear fission spread throughout the scientific world. Unfortunately, Bohr's announcement at the January conference in Washington, DC had created a sensation in the US, setting off publication frenzy. Bohr had attempted to persuade colleagues, who rushed to confirm experiments, *not* to give their stories about fission to the press until Meitner and Frisch had had time to publish *their* results. He also wrote to his wife on the same night that he spoke before the American Physical Society, "I was immediately frightened, as I had promised Frisch I would wait until Hahn's note was published and his [Frisch's] own was sent off."[1]

But priority races had already created a flurry of activity, publications, and resentment. Léon Rosenfeld, who had prematurely disclosed what he knew to the Princeton Physics Club, later admitted, "I was sorry I had unwittingly let it all out."[2] Alone in Stockholm, Lise Meitner worked on, unaware of the furor in America.

Events had already gone too far to stop other scientists from entering the field. Fermi and his new physics team at Columbia University were preparing to release reports of *their* January experiments, confirming fission to newspapers. The American educational journal *Science Service* published the Hahn–Strassmann data without even referring to Meitner and Frisch. By late February, when the Meitner–Frisch articles in *Nature* finally arrived in American laboratories and libraries, people had already read about "Otto Hahn's discovery" of fission. Many completely overlooked Meitner and Frisch's correct interpretation of the fission results and their careful experiments published that spring. As Bohr had feared, the "public story" of the discovery of fission and atomic energy had entirely neglected to cite the decades of work and key insights of Lise Meitner, let alone the contributions she and Frisch had made to understanding the fission process.

At first, no one in the Research Institute for Physics in Stockholm was aware of this storm of activity surrounding fission. Lise Meitner continued to struggle in isolation at Manne Siegbahn's Institute (which would always list her separately from the Institute personnel). She worried about Hahn's son Hanno, who had been drafted into the Hitler Youth Corps at age 16, and wrote to Edith Hahn as well. Cut off from family ties and the satisfaction of work, she was dismayed that Siegbahn showed more interest in his son's doctoral research than the international sensation caused by fission.

The Stockholm labs were being equipped slowly despite funding from the Swedish Academy of Sciences in 1938 and 1939. Siegbahn had lobbied for a cyclotron since the creation of the Institute for Physics plans in 1937; hence in January 1939 there were only few pieces of the smaller experimental apparatus available for the type of nuclear-fission related experiments Meitner wanted to perform. She wrote in a gloomy tone to Hahn that things were *not* harmonious in Siegbahn's new institute:

> ... I don't feel at all happy. Here I just have a workplace no position that would entitle me to anything. Try to imagine what it would be like if, instead of your nice private institute, you had a room in an institute which is not your own, without any help, without any rights, and with the attitude of Siegbahn who only loves big machines and

who is very confident and self-assured — and there am I with my inner shyness and embarassment! And that I have to do all the petty work which I haven't done for 20 years. Of course, it's my fault: I should have prepared my departure much better and much earlier, I should at least have had drawings of the most important apparatus, etc. Siegbahn also once told me that Debye hadn't written anything about collaborators or assistants (several times I had asked Debye for that) and he had only little room. Yet, that's not true. I think. But the essential thing is that I have come here so empty-handed. Now Siegbahn will soon believe – especially after your excellent results – that I didn't do anything and that you both did all the physics too at Berlin–Dahlem. I am losing all my courage. . . .[3]

Around this time 26-year-old Josef Mattauch, the Austrian physicist Hahn had been pressured to select to assume Meitner's former position at the KWI, was on a lecture tour in Scandinavia. Undaunted, Otto Frisch met him at the Copenhagen train station, and the two chatted about the fission process and the energy calculations Frisch and his aunt had made. Mattauch boasted that *he* had just proven by physical means what Hahn had proven chemically, recording the pulses of ionization released as each uranium nucleus was split. Dumbfounded, Frisch showed the new "head physicist" of Berlin–Dahlem his own parallel findings *he* had worked so hard on at Bohr's Institute.[4] Mattauch claimed he had thought up similar experiments!

Frisch hardly realized that the work he and Meitner had labored over, by telephone and through many late nights in the laboratory, was already creating another major stir in the U.S., and would now be quickly transported back to Nazi Germany by Mattauch. Unaware of priority races in the U.S., he unwittingly let the fission cat out of the bag to the very man selected to replace Lise Meitner.

In Berlin, Hahn was troubled by the politics inside (and outside) his own Institute. Throughout January and February he corresponded frequently with Meitner. Initially, while the new results concerning fission were being formulated, he had written almost pleadingly to Lise to find some "fantastic explanation" for the results he and Strassmann labored over. Still embarrassed by the discrepancies posed by their former publications on the transuranics, it would be some months before Hahn began to distance himself from this work — and that of Lise Meitner. For years Hahn would stress that the December 1938 discovery had been the "work [of] the three of us." But in the winter and spring of 1939, he found himself in a minority position in an increasingly Nazified Kaiser Wilhelm Society and its institutes. He had avoided talking to certain physicists, and wrote to Meitner in a humble tone, which would soon turn to arrogance in an "about face" excluding her from "his" discovery" February 7, 1939: "Lise . . . we absolutely never did physics, but instead

we did chemical separations over and over again. We know our limits. . . .[5] He could no longer discuss fission with his colleagues without being attacked for having "shared" his insights with his longtime partner.

Hence, fearing his own professional status and the tenuous positions of his younger non-Nazi staff members, including Strassmann, Otto Hahn began to separate himself from Meitner, and even her major contributions to the discovery of nuclear fission. It seemed through his own myopic vision that the limited "fame" fission was bringing him *might* keep him out of harm's reach, and he intended to ward off his detractors by publicly priding himself on his published results. His headaches with shipping Meitner's furniture and library, dealing with the bureaucratic policy procedures, and his own wife's mental illness and confinement in a sanatorium, had taxed him to the breaking point. He refused to take a stand on the politics in and out of his institute: the Third Reich was blinding Hahn and he began to discount Meitner's insights and contributions he had frantically *sought out* months earlier. We witness here appeasement, professional cowardice, and worse.

It was not until Meitner and Frisch's physical explanations published in February and March reached Hahn in Berlin that he and Strassmann would be prompted to begin their next series of experiments. It was Meitner who predicted that the fission fragment accompanying barium ($z = 56$) must be krypton ($z = 36$), which would decay to rubidium, strontium, yttrium, etc.[6] As he and Strassmann pursued this theory, Hahn was increasingly on the defensive, even scolding Meitner that the discovery "owed nothing to physics!" This denial would create a gulf in their relationship that would never be fully bridged.

Back in Stockholm, spring brought little change in the Siberian-style weather, or Lise's situation. Still unhappy, she kept up her correspondence, hoping for another job offer elsewhere, or even a brief opportunity to travel. An offer to work at the women's college, Girton, in Cambridge, England made months before would be reconsidered; she later regretted that she had not taken it seriously. She wrote to the head of the Cavendish Laboratory, famed experimentalist Lawrence Bragg, of some major difficulties involved in traveling that spring:

> My Austrian passport is invalid and it will probably take some months before I get a German one, if at all. So it seems quite doubtful whether I will be able to come as early as March. We better postpone the visit until the late term in July. By that time, I shall probably have a passport and be able to accept your kind invitation.[7]

Lise Meitner did travel to England months later, but decided to delay her decision to work at the Cavendish Laboratory. No one, of course, could have predicted the eventual bombing of London, or the reality that few passports would be given once war was declared.

But even without a legal passport, Meitner could travel within the Scandinavian countries with her current Swedish guest visa. She decided to go to Copenhagen for a much-needed visit with her nephew and a chance to work using the up-to-date equipment at Bohr's institute. She remained there a month, conducting detailed results concerning the long transuranic series, concluding that for elements to be truly "above" uranium, scientists would have to preclude fission fragments. As soon as she had time in the following week, Lise wrote to Otto Hahn, once and then twice the same day on March 10, 1939. She was happy to rest assured that the transuranics did indeed fall: lighter elements *were* the products of nuclear fission. But she insisted on priority credit as well — she scolded and reminded Hahn that "you and Strassmann" could not have made their important and "beautiful discovery" if "*we* had not done" research as a team with earlier uranium experiments.[8]

This letter is revealing: Meitner is still clearly preserving an outward deference to Hahn, and *sharing* results. She concedes the "beautiful discovery" of fission to him, yet still makes the point that their joint previous uranium work formed the *basis* of this discovery. Unlike Hahn, who had begun to deny the very interpretations that would lead to the dramatic possibility of nuclear energy being harnessed, Meitner embraced the complex interdisciplinary challenges posed by the "abundance of new problems." Yet here we find the heart of their growing distance as colleagues and friends: Hahn would continue to triumph his "discovery" while Lise Meitner pressed on in the interpretation of fission.

March 1939 was a dark month politically in Europe. Franco seized dictatorial power in Madrid, Italy planned to conquer Albania, and Roosevelt began to draft his ultimately futile appeal to Hitler to "respect the independence of European nations" despite the news from underground sources that Hitler was planning a dramatic offensive with his armies. The Czechoslovakian president, Emil Hacha, was forced to accept the German "protection" of his shaky government, and soon Nazi forces brutally occupied this formally independent country. Resistors were jailed, or worse. War in Europe seemed imminent.

During winter evenings in Copenhagen, in troubled discussions with other emigrants also looking for funds or positions outside of the growing Nazi empire, Meitner began to sense that her time in Stockholm may become more permanent than she had envisioned. Niels Bohr and his son Erik were still in America, so Lise enjoyed visiting with Bohr's wife.

The Meitner children's family portrait, 1887: nine year-old Lise with doll in center
(her mother Hedwig, pregnant, is seated)
(Courtesy Master and Fellows of Churchill College, Cambridge, England)

Shy Lise the doctoral candidate, 1906, Vienna
(Courtesy Master and Fellows of Churchill College, Cambridge, England)

Ludwig Boltzmann, Meitner's mentor from 1905-1907 at the University of Vienna
(Courtesy American Institute of Physics Emilio Segrè Visual Archives)

Lise Meitner, Ph.D., 1907
(Courtesy Master and Fellows of Churchill College, Cambridge, England)

Dr. Otto Hahn and Dr. Lise Meitner in the woodworking-shop-turned-laboratory, 1908
(Courtesy Archiv zur Geschichte der Max-Planck-Gesellschaft, Berlin)

Kaiser Wilhelm Institute (KWI) for Chemistry, completed 1913
(Courtesy Archiv zur Geschichte der Max-Planck-Gesellschaft, Berlin)

The Meitner-Hahn team in their laboratory in the new KWI for Chemistry, Dahlem, 1913
(Courtesy Archiv zur Geschichte der Max-Planck-Gesellschaft, Berlin)

Meitner served on the Austrian front as a radiologist/X-ray technician, 1915-1917
(Courtesy Master and Fellows of Churchill College, Cambridge, England)

Niels Bohr visits Berlin, 1920
(Courtesy Archiv zur Geschichte der Max-Planck-Gesellschaft, Berlin)

Four Nobel laureates: Einstein, Planck, Millikan, von Laue, 1920
(Courtesy Archiv zur Geschichte der Max-Planck-Gesellschaft, Berlin)

Fräulein Prof Meitner
freundschaftlich zugeeignet von
M. Planck. Jan. 192

DIE
RELATIVITÄTSTHEORIE
EINSTEINS

UND IHRE PHYSIKALISCHEN GRUNDLAGEN

GEMEINVERSTÄNDLICH DARGESTELLT

VON

MAX BORN

MIT 129 TEXTABBILDUNGEN
UND EINEM PORTRÄT EINSTEINS

BERLIN
VERLAG VON JULIUS SPRINGER
1920

Title page of Max Born's EINSTEIN'S THEORY OF RELATIVITY; Lise Meitner's copy, inscribed
"given to you by your friend, M. Planck, January 1920"
(Courtesy Nicholas Pinfield)

A gathering at the KWI for Physical Chemistry, 1920
Seated (l-r): Hertha Sponer, Albert Einstein, Ingrid Franck, James Franck, Director Fritz Haber, Otto Hahn
Standing: Hugo Grotrian, Wilhelm Westphal, Otto von Baeyer, Peter Pringsheim, Gustav Hertz
(Courtesy Archiv zur Geschichte der Max-Planck-Gesellschaft, Berlin)

Lise Meitner, Head of the Physics Department, KWI for Chemistry, 1928
(© Bildarchiv Preussischer Kulturbesitz, Berlin)

Otto Hahn, Lise Meitner, and Sir Ernest Rutherford
German Chemical Society Meeting, KWI for Chemistry, 1929
(© Süddeutscher Verlag, courtesy Archiv zur Geschichte der Max-Planck-Gesellschaft, Berlin)

Summer seminar at Niels Bohr's Institute for Physics, Copenhagen, 1930
Bohr is at far left (standing above 2nd row)
Front row includes Werner Heisenberg (2nd from left), Max Born, Lise Meitner, Otto Stern, and
James Franck (far right); Otto Frisch is in 2nd row (2nd from right)
(Courtesy American Institute of Physics Niels Bohr Library)

Solvay Conference, Paris, 1933
Standing (l-r): E. Henriot, F. Perrin, F. Joliot-Curie, W. Heisenberg, H.A. Kramers, E. Stahel, E. Fermi,
E.T.S. Walton, P.A.M. Dirac, P. Debye, N.F. Mott, B. Cabrera, G. Gamow, W. Bothe, P. Blackett, M.S.
Rosenblum, J. Errera, E. Bauer, W. Pauli, J.E. Verschaffelt, M. Cosyns, E. Herzen, J.D. Cockcroft, C.D. Ellis,
R. Peierls, A. Piccard, E.O. Lawrence, L. Rosenfeld
Seated (l-r): E. Schrödinger, I. Joliot-Curie, N. Bohr, A. Joffe, Mme. Curie, O.W. Richardson, P. Langevin,
E. Rutherford, T. deDonder, M. deBroglie, L. deBroglie, L. Meitner, J. Chadwick
(Courtesy Lawrence Berkeley Laboratory, University of California)

Max Planck, Meitner's mentor in the University of Berlin Physics Department
(Courtesy Archiv zur Geschichte der Max-Planck-Gessellschaft, Berlin)

Lise Meitner and Otto Hahn before her secret emigration out of Berlin, Summer 1938
(Courtesy Archiv zur Geschichte der Max-Planck-Gesellschaft, Berlin)

Meitner's nephew, physicist Otto Robert Frisch, age 34, Copenhagen, 1938
(Courtesy American Institute for Physics Emilio Segrè Visual Archives, Physics Today Collection)

John Archibald Wheeler, who met Niels Bohr at the pier in New York, January, 1939,
and relayed the news of fission to the Princeton Physics Club
(Courtesy American Institute for Physics Emilio Segrè Visual Archives, WheelerCollection)

Lise Meitner, 1940, after her escape from Germany
(Courtesy American Institute for Physics Emilio Segrè Visual Archives, Herzfeld Collection)

Manne Siegbahn and his physics staff and cyclotron engineers, Research Institute for Physics, Stockholm, 1943
(Photo: B. Aurelius, courtesy Hugo Atterling private collection)

J. Robert Oppenheimer, Los Alamos test site, discussing Hiroshima bombing with
U.S. Army advisors, Los Alamos, New Mexico, 1945
(Courtesy American Institute for Physics Emilio Segrè Visual Archives, Physics Today Collection)

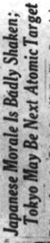

Newspaper article, "Woman Refugee Silent on Aid: Aided Bomb Research," about Lise Meitner, who ironically never worked on bomb research; August 7, 1945 (day after the bombing of Hiroshima)

Lise Meitner, the 1946 "Woman of the Year," U.S. Women's National Press Club Awards Ceremony, Washington, DC
Meitner is seated with ten women leaders chosen as "Makers and Promoters of Progress"
Seated (l-r): Dr. Esther Richards, award in science; Agnes deMille, award in ballet and music; Dr. Lise
Meitner, Woman of the Year; I.A.R. Wylie, award in literature; Anne O'Hare McCormick, award in journalism
Standing (l-r): Margaret Webster, award in drama; Georgia O'Keefe, award in art; Ruth M. Leach,
award in business; Margaret Cuthbert, award in radio
(Courtesy U.S. Women's National Press Club, Washington, DC)

Lise Meitner with President Harry S. Truman, Women's National Press Club Awards Ceremony dinner,
Washington, DC, February 9, 1946
(Courtesy The Catholic University of America Physics Department)

Otto Hahn receiving the 1944 Nobel Prize in Chemistry for the discovery of nuclear fission
(Swedish King Gustaf leaning forward in front row), December 10, 1946
(Courtesy Swedish Television Archives)

Fritz Strassmann, Lise Meitner, and Otto Hahn, Max-Planck-Institute for Chemistry, Mainz, West Germany, July 1956
(Courtesy Archiv zur Geschichte der Max-Planck-Gesellschaft, Berlin)

Meitner talking with students at Bryn Mawr College, Pennsylvania, 1959
(Courtesy American Institute for Physics Emilio Segrè Visual Archives, Photo by Heka)

Gottfried Frieherr von Droste Vischering and Lise Meitner, 1959
(Courtesy Archiv zur Geschichte der Max-Planck-Gesellschaft, Berlin)

Otto Hahn and Lise Meitner (both 81 years old) share a private moment at a ceremony for Max von Laue, 1959
(Courtesy Archiv zur Geschichte der Max-Planck-Gesellschaft, Berlin)

Grand opening of the Hahn-Meitner-Institute for Nuclear Research, Berlin, March 14, 1959
(© Hahn-Meitner-Institute, Berlin)

Max Born, Lise Meitner, Werner Heisenberg, and Otto Hahn, at the Lindau Conference of Nobel Laureates, 1962
(© Franz Thorbecke, Courtesy Archiv zur Geschichte der Max-Planck-Gesellschaft, Berlin)

Lise Meitner and Otto Hahn at the "Order Pour le Mérite" Awards Ceremony
West German President Heinrich Lübke at right, June 28, 1963, Bonn, West Germany
(© Bundesbildstelle Bonn)

Lise Meitner commemorative postage stamp, West Germany, 1988

Meitner and Otto Robert also worked on a joint article about their latest experiments; "Products of the Fission of the Uranium Nucleus" appeared in *Nature* on March 18.[9] In this article, Meitner and Frisch carefully explained their findings and attempted to set the record straight on the priority question. They stated:

> It can be shown by simple considerations that this type of nuclear reaction may be described in an essentially classical way like the fission of a liquid drop, and that the fission products must fly apart with kinetic energies of the order of a hundred million electron volts each.[10] Evidence for these high energies was first given by O.R. Frisch[11] and almost simultaneously by a number of other investigators.[12] The possibility of making use of these high energies in order to collect the fission products in the same way one collects the active deposit from alpha-recoil had been pointed out by L. Meitner.[13] In the meantime, F. Joliot has independently made experiments of this type.[14] We have now carried out the experiments, using the recently completed high-tension equipment of the Institute of Theoretical Physics, Copenhagen.

And experiment they did. By late March, Meitner published another paper, "New Products of the Fission of the Thorium Nucleus," utilizing Bohr's cyclotron as a source of irradiation. This article was published April 15, 1939 in *Nature*.[15] However, something essential was eluding them. Frisch later remembered:

> Lise Meitner felt that probably most of the radioactive substances which had been thought to lie outside uranium were also fission products...we proved that point by using a technique of "radio-recoil," which she had been the first to use about thirty years previously. And yet in all this excitement we had missed the most important thing — the chain reaction.[16]

Meitner and Frisch were excellent at experiment. Science fiction writers as well as professional physicists had speculated for years about the enormous potential energy waiting to be tapped within the atom, searching for a way to release this inherent "nuclear energy." As early as 1923, H.A. Kramers, a young physicist whom Lise Meitner had met at Bohr's Institute, wrote in his first book *The Atom and the Bohr Theory of its Structure*:

> One interested in speculating about what would happen if it were possible to bring about artificially a transformation of elements propagating itself from atom to atom with the liberation of energy would find food for serious thought in the fact that the quantities of energy that would be liberated in this way would be many many times greater than those which we now know in connection with chemical processes. There is then offered the possibility of explosives more extensive and more violent than any mind can conceive.... But this is, of course, mere fanciful conjecture.[17]

"Conjecture" or not, by 1939, what had once been "food for serious thought" was now on many minds. Hans Dominik, a popular novelist in Germany during the twenties, had written a book set in an "atom powered city." Although seen as complete fantasy during the Great Depression, Dominik was one of the first science fiction writers to become popular in Berlin since the work of Jules Verne and H.G. Wells.[18] It was a book that fueled the imagination of younger scientists. An earlier generation of physicists, people like Ernest Rutherford, had declared as late as 1933 that "anyone who expects a source of power from the transformation of these atoms is talking moonshine," and even Einstein, in a newspaper interview in America, had thrown a damp towel over speculations about the release of energy from the atomic nucleus.[19]

But not all scientists hesitated to predict the coming reality of atomic energy. One of Lise Meitner's younger colleagues, Hungarian physicist Leo Szilard, talked to anyone in the late 1930s who would listen to him about his own ideas on the implications of fission. Szilard, you recall, had left Berlin immediately after being denied lecturing privileges in 1933 when Meitner's classes were canceled at the university. For six years thereafter, he lived in various hotel rooms in America and met with physicists wherever he could find a friendly ear. He followed Meitner and Hahn's work carefully through the journals; even before Hahn and Strassmann's 1938 experiments, Szilard had foreseen the potential release of nuclear energy. He recalled that in the winter of 1938 "I was still intrigued with the possibility of a chain reaction, and for that reason I was interested in elements which became radioactive when they were bombarded by neutrons and where there were more radioactive isotopes than there should have been."[20] When the Meitner–Frisch *Nature* article was published on February 11, 1939, he jumped at the chance to share his own insights about fission.

Szilard was still seeking a job in America. He carefully followed the news of fission and devoured the various publications coming from Europe. Despite the absence of a university lab or position, he also sought the support of industrialists and other financial backers for his own experiments with uranium as a potential source of energy. Szilard would be one of the key emigrant physicists to trigger the chain reaction of events in America's race to harness nuclear fission.[21]

Szilard approached Rabi at Columbia University. Once the announcement of nuclear fission was made, enthusiasm for Szilard's "pet project" started to grow. The still skeptical older generation of salaried physicists continued to doubt any "wild speculations" about energy released from the nucleus. So Szilard sought out Rabi, another emigrant friend, who vividly recalled his own skepticism about a "nuclear reaction" being triggered by the fission process. Szilard recalled:

Rabi told me that Fermi had similar ideas [about a chain reaction] and that he had talked about them in Washington. Fermi was not in, so I told Rabi to talk to Fermi and say that these things ought to be kept secret because it was very likely that neutrons are emitted, that this might lead to a chain reaction, and this might lead to the construction of bombs.... A few days later I got up to see Rabi and asked, 'Did you talk to Fermi?' Rabi said, 'Yes, I did.' I asked, 'What did Fermi say?' and he said Fermi said, 'Nuts!' So I asked, 'Why did he say nuts?' and Rabi said, 'Well, I don't know, but he is in and we can talk to him.' So we went over to Fermi's office, and Rabi said to Fermi, 'Szilard wants to know why you said Nuts!' So Fermi said, 'Well, there is the remote possibility that neutrons may be emitted in the fission of uranium and then of course that a chain reaction can be made.' Rabi said, 'What do you mean by remote possibility?' and Fermi said, 'Well, ten percent.' And Rabi said, 'Ten percent is not a remote possibility if it means we may die of it. If I have pneumonia and the doctor tells me that there is a remote possibility that I might die, and that it's ten percent, I get excited about it.'[22]

From the very beginning, Szilard would later tell his biographers, "the line was drawn" — the difference between Fermi's position and his "was marked the first day we talked about it." Fermi initially played down the possibility of a chain reaction. Yet Szilard was more concerned with the social and political implications of nuclear energy than the processes of experimentation: in an era dominated by fascist leaders bent on racial superiority and conquest, the control and use of nuclear energy could change the fate of the world. "I thought the conservative thing was to assume that [a chain reaction] *would* happen, and we should take all necessary precautions," he later recalled.[23] Construction of atomic bombs was definitely on Leo Szilard's mind.

In February 1939 Szilard took action: he voiced his concern that an international scientific moratorium be created in order to keep information about potential explosive results of fission away from fascist countries. Based upon the research of Meitner and Frisch, he wrote to Frédéric Joliot–Curie in Paris, stating:

I personally feel that these things should be discussed privately among the physicists of England, France, and America; and that there should be no publication on this topic if it should turn out that neutrons are, in fact, emitted, and that a chain reaction might be possible.[24]

Meanwhile, Niels Bohr continued to work throughout February and March with John Archibald Wheeler on his own physical explanation of the fission process.

Others had insights as well concerning the potential release of nuclear energy. As Otto Frisch recalled, it was the Danish physicist Möller who "first suggested fission fragments might contain enough energy to send out a neutron (or even two)."

My immediate answer was that in this case no uranium deposits could exist; they would have blown up long ago by the explosive multiplication of neutrons in them! But I quickly saw that my argument was too naive — ores contained lots of other elements which might swallow up the neutrons; and the seams were perhaps thin, and then most of the neutrons would escape. With that, the exciting vision arose that by assembling enough pure uranium (with appropriate care!) one might start a controlled chain reaction and liberate nuclear energy on a scale that mattered. Of course, the specter of a bomb was there as well; but for a while, anyhow, it looked as though it need not cause us much fear. That complacency was based on an argument by Bohr, which was subtle and appeared quite sound.[25]

Bohr had attributed fission by slow neutrons to the rare isotope U235, concluding that "the only chance of getting a chain reaction with natural uranium was to arrange for the neutrons to be slowed down, whereby their effect on U235 is increased."[26] Yet the challenge remained: how could neutrons trigger a chain reaction? Otto Frisch recalled their explorations:

> ...one could not get an efficient explosion; slow neutrons take their time, and even if one did get the conditions for rapid neutron multiplication, this would at best (or at worst!) cause the assembly to disperse itself, with only a minute fraction of its nuclear energy liberated...the development of nuclear reactors followed closely the lines which Bohr foresaw within a few months of the discovery of fission.[27]

Meanwhile, back in Stockholm Meitner despaired at her unhappiness and lack of news from Berlin. All that winter of 1939 she heard second-hand reports of friends who faced greater dangers in Vienna under the Nazis than Berliners faced on a day-to-day basis. Adolf Eichmann, a Gestapo lieutenant, was systematically implementing a reign of terror on Jews in Austria. And soon, Hitler invaded Czechoslovakia. The world watched with embarrassment as first Neville Chamberlain, then other European countries, compromised with Hitler.[28]

Hahn was extremely worried about his own priorities. Paris was not paying attention to his fission findings, and Berkeley, Copenhagen, Cambridge, and Princeton researchers continued their latest fission research where strong neutron sources from the cyclotrons were available. Hahn was angry, and pressed forcefully for recognition when, in the same March 18 issue of *Nature* in which Meitner and Frisch published, the French team did not even cite the December work of Hahn and Strassmann.[29] As spring drew to a close, Lise wrote to him, urging a meeting in Copenhagen. They would only have this final reunion before a dark curtain fell over the Third Reich in April 1939.

Meitner's financial dependence on Siegbahn's Institute would curtail her Copenhagen visit, but she was excited about possible discussion with Hahn, so isolated and sidelined did she feel from their "important work." Otto Robert was conducting slow neutron experiments in Bohr's Institute all winter. She too longed to participate directly in the on-going research that was central to her existence. She was unaware of Bohr's work with Wheeler in Princeton, or Einstein's interest in her publication. It was difficult for her to go back and forth between an empty hotel room and a barely equipped lab in subzero temperatures. Soon her sister and emancipated brother-in-law (Frisch's parents) would arrive with few belongings from Vienna; meanwhile, her furniture, dishes, and library were shipped. Most of it had been systematically broken or destroyed by the Nazis, leaving her little from her past. She wrote to Hahn:

> ...Probably you can't really imagine what it means for a person my age to have been living in a small hotel room for *nine months*, without any comfort, without any scientific aids, and with the fear that nobody has the time required to advance my affairs [at Berlin].... And here in the institute, too, I'm without any help, and I don't think that Siegbahn is much inclined to assist me....[30]

Throughout the spring and summer of 1939, dozens of articles on fission and fission products appeared in *Nature, Physical Review*, and other publications.[31] Bohr and Wheeler's mid-March article, the 19-page "The Mechanism of Nuclear Fission," sparked an international debate about U235, but it also assured priority credit for Meitner and Frisch's work. It was a careful and systematic physics paper that remains a classic to this day.

Jealousies, however, remained in Berlin. Meitner did what she could to keep actively abreast of the news surrounding fission, discussing each new finding with her German-speaking colleague and friend in the Stockholm University's Physics Department, Oskar Klein. He and his wife did all they could for her, and urged her to reunite with Hahn in Copenhagen if possible. At the Royal Institute for Technology (KTH), Swedish physicist Gudmund Borelius also took a friendly interest in her and tried to discuss current work with uranium.[32] He was an influential member of the Nobel Prize-granting Swedish Royal Academy of Science, and would become a dear colleague to her as years passed in Sweden. Their friendship lasted nearly thirty years; but in 1939, Meitner was not interested in remaining in Sweden forever. She was actively contemplating the offer from England, where her nephew had decided to take a position, and began planning a trip there in June. Little did the British foresee that they would soon be building bomb shelters, not new scientific labs, in Cambridge, Oxford, and London.

In the fall of 1939, Hitler's army invaded Poland. What had once been an uneasy truce now turned into active aggression. In September, England declared war against Germany. At this time, the members of the Swedish Academy of Sciences began meeting to discuss the Nobel nominations for 1939. It is unthinkable that the Physics as well as Chemistry Committee would not have discussed nuclear fission, an undeniable milestone in the history of science. But in Sweden throughout that year, Siegbahn's focus on the cyclotron upstaged the news of fission. Manne Siegbahn played an influential role on the Physics Committee. His disregard for Meitner's achievements was evident. An argument may have been put forth that her 1938–39 contributions were not "empirically based," a well-known bias in the Swedish Academy members' voting for the Nobel Prize scientific awards.[33] Whatever the deciding factors, it was the American Ernest O. Lawrence who was awarded the 1939 Nobel Prize in Physics for his development of the cyclotron in Berkeley, California.[34] Despite a strongly-worded nomination made in 1938 by the respected Swedish chemist Theodor [Thé] Svedberg of Uppsala University for the Hahn–Meitner team to receive the 1939 Nobel Prize in Chemistry, the volatile political situation led the five-member Chemistry committee to a highly controversial stand: *not* to award a Nobel Prize in Chemistry at all in 1939 or 1940.[35]

Alone in her hotel room, Lise Meitner felt the world collapsing around her. Why hadn't the discovery of fission been duly recognized by her peers, particularly by those with whom she had daily contact? Her long years of effort and those of Otto Hahn had again been overlooked by the Nobel committees. Or did the politics of Nazi Germany enter into the committees' decisions? In despair she wrote to Hahn in October, "my work is equivalent to zero. . . ."[36] Her private grief was compounded by the encroachments of the war. That summer, with Denmark in danger of invasion, her nephew Otto Robert accepted an invitation to work at the University of Birmingham, England. He was there when Great Britain declared war against Germany: there would be no more appeasement. Frisch never dreamed that he would soon be drawn into a "top-secret" project concerning a new discovery, the discovery that had its genesis in the winter of '38 in talks between him and his aunt in the Swedish woods.

CHAPTER XII

Secrecy and Code Names: War Research Surrounding Nuclear Fission, 1939–1942

Terror was a weapon that Hitler especially prized, the destruction of what he called the enemy's 'will to resist.'

Richard Rhodes
The Making of the Atomic Bomb

The representative from the Army told us that it was naive to believe that we could make a significant contribution to defense by creating a new weapon. He said that if a new weapon is created, it usually takes two wars before one can know whether the weapon is any good or not.

Leo Szilard
Leo Szilard: His Version of the Facts

A TINY OFFICE for nuclear research was established in the German War Department about six months after Hahn and Strassmann's research was published. Most Nazi officers, however, knew little or nothing about the potentials of "atomic power," nor did they discuss such possibilities with their military superiors. Fortunately, few read the latest scientific articles or knew of Hahn's international notoriety in 1939.[1]

After celebrating his sixtieth birthday in grand style on March 8, 1939 with Institute members and university friends, the Third Reich finally gave Otto Hahn his visa papers so that he could go on an "academic lecture tour" in the Scandinavian countries of Denmark, Sweden, and Norway.[2] In April, he traveled first class to the

University of Oslo, where he was received with fanfare by the chair of the Chemistry Division, Ellen Gleditch, who, in front of Norwegian students and faculty, heartily congratulated him on "the discovery of nuclear fission." One of her younger staff members, Elizabeth Rona, recalled Hahn's strong opposition to Hitler:

> He started his lecture with the words, 'I am sorry to announce that the artificial radioactive elements are not transuranium elements, as I thought before, but products of the splitting of the uranium atom. He then went on to describe his [December] experiments which had led to the production of barium as a consequence of the splitting of the uranium atom. Hahn had obtained permission from fascist Germany to give a talk in Norway under two conditions: that he would speak only in German, and that he would present his compliments to the German ambassador. Hahn was a fierce liberal, opposed to fascism and Hitler's phony philosophy. He talked mostly in English and went to see the ambassador at noon, knowing that he would be out for lunch; he only left his card. One day, Hahn and I were on the porch of the University of Oslo, which faced the road. A battalion of Norwegian soldiers passed by. Hahn exclaimed: 'What a beautiful sight! Each soldier holds his gun in a different way, and they do not march in goose steps!'[3]

Hahn's trip was a brief window out of his despair over the priority credit controversy surrounding fission. He was bitter that spring, and it showed in his correspondence to Lise Meitner. These crucial months, between December and March, would solidify Hahn's resentment and unfortunately, color all relations between him and others regarding "credit" for fission. Years later Meitner would painfully recall the sting of his rejection, since, during this period, with the clouds of war hanging over their every day affairs, Lise looked forward to his arrival in Copenhagen.

Niels Bohr's family and the Bohr Institute offered a welcome haven for Meitner. The guest book shows that Frisch was also there for the animated discussions that ensued after Hahn's arrival. Talk obviously also turned to options for Meitner and Frisch. Two of Lise's sisters and beloved older brother Walter had emigrated to America, but Lise still maintained that Siegbahn's Institute would provide a good research environment for her. Offers from Bragg and Cockcroft in England also looked promising; later on she regretted that she had not taken them up on their generosity.[4] Also, since radioisotopes had been a key part in the treatment of cancer during the late 1930s, Meitner's expertise was of interest to American scientists as well. Esther Caulkin Brunauer and Florence Sabin, upon hearing of Meitner's plight, had lobbied colleagues to raise funds to bring her to America, but Lise held to the belief that she needed to remain in Scandinavia both for Bohr's Institute and for her correspondence with Berlin.[5]

It was Frisch who needed a new home. At last an offer came: the respected English experimental physicist, Mark Oliphant, invited Otto Robert to spend the summer in Birmingham. Frisch later recalled:

Hitler had occupied the Rhineland, and even I began to realize that the whole balance of Europe was tottering; that we could not expect that precarious peace to last much longer, and that sooner or later Hitler would occupy Denmark, and I would be back in the frying pan. Placzek said coolly, 'Why should Hitler occupy Denmark? He can just telephone, can't he?' It was a cruel joke, but not far from the truth.... From then on, whenever an English visitor came, I put out feelers whether a job could be found in England where I would be out of the real danger of being sent to a concentration or extermination camp, and where I might even have a chance to help fight against that menace to civilization.[6]

Another young German emigrant, Rudolf Peierls, was also relocating to England with his Russian emigré wife. That summer, as Frisch traveled through the British countryside, he reflected how unlike Nazi Germany this nation was: no sirens, no radio broadcasts, no marching soldiers . . . yet. Soon thereafter, in June 1939, Lise Meitner also visited England to meet with Bragg and survey prospects of working at a women's college. Frisch, in high spirits, seemed to adjust much easier than his aunt to speaking English.

As Frisch and Peierls compared notes over coffee and conversation while they settled into industrial Birmingham, they admitted quietly to one another that if war did come, they might once again be on the "outside" of official university research.[7] Everyone's position was tenuous, but the young Germans in their mid-thirties realized that, safely out of reach of fascism, they might collaborate on several interesting issues presented by uranium.

Secrecy soon surrounded research on fission around the world. Alerted by Leo Szilard, many younger emigrants in America grew more worried as the Joliot–Curie team in Paris continued publishing articles on a potential "chain reaction." From his hotel room in New York City, Szilard contacted Fermi, Wigner, and would soon meet with Einstein concerning the dangers of permitting fascist nations to learn about a potential chain reaction. If such results were indeed applied, he argued, by competent Nazi scientists, imagine what a "fission bomb" would be like in the hands of Hitler's air force!

The scientific community in New York split over what course of action to take. Szilard's argument for "self-censorship" seemed convincing: as he recalled, "If *we* persisted in *not* publishing, Joliot [in Paris] would have to come around; otherwise, he would be at a disadvantage, because we could know *his* results and yet, he would *not* know ours."[8]

By the time Churchill declared war against Hitler in September, Frisch was comfortably settled in England, a bachelor renting from a landlady. Yet he and Peierls were immediately declared "enemy aliens" by government officials, and curfews were imposed upon their coming and going between their flats and the university. With their German accents, older British scientists shied away from them. Frisch sorely missed the openness of Bohr's institute.

Churchill's chief science advisor had been urging him since 1935 to fund research on radar, stressing the importance of this "applied physics" to help realize potential technologies and make them available to the Royal Air Force and Navy. By the late fall of 1939 "scientific mobilization" had begun. Government funds began to flow for physics research on radar, and most British scientists were drafted into this work.

Since the staff and busy British scientists were now concentrating on "top secret" work, Frisch and Peierls had time on their hands. Bohr's subtle arguments concerning U235, which is only 0.07% of natural uranium on earth, did not convince everyone. A conclusive test would require separation of uranium isotopes. Frisch recalled: " . . . after studying the theory of thermal diffusion, I decided to build a Clusius separation tube to produce a small sample of enriched uranium."[9] His results surprised many.

Meanwhile in America, Victor Weisskopf, an Austrian and distant Jewish cousin of the Meitners and friend of Frisch's who had emigrated to New York City, became convinced that Szilard's campaign for self-censorship among physicists was the right one to pursue. He telegraphed editors of leading scientific journals, including *Nature*, pleading with them not to publish articles relating to fission or a potential chain reaction.[10] Weisskopf had also spoken passionately to Bohr and Fermi about the potential military applications that might be developed if the German Army literally saw the potential of a "super bomb" based on a chain-reaction triggered explosion.

However, censorship runs against the core beliefs of many scientists. The French team remained unswayed during the spring of 1939. On April 7, Joliot sent *Nature* the results of their experiments and calculations which estimated the number of neutrons emitted per fission; the article was published on April 22, 1939.[11] The French had the full support of their government in their continued nuclear research, and soon, the Germans were reading their results.[12]

That same spring, in the Soviet Union, I.V. Kurchatov and others sent the Soviet Academy of Sciences a series of letters on the possibility of harnessing uranium fission.[13] In Great Britain, George P. Thomson warned his government of the dangers of an unprecedented "nuclear explosion," while he himself began a series of new experiments on uranium in his own university physics lab. German physicist Georg Joos wrote an official letter to the Reich Ministry of Education about the potentials of nuclear fission, whiel almost simultaneously, Paul Harteck and William Groth sent their speculations concerning the applications of nuclear energy to the Nazi War Office. [14]

Emigrants in America who felt that "the cat was nearly out of the bag" were frantic: by summer, nothing seemed to be moving the Pentagon bureaucracy to alert those in power to the very real dangers of atomic energy development in Germany.

Finally, Leo Szilard decided to take matters a step further. Enlisting his young friend and fellow Hungarian emigré Eugene Wigner, they drove together to the Institute for Advanced Study in Princeton, where they learned that Albert Einstein was on vacation.[15] Not to be sidetracked, they decided to track down their pacifist hero in his hideaway on Long Island. After getting lost and asking a young child for directions, they at last met with Einstein. Einstein listened carefully to their pleas and clearly realized the seriousness of research in nuclear energy by von Weizsäcker, Heisenberg, Hahn, and others.

Long afterward Szilard recalled Einstein's surprise when he learned of the possibility of a chain reaction being triggered in uranium: *"Daran habe ich gar nicht gedacht*! I never thought of that!" came his reply.[16] The famous letter drafted by Szilard and Einstein was, and still is considered by historians around the world, the impetus that compelled President Roosevelt to take seriously the possibilities of a German nuclear threat. It reads, in part:

August 2, 1939

Sir:

Some recent work by E. Fermi and L. Szilard, which had been communicated to me in manuscript, leads me to expect that the element uranium may be turned into a new and important source of energy in the immediate future. Certain aspects of the situation which has arisen seem to call for watchfulness and, if necessary, quick action on the part of the Administration. I believe therefore that it is my duty to bring to your attention the following facts and recommendations:

In the course of the last four months it has been made probable – through the work of Joliot in France as well as Fermi and Szilard in America – that it may become possible to set up a nuclear chain reaction in a large mass of uranium by which vast amounts of power and large quantities of new radium-like elements would be generated. Now it appears almost certain that this could be achieved in the immediate future.

This new phenomenon would also lead to the construction of bombs, and it is conceivable – though much less certain – that extremely powerful bombs of a new type may thus be constructed. A single bomb of this type, carried by boat and exploded in a port, might very well destroy the whole port together with some of the surrounding territory. However, such bombs might very well prove to be too heavy for transportation by air.[17]

Einstein went on to outline why the Belgian Congo and Czechoslovakia, as important sources of uranium (which could fall into Nazi hands) should be observed carefully by the Administration. Then he urged President Roosevelt: "You may think it desirable to have some permanent contact maintained between the Administration and the group of physicists working on chain reactions in America."[18]

In a moving chapter in the history of modern science, Roosevelt began to mobilize resources and personnel, responding to Einstein formally two months later, on October 19, 1939, from the White House:

My dear Professor:

I want to thank you for your recent letter and the most interesting and important enclosure. I found this data of such import that I have convened a board consisting of the head of the Bureau of Standards and chosen representatives of the Army and Navy to thoroughly investigate the possibilities of your suggestion regarding the element uranium. I am glad to say that Dr. Sachs will cooperate and work with this Committee and I feel this is the most practical and effective method of dealing with the subject.

Please accept my sincere thanks.[19]

This was the beginning of the U.S. atomic research program.

Soon after the formation of the American-based "Manhattan Project," Stalin's master spy, Pavel Sudoplatov, was pressing to form a commission of Soviet scientists to work on atom bomb research.[20] And two years after Einstein's letter to Roosevelt, a Soviet agent, who had been a covert member of the spy ring infiltrating British nuclear efforts, reported to Stalin that a uranium bomb could technically be constructed in two years and that "the British government was seriously interested in developing a bomb with unbelievable destructive force based upon atomic energy."[21] Before Stalin would arrive in Washington D.C. in 1941 to press Roosevelt for Allied support, he ordered his intelligence agents to monitor reports on the development of nuclear weapons from Britain, Scandinavia, Germany — and America. The U.S.S.R. had covertly entered the nuclear age.

The German invasion of Poland had set the stage for Great Britain and France to declare war against the Third Reich. Scientists in both countries were officially ordered to withhold any further publications on nuclear fission or uranium. Through the eloquent appeals of economist Alexander Sachs to President Roosevelt, there was also a call for action in America. With a representative from the Navy, the U.S. Army, and the Director of the Bureau of Standards, Dr. Lyman Briggs, the American "advisory committee on uranium" met with Eugene Wigner, Leo Szilard, Edward Teller, and Richard Roberts (who stood in for Fermi): clearly, this emigré team was concerned as much about world peace as the potentials of uranium-based weapons in fascist hands. Under Roosevelt's orders, the U.S. Executive Branch thus budgeted for this secret advisory committee under the auspices of the nation's Bureau of Standards physics laboratory, which was in charge of "applying science and technology in the national interest and for public benefit."[22]

"If there *was* something to [a potential atomic bomb]," Alexander Sachs recalls, "there was danger of our being blown up. We *had* to take time by the forelock. . . ."[23] The Germans were also forming a top secret committee on uranium: Otto Hahn was one of its senior members. Werner Heisenberg provided another brilliant theoretical insight during this time: ". . . although energy could be generated from natural uranium, if it was used in conjunction with heavy water," he reported to Nazi superiors, "or graphite to slow down the neutrons," the U-238 would have to be "enriched."[24] This process presented a technical problem for all involved. Certainly the small German group did not yet have the resources at their disposal to follow up on Heisenberg's idea and would not fund large scale research for it in 1939. The German Army was not interested in scientific speculations. A report was filed with the German Army, and the Berlin-based "Uranium Committee" lobbied for more funds.

Back in Stockholm, Lise Meitner, alone, still reflected on her unexpected emigration. The Reich Ministry had tied her shipment of clothing and furnishing up in endless bureaucracy, placing further demands upon Hahn. A year earlier, she had written to Hahn about her former assistants at the Institute for Chemistry: none had contacted her, and several were even vying for her former apartment. News was so censored that she had only Max von Laue, a loyal correspondent, who conscientiously numbered each letter sequentially during their six-year exchange of letters, all of which were subjected to the Nazi censors. Still, they communicated cryptically and sympathetically, and it was von Laue who kept her informed about Planck and others close to her throughout the war.[25]

In between continued frustrations with Manne Siegbahn, Meitner continued her search for the mysterious elements that theoretical physicists insisted *were* transuranic. In mid-1940, news reached her from Berkeley, California, that Philip Abelson and Edwin McMillan had identified a decay product of U-239, with a 2.3-day activity, later called neptunium: the first actual element so identified above uranium.[26] It depressed Lise Meitner no end that she was not the one, after so many years of careful analysis, to announce such a discovery. Neptunium, a beta emitter, would lead to the discovery of yet another transuranic element not known in 1939, but the whole complex new field was one Meitner, without access to the Bohr Institute cyclotron and without a qualified chemistry assistant, just could not keep up with.

Depressed as she was in the face of personal and professional disappointments, she did the best she could in Stockholm. The plight of Jews worsened and all of Europe braced for war. Meitner reached out to other refugees in need of help. There is a voluminous correspondence between her and Max Born in Göttingen about the Breslau physicist Hedwig Kohn. Meitner lobbied international women's organizations, Jewish aid funds, and countless colleagues to obtain a small sum of money to bring Hedwig Kohn, one of the first women lecturers in any German university, to neutral Sweden. She waited, wrote, and waited again. On edge all that year, Kohn finally escaped from Germany, and traveled to Stockholm, alone. Meitner then assisted her in obtaining the coveted American visa.[27]

Lise's oldest brother Walter was living in England, and in June 1939, Otto Hahn was able to travel to London to lecture there, accompanied by an undercover "chaperone," a Berlin chemist who was a known Nazi Party member surveilling Hahn. Meitner openly warned her brother about a change in Hahn's attitude: that he was not "completely free because of a colleague. And perhaps inwardly he is also not completely free."[28] This honest appraisal of Hahn's catch-22 within the Third Reich is essential to note. To her own family, Lise could communicate directly about the very real moral shifts in her colleague. Hahn had written to Lise in careful terms during this trip out of Germany, one of his last for five years: "Together with a Party member. I cannot hence tell when I will have the free time to visit your brother, but I hope to do so anyway.... My 'colleague' is very nice but it is best that I am careful." (This so-called chemistry colleague later, unbeknownst to Otto Hahn, wrote up an official report to his Reich superiors noting 'disgust' at meeting German refugees living in England, the majority of them of Jewish descent.)[29]

Refugees were fleeing to England. Only a handful understood the complex challenges presented by nuclear fission. By early 1940, Otto Robert Frisch and Rudolf

Peierls realized, after many late night discussions, that only 100,000 or so Clusius tubes would be needed to separate the rare isotope U-235 from the more abundant U-238.[30] They also came to the realization that only a few pounds of U-235 could make a highly explosive bomb. Little did they foresee how quickly their insight would become reality.

Frisch realized he had a social responsibility to inform others about this potential: but what to do? Frisch later recalled that both he and Peierls felt that large-scale government funding was needed — immediately. The cost of even a factory to produce U-235, they reasoned with one another, would be "insignificant compared to the cost of war," meaning that governments should mobilize large-scale industrial forces and funds to separate as much of this rare isotope as possible before Hitler and Mussolini did.[31] Two years later, forces would be mobilized to do just that in America. Many years later, Sir Rudolf Peierls recalled:

> My rather abstract paper was exactly applicable. . . . We foresaw that a bomb made from U-235 would not be excessively large . . . and that if you got a chain reaction started . . . a great portion of the energy could be released. . . . Frisch and I discussed that a weapon made out of U-235 which was this powerful could repay the expensive separation of isotope. . . . Even if a separation plant would cost as much as a battleship, it would be worth it.[32]

But wartime conditions for German speaking scientists, very much on the lonely periphery of lab activity in those years, was not always supportive of such grand ideas. Restrictions on Austrians and Germans in Great Britain were numerous: they were not allowed to drive cars, or be out after dark under the local curfew laws. Frisch recalls his "difficult times as an enemy alien" in 1939–1940. Some of his friends had even been interned in British prisoner of war camps, a fact that few British citizens knew about.[33] But he kept his humor, recounting stories to his Aunt Lise in later years:

> Around May 1940 I was asked to go to the police station and was quizzed on a number of points, all adding up to the question: Was there any reason not to intern that guy? Decisions were made in a humane way: internment and release were to some extent governed by the hardship that might be inflicted upon dependents, of which I had none in England. After that police interview, I spoke to Phillip Moon who, when Oliphant was away, was his deputy [in the Birmingham University Physics Department] and he presumably told the police that 'the work [Frisch is] doing was important.' I never heard from them again.[34]

Because of German bombing raids, all lights had to be out soon after dusk, but during many nights Frisch and Peierls continued their discussions over candles well after midnight. Finally, they decided to formalize their critical insights in an official memo to their university leader, Oliphant, informing him about their findings. Oliphant was impressed, and suggested they write to Henry Tizard, chairman of the United Kingdom Committee on the Scientific Survey of Air Defense. So once again they carefully constructed their thoughts in writing, which became known historically as the famous Frisch–Peierls memorandum. In it, they stated in part:

> The possible construction of 'super bombs' based on a nuclear chain reaction in uranium has been discussed a great deal, and arguments have been brought forward which seemed to exclude this possibility. We wish here to point out and discuss a possibility which seems to have been overlooked in these earlier discussions.... In any case, no arrangement containing hydrogen and based on the action of slow neutrons could act as an effective super-bomb, because the reaction would be too slow.... The behavior of U-235 under bombardment with fast neutrons is not known experimentally, but from rather simple theoretical arguments, it can be concluded that almost every collision produces fissions and that neutrons of any energy are effective. Therefore, it is not necessary to add hydrogen, and the reaction, depending on the action of fast neutrons, develops with very great rapidity so that a considerable part of the total energy is liberated before the reaction gets stopped in the expansion of the material....[35]

In this document, they go on to outline the essential points and calculations regarding fast and slow neutron reactions, diffusion theory, and the expression for the energy "liberated before the mass expands so much that the reaction is interrupted." In short, Frisch and Peierls foresaw that the liberation of nuclear energy through a chain reaction would be possible after all, and outlined the technical details necessary to sustain such a nuclear reaction months before American scientists were mobilized. In the six months between Meitner's insights and this memorandum, all of Europe, Russia, and Japan would begin to mobilize for world war.

The suggestions in the Frisch–Peierls memorandum were taken seriously by the Royal Air Defense Council, forwarded by the respected Oliphant. The result was the British government's formation of a top secret committee, code named "Tube Alloys."[36] Frisch and Peierls would be sent to Liverpool, where a cyclotron made their ideas about separating uranium isotopes possible. War research had begun in earnest.

In a short year and a half, dramatic changes concerning fission had taken place around the world. On January 31, 1939, Swedish chemist and Nobel Laureate Thé Svedberg had nominated Hahn and Meitner for the Nobel Prize in Chemistry. In concluding the required report necessary for review by the committee, Svedberg summarized the reasons for his nomination:

It seems that sharing the Prize between Hahn and Meitner for the discovery of uranium fission or in common for their work with uranium fission products should not be questioned. Therefore, the sharing of the Prize could also possibly be proposed to a great extent for the totality of their common work in the field of radioactivity.[37]

But politics intervened during the 1939 deliberations. Hitler had been incensed when the 1936 Nobel Prize for Peace had been awarded to the outspoken German journalist, Carl von Ossietzky, who had exposed secret financial collaborations between the Nazi Party, the Reich government, and major German industries. Obviously, the award of the Nobel Prize gave even more publicity to such "under the table" deals. To get even with the Nobel process, Hitler forbade any German citizens' acceptance of – or even nomination for – the Nobel Prizes after 1938. This put Laureates Planck, Heisenberg, and others in a quandary, but effectively silenced them internationally until the end of the war. The entire international academic and scientific community was shocked.

The Nobel Committee for Chemistry hotly debated what to do, and after long deliberations, decided not to award any Prize in 1939. The next year, the Nobel Committee for Physics would follow suit.[38] Hence, despite over 15 nominations during their lifetime, Hahn and Meitner were passed over again and again for the Nobel Prize. And ironically, Thé Svedberg would write up a very slanted report to the Nobel Chemistry Committee as its chair in 1941 making the argument, while Hahn's work was essential for the "discovery" of fission, Meitner and Frisch's work was not extraordinary.[39] However, between 1940 and 1943 several Nobel laureates in physics proposed in writing that Hahn and Meitner be recognized by a Nobel Prize in Physics, but the Physics Committee, reporting to the Royal Swedish Academy of Sciences, stressed that their work "belonged" to chemistry. This committee was chaired by Manne Siegbahn who continued to champion his own experimental biases and ignore Lise Meitner and her major contributions.[40]

Despite years of oversight by the Nobel Committees, others *had* taken the news of fission and a potential chain reaction seriously. As early as September 1939, Third Reich army officer K. Diebner and a physicist named E. Bagge chaired a second major physics conference in Berlin concerning potential applications of nuclear fission. Werner Heisenberg was named head of a new theoretical division of the Kaiser Wihelm Institute for Physics, which had just opened amid great fanfare, and his friend Carl von Weizsäcker met him there often while he continued a commute from his professorship in Leipzig. Years later, von Weizsäcker, young and idealistic, recalled his dramatic realization that the discovery of nuclear fission, if used for making bombs, "could radically change the political structure of the world."[41] They talked about this in hushed tones and yet advanced their own careers under the Nazi swastikas. Most foresaw that war would soon be declared. However, most German scientists at the 1939 conference agreed together that it was the harnessing of atomic energy for potential *power* sources that offered the greatest potential for research in the future.[42]

Others thought differently. Emigrant physicists from the Third Reich knew that it would just be a matter of time before the Germans moved into weapons research focused upon uranium. The British saw major problems in building a uranium separation plant as bombing raids began to intensify, and negotiations began with the United States about working jointly.[43]

The first six months of war in Europe were a living hell for many. Newspapers reported grim casualties, but Meitner was unaware of the military interest in her work. She was able to visit Bohr's Institute for the last time in early 1940, and published an interesting paper in *Nature* that year entitled "Capture Cross-Sections for Thermal Neutrons in Thorium, Lead and Uranium-238."[44] While Hahn complained bitterly about lack of access to "foreign journals" such as *Comptes Rendus*, *Nature*, and other "enemy publications" banned from the Third Reich, a veil of secrecy fell over Allied nuclear research. Few physicists would communicate openly about their uranium research for the next five years.

Meitner was as astounded as were others when Hitler's troops fought fiercely with the Norwegian underground, imprisoning the nephew of her dear friend Gerta von Ubisch and thousands of others.[45] With Norway under Hitler's control, life in Sweden, a "neutral country," was a strange drama for Lise Meitner. After December 1941 and the Japanese bombing of Pearl Harbor, the entire globe became engulfed in war. Still lukewarm to her presence, Manne Siegbahn, the busy Director of the Stockholm Research Institute for Physics, tolerated Meitner's work while he himself was involved in responsibilities she knew nothing about: a Swedish military board had been called together to study the military "applications" of nuclear fission.[46] Under the domain of aggressive military leaders, few realized that the Japanese were also forming a top secret nuclear weapons research team as well.[47]

1942 was a terrible year. Millions died fighting on the Russian and Western fronts and scores of others bore the atrocities of the Hitler and Mussolini regimes. Early in the war, Hitler had basically overlooked the organized nuclear research team in Berlin, deciding that Nazi scientific efforts needed to be utilized for aerial bomb development.[48] Germany's Munitions Minister Albert Speer, however, recorded memoirs of his own for several years regarding nuclear matters:

> American technical journals suggested that plenty of technical and financial resources were available there for nuclear research.... In view of the revolutionary possibilities for nuclear fission, dominance in this field was fraught with enormous consequences. I asked Heisenberg how nuclear physics could be applied to the manufacture of atomic bombs. His answer was by no means encouraging. He declared, to be sure, that the scientific solution had already been found and that theoretically nothing stood in the way of building such a bomb. But the technical prerequisites for production would take years to develop, two years at the earliest, even provided that the program was given maximum support.[49]

Historians are still debating Heisenberg's role in the research of nuclear weapons. Hahn, Gerlach, and senior members of the "Uranium Club" would regularly be called together for top-secret meetings, and although few were Nazi party members, most wanted fission research to continue in Germany.[50] By 1942, when Heisenberg was named the Director of the Kaiser Wilhelm Institute for Physics, uranium research intensified. With Paris occupied, the French team of Joliot–Curie had been silenced, and their labs confiscated. French equipment was crated and shipped to the Third Reich for use. Also, little work was being done in Rome without Fermi's leadership. Without the Paris team to criticize Berlin, the German physicists and fear of a Nazi-controlled atomic bomb were on many emigrant physicists' minds. Speer later recalled:

> Hitler had sometimes spoken to me about the possibility of an atom bomb, but the idea quite obviously strained his intellectual capacity. He was also unable to grasp the revolutionary nature of nuclear physics. In the 2,200 written points of reference from my conferences with Hitler, nuclear fission comes up only once, and then it is mentioned with extreme brevity. Actually, Professor Heisenberg had not given any final answer to my question whether a successful nuclear fission could be kept under control with absolute certainty or might continue as a chain reaction. I am sure that Hitler would have not hesitated for a moment to employ atom bombs against England.[51]

Given Hitler's attitude, Speer decided to shelve further research on the explosive aspects of atomic energy during the war in favor of work on a "uranium engine," an atomic reactor the German team envisioned would be fueled by the fission process. In Heisenberg's words, U-235 seemed to have lost its attraction as a bomb material "owing to the difficulties of separating sufficient quantities." Yet this ambitious physicist continued to interact with his Reich superiors concerning fission potentials.[52] Years later Speer wrote:

No method was known that would have allowed production of an atomic explosive without enormous and therefore impossible technical equipment ... the work was to go forward as before on a comparatively small scale. Thus the only attainable goal was the development of the uranium pile.[53]

Aerial bombing intensified over England, and yet Churchill had "only begun to fight." While spies brought back reports from Heisenberg's labs, few knew what Hitler's plans were for atomic research. For two tense years, Otto Robert Frisch, Rudolf Peierls, and others once considered "at risk" to work on government science research, now worked cooperatively with the British atomic project, code named the "Maud Committee." Then, in early 1943, Frisch moved southeast to Oxford to work on diffusion techniques, supported by equipment he had only envisioned in 1940. Of course, he was not permitted to inform Lise Meitner of his work or whereabouts, but at least she knew he was safe — and employed. He looked back at his brief period in Oxford:

> This was a peaceful interlude for me, as Liverpool was also being bombed.... On one occasion, the whole laboratory spent a day in replacing most of the windows with cardboard because a land mine (actually, a large parachute bomb) had exploded in the courtyard.[54]

Meanwhile, work on fission in America was also speeding up. On December 2, 1942, four years after the profound ideas Meitner and Frisch had arrived at, based on Hahn and Strassmann's experimental findings, another historic juncture in the history of physics would be crossed. Known only to the U.S. government, which financially sponsored several teams of experimental physicists, Fermi had been working steadily with a team at the University of Chicago. The government-sponsored team's goal was clear: to construct an enormous lead-shielded "pile" to test whether the fission of U-235 isotopes would really become a "self-sustaining" chain reaction.

Located deep in the university's squash courts, away from science labs and prying eyes, this huge experiment held a core of uranium and was crossed by neutron-absorbing cadmium rods.[55] Graphite bricks interspersed at regular intervals with small bricks of pure U-235 had been constructed by crews who did not know the meaning of Fermi's secret experiment. Control rods of cadmium would be pulled out when Fermi was ready. At last the test date arrived. A nuclear chain reaction, which a gifted few had speculated about as early as 1939, was started.

When the last cadmium rod, nicknamed "Zip" was pulled out, the chain reaction became "critical" — it became self sustaining. The naturally radioactive U-235 was fissioning and emitting neutrons fast enough to create an amazing result.[56] Fermi's pile was a classically designed experiment — no one knew for sure whether the entire city of Chicago would be affected by such a nuclear reaction, but the team had taken the risk. Chianti wine was served as a congratulatory toast to the exhausted physicists. After 28 minutes of the reaction, the "Zip" rod was reinserted and the fissioning slowed down. This was the first time that a self-sustaining nuclear reaction had occurred. The "nuclear age," with all its prospects and apprehensions, had begun.

The news of Fermi's success spread quickly in select scientific circles across America. Arthur Compton telephoned the president of Harvard University, James Conant, that historic day and cryptically reported the news to him. "You'll be interested to know," he announced, "that the Italian navigator has just landed in the new world."[57]

CHAPTER XIII

The Dark Days of War: 1941–1945

BRIDGES TO THE PAST were annihilated as the war continued. Nazi victories penetrated into the Soviet Union. Stalin called upon the Allied forces for support when the non-aggression pact with Hitler was broken. Beating Stalin's attack by a mere three days, despite Russian troops amassed on the border, Hitler rolled tanks and the Army into Russia. By the winter of 1941, Germany's eastern offensive met fierce Russian counterattacks, and German troops began to suffer large losses. Four years of bloody fighting would drain millions of lives from both armies.

Although "uranium machines" were still seen as a potential source of weapons by German scientists, the technical problems concerning the production of heavy water and separation of U-235 presented major hurdles that were not solved during these years.[1] Recently published memoirs of Stalin's spymaster Sudoplatov reveal that as early as 1940, a commission of Soviet scientists supported by the Russian government began to investigate the possibility of creating a weapon based on uranium fission.[2] A year later, in September 1941, British "double agent" Kim Philby, who had been part of the Soviet spy ring, reported from London that he had uncovered "proof" that the Allies might construct a uranium bomb within two years. The report ended in revealing what another top secret department was also working on: namely, that "the British government was seriously interested in developing a bomb with unbelievable destructive force based on atomic energy."[3]

By December 1939, all fission research in Germany was under the authority of the War Office and military commander Kurt Diebner. Diebner ordered theoretical physicist Erich Bagge to leave his university post in Leipzig and come to Berlin. Meitner was well aware that the "younger generation" of German physicists were shifting towards applied research for the military, but she was generally unaware that key theoretical questions regarding uranium fission would occupy Reich scientists for years. The new fission project director Bagge was ordered to summon senior scientists Hans Geiger, Walther Bothe, Walther Gerlach, Otto Hahn, and others to a meeting in the War Office headquarters. Hahn recalled being very nervous; were they being invited for scientific consultation or conscription, ordered back into military service? Most had never joined the Nazi Party, and yet all had considerable knowlege about the fission process.[4]

Under the leadership of Bagge and later Heisenberg, this elite group met periodically to discuss, among other agenda items, Paul Harteck's research on the importance of "layering" uranium to create a chain reaction, and the uses of heavy water as a moderator.[5] While domestic surveillance as well as brutal repression by the SS and Gestapo intensified in occupied lands of former Poland, Austria, France, Netherlands, Belgium, Czechoslovakia, and soon Norway and Denmark, this group continued to meet to discuss potential applications of nuclear fission. Some German scientists later insisted that they were "stalling" Reich officials for many years,[6] while many historians call these "memories written with hindsight."[7]

Meitner had been splitting uranium for about four years without realizing it during the 1930s. Once the explanation she and Frisch had come to was published internationally, research efforts intensified around theoretical challenges. None knew better than Otto Hahn what that might entail: it was just a matter of time before natural uranium, 99% of which is in the isotope U-238, would be separated into the more highly fissile isotope U-235, which Bohr had predicted would be the basis of a sustained chain reaction. But separation *was* the issue during the 1940s. Factories were soon to be built for this in America; in Germany, however, such funds were not available.

A major cause for alarm was a meeting between Bohr and his protegé Heisenberg in mid-September 1941. The German Army had been occupying Denmark for over a year, and a so-called "German Cultural Institute" had been established in Copenhagen, basically to generate propaganda. At an official meeting of astronomers, Heisenberg arrived, funded by the German Institute; he met several times with Bohr. Certainly memories differ regarding this occasion. Bohr recalled that in a private conversation held in Tivoli, Heisenberg alluded to the German uranium project team, and "brought up the question of the military applications of atomic energy."[8] Bohr's son Aage later recalled:

> My father was very reticent and expressed his skepticism because of the great technical difficulties that had to be overcome, but he had the impression that Heisenberg thought that the new possibilities could decide the outcome of the war if the war dragged on.[9]

Others, including Hans Bethe, clearly recall that Bohr's alarm was based on a drawing Heisenberg showed him: a diagram of a reactor to produce plutonium.[10] The most secret project of all alarmed Bohr enough to alert others in his own government that the Germans *were* working on atomic weapons.

In January 1942, a meeting took place between the new President of the Kaiser Wilhelm Society, Dr. Vögler, and Army General Leeb, head of the weapons testing office. The former President of the Society, Karl Bosch (who had aided Meitner with her escape) had died suddenly in 1940. Vögler, his successor, now the administrative leader of all the Kaiser Wilhelm Institutes, was an ardent supporter of Nazi policies and a wealthy industrialist. General Leeb, proud of his division's aggressive weapons development, felt that since the German nuclear power research surrounding potential electrical generation and production had been "formally concluded" in 1941, the Kaiser Wilhelm Society should now take over uranium research, shifting potential war-related experiments to "longer-term scientific inquiry."[11] And because the Army already had its hands full with the scientists in the rocket research division, the general argued that an "upgrade" of the uranium project to control by industry was not logical during the war and hence, he concluded, it should fall upon civilian scientists to carry on basic research. In effect, Leeb was washing his hands of nuclear research.

But the President of the Kaiser Wilhelm Society would not let this research be swept aside. Vögler, formerly head of the largest steel company in Germany and highly influential within the Nazi Party, brought Speer's attention to this "neglected" field; he complained of "the inadequate support for fundamental research from the Ministry of Education and Science."[12]

Soon Speer decided to speak to the key decision maker, and recalled in his memoirs: "On May 6, 1942, I discussed this situation with Hitler." It was proposed that Goering become the new head of the Reich's "Research Council," to which he was appointed on June 9, 1942, thereby "emphasizing its importance."[13] The Allies, including Soviet spies, had been observing this reorganization of the uranium research program, and feared – correctly – that the German nuclear project was picking up steam. Speer later recounted:

> Around the same time, the three military representatives of armaments production – Milch, Fromm, and Witzell – met with me at Harnack House, the Berlin center of the Kaiser Wilhelm Society, to be briefed on the subject of German atomic research. Along with scientists whose names I no longer recall... Otto Hahn and Werner Heisenberg were present. After a few demonstration lectures on the matter as a whole, Heisenberg reported on "Atom Smashing and the Development of the Uranium Machine and Cyclotron." Heisenberg had bitter words to say about the Ministry of Education's neglect of nuclear research, about the lack of funds and materials, and the drafting of scientific men into the services.... I was familiar with Hitler's tendency to push fantastic projects by making senseless demands, so that on June 23, 1942, I reported to him only very briefly on the nuclear-fission conference and what we had decided to do.
>
> Hitler received more detailed and glowing reports from his photographer, Heinrich Hoffmann, who was friendly with Post Office Minister Ohnesorge. Goebbels too may have told him something about it. Ohnesorge was interested in nuclear research and was supporting – like the SS – an independent research apparatus under the direction of Manfred von Ardenne, a young physicist. It is significant that Hitler did not choose the direct route of obtaining information on this matter from responsible people but depended instead on unreliable and incompetent informants to give him a Sunday-supplement account. Here again was proof of his love for amateurishness and his lack of understanding of fundamental scientific research.[14]

However, German physicists younger than Hahn and Heisenberg continued their intense focus on nuclear issues. When von Weizsäcker organized meetings for a "German Scientific Congress" in occupied Copenhagen, Lise Meitner was furious. She never forgave many physicists, including Heisenberg, who arrived spouting Nazi propaganda, and whose "unpsychological ideas" touted rhetoric instead of reality.[15] She wrote to many of her colleagues about their shameless nationalism, and scholars are still debating Heisenberg's contentions that he tried to "warn" Bohr about the German atomic research project.[16] Ten years after the war, Lise Meitner would take post-war historians such as Robert Jungk to task for misinterpreting Heisenberg's active role in supporting German supremacy throughout the 1940s.[17]

In 1943 Hitler continued to insist on the expansion of the Third Reich, despite casualties in the millions. On September 28, 1943, a diplomat visited the Bohr's Carlsberg residence, emphasizing that many people were "quickly departing" from Denmark — even professors. "He did not say so directly," recalled Margrethe Bohr, "but we knew that the time had come when we must go."[18] Despite years of occupation, few arrests had been made. Bohr was half Jewish from his mother's lineage, and no one could predict his fate under the Nazis. News spread quickly through the city of Copenhagen concerning pending arrests. Meitner had been in and out since 1940 when the German occupation began, but few could guess Bohr's fate. While the underground resistance movement intensified its resistance, scientists mobilized to aid Bohr.

George von Hevesy went directly through papers in the Institute for Physics that recorded Bohr's assistance with refugees for many years. These were not to fall into the hands of Nazi soldiers! But a puzzle presented itself — what was he to do with the solid gold Nobel Prizes that his friend in Berlin, Max von Laue, had left with Bohr for safekeeping? If the Nazis found them in his possession, it would also endanger von Laue, already a black sheep in Berlin. Bohr's own gold medal had been donated to a Finnish war aide appeal. But with a stroke of insight, always the scientist, von Hevesy realized that gold was just gold — and dissolved the medals in a strong chemical solution of aqua regia that was stored safely in a small bottle on a shelf, to be recast respectfully years later.[19]

The next day, September 29, 1943, an anti-Nazi in the German diplomatic corps stationed in Denmark confirmed that orders had been issued for the "immediate arrest" of Niels and his brother Harald Bohr, including their "removal" to Germany. No time could be lost. Husband and wife escaped to Sweden that night, dramatically hidden in an old fishing boat that traversed the rough autumn waters. Their grown boys would soon follow with their families.

Hitler's signals were unmistakable. Days later, on October 2, rumor became reality. Scores of Jewish citizens were arrested and forced to board a cramped ship in Copenhagen which transported them to work camps.[20] Danes would never forgive that horrible day.

Meitner and Klein met the Bohr family and made them as comfortable as possible. Bohr acted quickly in Sweden. He petitioned the aging King through the assistance of members of the Swedish Academy of Sciences and diplomatic corps to "assist" his fellow Scandinavians. King Gustav Adolf pointed out that Sweden had three years before made an offer of asylum for any and all "citizens or refugees" directly to the Reich government when the Nazis first began their mass deportation of Norwegian "resisters." The King's veiled message to Hitler again in1943 was that instead of transporting Danish Jews, other innocent victims, and resistance fighters to concentration camps, the Swedes, as a neutral nation, would take them in.

This plea was abruptly rejected by Hitler as "unjustifiable interference in the affairs of Germany." Bohr, now himself a refugee in Stockholm, again asked the King of Sweden to help, stressing that the desperate situation of neighboring Danes threatened "the very survival of all Scandinavians" during wartime. If all else failed, perhaps a personal audience with Hitler would change the fate of nations and save thousands of lives. He also urged Sweden to make a public announcement of its "willingness to accept responsibility" for Jewish refugees.

Several hours after Bohr's meeting at the palace, Bohr received word that the King would broadcast an announcement that night.[21] Meitner and her friends rejoiced. Bohr and his wife, Oskar Klein, his father (a rabbi in Stockholm) and his wife, Stefan Rozenthal and Lise Meitner all sat intently around the radio in Klein's home that night and listened as the King of Sweden offered formal asylum to refugees from Denmark, and urged the Nazis to reroute their prison ships to Swedish ports. As one of Bohr's biographers relates:

> The Swedish offer greatly facilitated the full-scale rescue operation already in effect within the Danish underground. Swedish ships went out to the limit of the territorial waters to embark refugees from the Danish boats. Every night a fleet of small boats shuttled back and forth on their dangerous mission.
>
> Within a few weeks, almost 6,000 reached safety and sanctuary in Sweden. About 300 refugees were captured or lost when the Nazis tried to halt the great rescue. Altogether 472 Jews, including some who were unable to attempt the hazardous escape, were sent to the Nazi concentration camps. Forty-three died there.[22]

The scientific community in Stockholm extended their love and support to the Bohrs and newly-arriving refugees. Erik, who had accompanied his father during the 1939 trans-Atlantic journey in which Niels Bohr had reconfirmed Meitner and Frisch's explanation of fission, had married and was working on his Ph.D. in physical chemistry at Siegbahn's Institute. He later recalled that their baby son would run into Lise Meitner's arms as they routinely met in his mother's hotel room.[23] Anxious months would pass before the 58-year-old Niels Bohr would be off again, summoned by Lord Cherwell (known as Dr. Lindemann, Churchill's advisor on the "atomic project") to work in London on a top-secret project no one had yet dreamed about in Stockholm.

Over a year had passed since Goering was appointed head of the Reich Research Council. The top secret "Manhattan Project" would not be mobilized on a large-scale until 1943. On February 11, 1943, Stalin signed a decree "organizing a special committee to develop atomic energy for military weapons" with Soviet Foreign Minister Molotov in charge.[24] Physicists ordered to project headquarters in Lubyanka were shown scientific papers and clandestine reports gathered by secret agents, particularly those describing the first nuclear chain reaction in Chicago.

Fermi's brilliant work was recognized immediately: "he is the only one capable of producing such a miracle," physicist Issac Kikoin was said to have exclaimed. Abram Ioffe identified Otto R. Frisch as the author of another document;[25] Frisch, now 39, had been working on the British atomic project for three years.

During mid-1943, Churchill agreed to merge the British atomic project with the top secret U.S. Army operation, the Manhattan Project. And on October 6, a British Mosquito bomber plane landed at the Stockholm airport (unarmed in order not to violate Swedish neutrality) to pick up a valuable passenger. Her Majesty's Secret Service had been sent to transport Niels Bohr into an international physics chess game.

However, the young pilot, who secretly flew across enemy airfields in Norway en route back to London, had a nasty shock when he learned that his "valuable cargo" had passed out flying at high altitudes in the empty bomb bay prepared for him! Bohr's famous "big head and Danish ears" were too big for the aviator's helmet and he had consequently not heard the pilot's instructions to turn on his oxygen supply. Hence, unconscious, Bohr made the passage safely to Great Britain, and only upon his arrival was he briefed by the British Secret Service: scientists had for years been involved in the top secret efforts of the "Tube Alloys" project.[26] The real surprise would come soon: the physicists learned they would all be departing to join the Americans in an unrevealed region, where other emigrants were already working on something so large it amazed even Bohr — a "classified" atomic weapons project with factories scattered across the United States.

In Great Britain, concern turned towards the Nazi atomic efforts. By late 1943 German uranium efforts had slowed down, but there are varying accounts of why — or if key theoretical mistakes rather than the availability of funds was the problem. As Speer recalled:

> Wolframite imports from Portugal were cut off, which created a critical situation for the production of solid-core ammunition. I thereupon ordered the use of uranium cores for this type of ammunition. My release of our uranium stocks of about 1200 metric tons showed that we no longer had any thought of producing atom bombs.[27]

Others, however, recall working under Heisenberg to do just that. Although there were not ready to work with the Gestapo, the research team under Speer seemed willing to compromise their morals and work on uranium research to develop nuclear weapons for the Third Reich.

Niels Bohr and others became increasingly alarmed over the prospects of *any* world power developing a lethal atomic weapon, and he worked feverishly, sometimes around the clock, to alert other Allied scientists about the global dangers such "uranium projects" might lead to after the war's end. However, few but Einstein seemed concerned about international peace or the open exchange of scientific information in the midst of world war.[28] Bohr's prophetic glimmerings would remain unheard. When Churchill was challenged to join forces with the Soviet atomic project, he bluntly refused. Much later, he would even suspect Bohr of espionage and sent his spies after him.[29]

Cryptic messages from Bohr to his wife were sent to Lise Meitner and Oskar Klein to decipher. None in the Stockholm circle knew that the entire British team would soon be headed for New Mexico to join a growing community of U.S. Army engineers and their families, thousands of civilian workers, and an elite team of physicists under the direction of Army General Leslie Groves. Groves had chosen J. Robert Oppenheimer to lead the research and engineering efforts. The Manhattan Project was located in the New Mexico desert, high on a mesa, which had housed a ranch Oppenheimer worked on as a boy near the small town of Los Alamos. The project would receive millions of dollars in U.S. funds in the two-year race to construct the world's first atomic bomb.[30]

There was no time to lose. Hitler's bombing of London and industrial cities across England intensified. While the diplomatic perils of a world armed with atomic weapons loomed on the horizon, in the laboratories, other problems and paradoxes on the nuclear level presented themselves. Frisch had moved to Oxford to work on diffusion techniques. But with limited travel privileges, he went back to visit his mentor Chadwick one weekend and was surprised with the question, "How would you like to work in the U.S.A.?" Frisch had no idea what research was being pursued in America, but recalled the roundabout conversation:

> 'But then,' said Chadwick, 'you would have to become a British citizen!' 'I would like that even more,' I responded.
>
> Thereafter things moved with amazing speed. Within a couple of days, I was visited by a policeman for all my personal data required for a citizenship application. A week later I was asked to pack my things and go to London.[31]

Rudolf Peierls was also notified that he could change his refugee status immediately, and remembered: "We were keen to further the project . . . it was the official view that this was the best thing to do. And we didn't disagree . . . the question of whether it was our choice or someone else's choice really didn't arise."[32] America was foreign to them, but duty called. Nicholas Kurti, a Hungarian physicist and refugee, also went through a similar "speeded up" naturalization process before departing for a voyage across the rough Atlantic to a country whose language, English, was still new to him. An ironic scene unfolded the day of their departure, he recalled:

> There were no taxis available the day our boat was about to leave — they were often being called up as ambulances by this time, as the bombing of the city was getting worse. On that day – I will never forget it – we were just a group of young physicists, who were previously not permitted to work on government projects at all, preparing to work on an 'unknown project' in distant America — when outside, a big black hearse pulled up to transport us and our luggage to the docks.[33]

It was an ironic sign. They made the rite of passage, most seasick. Otto Robert Frisch recalled that there were mistaken impressions in later years about why British government officials would send so many refugees and Jewish emigrants to an unknown research project:

> This had been attributed to a 'wish for revenge' . . . but there was in fact a much simpler reason. Most British physicists had already been drafted into war work of some other kind, particularly on radar. Because of security reasons, refugees had been excluded from this work, so now they were almost the only physicists available for the most secret project of all.[34]

Hence, just months after Fermi demonstrated experimentally that a chain reaction was possible, U. S. government and Allied forces began to coalesce a team 100 times larger than the German "uranium project." By mid-1943, engineers, civilians and soon physicists were hard at work in Los Alamos. None of this was known to Lise Meitner; Frisch wrote his last letters to her from England. And he later recalled:

> It was not until Los Alamos that I saw Bohr again. By that time, it was clear that there were even two ways for getting an effective nuclear explosion: one through the separation of the highly fissile isotope U-238, and the other by using the new element plutonium formed in a nuclear reactor.
>
> Many of us were beginning to worry what the future might hold for a humanity in possession of such weapons. It was Bohr who taught us to think constructively and hopefully about that situation.[35]

Others were trying to do the same thing. Through behind-the-scenes work, in the fall of 1943, the chemistry division of the Swedish Royal Academy of Sciences invited Otto Hahn to lecture in Stockholm on the discovery of fission. Surprisingly, the Nazi government granted him a travel visa to deliver this lecture. This was the first reunion between Hahn and Meitner in three years. They caught up on family and professional news while "celebrating" Lise's 65th birthday. It was an ironic time to be so cheerful.

Hahn delivered his official lecture to the Royal Academy of Sciences, and Lise Meitner listened attentively as scientists from a variety of fields gathered in the hall. In his address, he referred to the "possibility" of splitting uranium by means of a "chain reaction" and in such a process, he conceded, enormous amounts of energy could "hypothetically be produced in a short instant." He also pointed out that such a process "might exceed any explosive phenomenon" then known. However, in this sobering Stockholm lecture, Hahn publicly stated that he "doubted whether it was possible to surmount the technical difficulties involved" in harnessing nuclear energy. "Providence has not wanted the trees to reach the sky," he obliquely affirmed, and the audience guessed from the passion in his voice that he may have wished that the conquest of atomic energy be approached at a much later date.[36] Then his departure came, much too soon for Lise Meitner. She was left depressed from final talks with Hahn in which he related the hardships, suicides, and deaths in Berlin. Yet they remained hopeful that after their own retirements, someday their Institute would continue.

Meitner remained disillusioned over the international shift toward "big science" based on expensive technological equipment and huge amounts of military funding. What is essential to understand from a sociological context is that her alienation in the face of the growing chasm between "basic" research and large-scale, technologically oriented "applied" research was not a loss of faith in the methods of science but in the drastically different "social contexts" in which scientific research was shifting.[37] She witnessed in Sweden what was occurring around the world during World War II: organizational and funding efforts redirected, and in many ways, redefined by government officials, not scientists themselves. The complex chemical, physical, and metallurgical issues of bomb construction and the scores of other scientific problems that arose between 1940 and 1945 surrounding the construction of nuclear weapons have been well documented by historians of science.[38] Scientific research would literally never be the same after World War II; in the race to counter fascism as well as Japan's warlords in the East, the scale and accompanying breakthroughs made by Allied organized efforts far surpassed most scientists' dreams. As sociologist of science Michael Mulkay asserts:

> Many scientists repeatedly rejected in principle the possibility that the form or content of scientific knowledge, as distinct from its incidence or reception, might also be in some way socially contingent.[39]

Lobbying efforts for further funding would replace publication records, and in many ways, under Manne Siegbahn, cyclotron research became dominant. Without command of the Swedish language, Lise Meitner was left on her own in the lonely Nobel-funded Research Institute for Physics. Yet thankfully, there were Swedish doctoral students who, exempt from military service, did seek her out, and her friend Gudmund Borelius, at the Stockholm Royal Institute for Technology, also lent her support and began the process to initiate a part-time pension for her retirement years.[40]

Lise Meitner, walking six to eight miles a day in her normal robust health, continued her avid experimentation throughout the war. With the assistance of those few male doctoral students she was able to mentor, she constructed creative apparatus, and true to her love of experimental work, was able to publish original results, demonstrating her new theories on fission as well as investigations on secondary electrons excited by gamma rays.[41] The psychology of invention, of pressing boundaries with new elements and their physical properties, engaged her mind, and her great gifts in this field made small experiments into original findings, worth publishing and contributing to the international community of science.[42]

Otto Hahn's research and health suffered during 1944: during the Allied bombings of Berlin, he had to finally evacuate his staff from the Institute for Chemistry, which was almost totally destroyed by major bombing that spring. The Institute – and its portion of the top secret uranium project – was then transferred to the south German town of Tailfingen, to which Hahn moved with his wife Edith. Tragedies continued to face the Hahns. Their son Hanno had lost an arm on the Russian front. He would be separated from his parents for over a year. Meanwhile, Hahn was called on by Heisenberg and urged to continue work on the uranium project.

The last secret research experiments were being carried out in an underground cavern in the Rhineland. They centered around a centrifuge developed to separate isotopes of uranium.[43] Research in a fortified bunker in Berlin had become impossible; in fact, radioactive contamination was a fact of life for the Institute for Physics staff before they left Berlin–Dahlem. When vital parts ran out, equipment that had been confiscated in occupied labs such as Paris, began to break down. Morale flagged.

Heisenberg had transferred the whole operation to the towns of Hechingen and Haigerloch.[44] Some manual laborers whispered that it was the government's last efforts to construct a "super weapon" but Hahn's close friend Walther Gerlach silenced these rumors. The "uranium machine" they had envisioned was close at hand, most thought, and von Weiszäcker and others were present for the final experiment.

It was a minor success. On March 1, 1945, Heisenberg sent a telegram to Berlin that a ten-fold increase in neutrons had been achieved with their "uranium machine" in the cellar laboratory. More heavy water and uranium were desperately needed to ignite a chain reaction.

However, the German team would be disappointed. On April 23, near the date of Hitler's suicide in Berlin, an Anglo-American intelligence team, code-named the ALSOS mission, staffed by physicists and supported by a special military force, arrived in Hechingen. Heisenberg, who returned to his family by bicycle, was captured in Bavaria over a week later.[45] Other members of the project were taken captive, and on April 25, Dutch–American physicist Samuel Goudsmit, scientific head of the ALSOS project, carefully searched Otto Hahn's laboratory for U-238 and arrested him in Tailfingen.

The ALSOS mission captured many German physicists, and transported them first to Belgium. At the end of the war in Europe, the German scientists were taken to a secret site outside of London. This team of top German scientists was imprisoned in a rustic estate house called "Farm Hall." Although under surveillance for six months, they were relatively comfortable. This veritable "Who's Who" of German physics included two Nobel laureates: von Laue and Heisenberg, and soon Gerlach, von Weizsäcker, Hahn, and others. Some of their conversations were taped by the British military and make interesting reading; their lack of scruples about Nazi atom bomb research and myopic views about the fate of Germany are still being debated today.[46]

In the summer of 1945, the war in Europe ended, but fighting in the Pacific Islands and Philippines continued. Now came a time of reckoning. President Roosevelt had died, and the new President, Harry Truman, supported by Generals Eisenhower, Patton, and others, approved Secretary of War Henry Stimson's suggestion of setting up an advisory committee to consider questions relating to post-war atomic energy policy.[47]

The Allies marched victoriously into Berlin; Stalin would not reveal his plans. The British were not consulted regarding atomic policy in the Truman administration, and Churchill would soon lose the elections at the end of the war in Europe. When Niels Bohr returned to Copenhagen, Soviet agents arranged a clandestine meeting with him;[48] American top-secret research continued and all were pressured about what was emerging in Los Alamos. At a meeting held in Washington, D.C. on May 31, 1945, General Marshall brought up the question of whether Russia should be "informed" about the development of the atom bomb, or if Russians should be invited to observe the first test, scheduled for July, to be held in a desert site near Los Alamos known as Alamagordo, "the valley of death."[49]

This suggestion was virulently opposed by several committee members, especially President Truman's representative, F. Byrnes. But fear remained that such a weapon could not be kept secret for long. O.E. Brewster, an engineer working on the Manhattan Project for the Kellex Corporation (subsidiary of the Kellogg Corporation), wrote a letter to the president warning that "other great powers would never permit the United States to enjoy a monopoly in nuclear weapons" and in order to avoid an inevitable arms race (which Niels Bohr had also been warning about since the 1930s) the U.S. should demonstrate its strength by using the bomb against Japan — then halt production "as evidence of good faith."[50] Several members of this interim committee were impressed by Brewster's logic. They decided that no information should be revealed to Russia. Some suggested that the bomb be "demonstrated" over an uninhabited island; others, that it be used directly against Japanese industrial cities, since the Japanese army would not surrender or release prisoners of war. From reports, they were aware that the Japanese were also working on wartime research surrounding applications of nuclear fission.[51]

Soon there would be more reminders of the war's irreversible effect upon humanity, and the harsh treatment of prisoners of war in both Japan and Germany. Count Bernadotte of Sweden, working tirelessly with the International Red Cross, at last secured the release and transportation of thousands of debilitated concentration camp victims to Sweden. Lise Meitner was horrified to witness their arrival. She knew relatives of friends and tried to assist them in any way possible.[52] Although she had been circumspect in her letters to Berlin all during the war, she had continued to send food packages and what little she could, writing petitions immediately after the war in order to assist former colleagues or friends.[53] But nothing prepared her – or the world – to face the realities of Hitler's death camps and the gaunt faces of those who were liberated by Allied soldiers.

Lise Meitner's full realization of the irreversible consequences of World War II, and the deep reckoning with her own identity, at last broke through. She decided to write a last letter to Otto Hahn, by then in captivity. Dated June 27, 1945, Hahn never received this letter, but a carbon copy of its contents was filed by his secretary and remained there for forty years after the war. In this letter, Meitner's moral stand against Hitler's "final solution" towards Jews and non-Jewish peoples rings forth clearly. Her letter read:

Dear Otto,

Your last letter is dated 25 March. You can imagine how much I want to hear reports about all of you. I follow events carefully in English war reports, and think I can safely assume that the area you and [Max von] Laue are in is an area which was occupied without a battle. From the bottom of my heart, I hope you didn't suffer personally. Naturally it will be very hard for you now, but that is unavoidable. On the other hand, I'm very much worried about the Plancks, since there was bitter fighting in their area. Do you know anything about them and our Berlin friends?

An American is going to take along this letter; he's going to come to get it right away, and so I'm writing in a great hurry, although I have so much to say that's so close to my heart. Remember that, please, and read the letter with the assurance of my unwavering friendship.

In my mind I've written you many letters in the last months, because it was clear to me that even people like you and Laue did not understand the real situation [in Nazi Germany]. Among other reasons, I noticed that when Laue wrote me once about Wettstein's death, saying that his death was a loss 'in a broader sense' since W. would have been very useful at the end of the war with his diplomatic abilities. How could a man who never opposed the crimes of the last few years be of 'use' for Germany? That has been Germany's misfortune: that all of you lost your standards of justice and fairness. As early as March 1935, you told me that Hörlein had told you that horrible things would be done to the Jews. He knew about all the crimes that had been planned and that would later be carried out; in spite of that he was a member of the Party and you still regarded him – in spite of it – as a very respectable man, and let him guide you in your behavior towards your best friend [referring to herself].

All of you also worked for Nazi Germany, and never even attempted passive resistance. Of course, to save your troubled consciences, you occasionally helped an oppressed person; still, you let millions of innocent people be murdered, and there was never a sound of protest.

I must write you this because so much of what happens to you and the Third Reich now depends upon your recognizing what all of you allowed to happen. Long before the end of the war, people here in neutral Sweden were discussing what ought to be done with German scholars at the end of the war. What will the Americans and the British have to say about this? I believe, as do many others, that one possibility would be for you to make a statement, namely that you know you bear a responsibility for the occurrences as a result of your passiveness, and that you feel it is necessary to help out in making reparations for the occurrences as far as that is even possible. Many, though, believe that it is already too late for that. They say that you all betrayed your friends at the outset, then your men and children by allowing them to risk their lives in a criminal war, and that finally, you betrayed Germany itself: when the war was totally hopeless, you didn't even oppose the senseless destruction of Germany. That sounds pitiless, but yet believe me, it is the truest friendship that makes me write you all this. You really can't expect the world to pity Germany. What we have heard recently about the unfathomable atrocities of the concentration camps exceeds everything we had feared.

When I heard a very objective report prepared by the British and Americans for British radio about Bergen-Belsen and Buchenwald, I began to wail aloud and couldn't sleep all night. If only you had seen the people who came here from the camps. They should force a man like Heisenberg, and millions of others with him, to see these camps and the tortured people. His performance in Denmark in 1941 cannot be forgotten.

Perhaps you remember that when I was still in Germany (and I know today that it was not only stupid, but a great injustice that I didn't leave immediately), I often said to you, 'As long as just we [the Jewish people] and not you have sleepless nights, it won't get any better in Germany.' But you never had any sleepless nights: you didn't want to see — it was too disturbing. I could prove it to you with many examples, large and small. Please believe me that everything I write here is an attempt to help you.[54]

The letter was sent off and never received.

Lise Meitner had reached an ethical turning point. She had at last expressed her deep sense of betrayal, exclusion, and most of all, moral separation from those who "didn't oppose the senseless destruction of Germany." Her sense of loss and personal responsibility for not leaving immediately were finally stated. She was able to remind Hahn truthfully that even her closest friends were also fully responsible and that their "loss of the standards of justice and fairness" and lack of "even attempted passive resistance" cut to her core — that they "didn't want to see" the fate in front of them, and hence, had *collaborated* with Nazi Germany.

Later, many would have their reputations resurrected, including Hahn, as Germany was rebuilt. However, Lise Meitner, more than most, realized fully that because they lacked a moral conscience, her friends had actively worked for the government that had brought about mass destruction for over twelve years. What Meitner was experiencing through this pain was as deep as her shock upon learning of the "unfathomable atrocities" of the concentration camps, which "exceeded everything we had feared." Never again would Meitner seek assistance from Hahn or others in Germany. Her attempt to "help" him face the truth was a commitment of love and healing for her friend, a gesture of moral awakening and personal responsibility. By admitting that it was also unjust that she too had not left "immediately" when Hitler seized power in Germany, she faced her own survivor guilt and began to be freed of its heavy burden. On her 67th birthday in November 1945, she filed papers to become a Swedish citizen.[55]

The Atomic Bomb, a Trip to Washington, and the Nobel Prize Controversy: 1945–1946

What still gives ground for anxiety, of course, is what mankind will make of this newly won knowledge, which could come to be used for destruction on a tremendous scale.

Lise Meitner
"Looking Back"
Bulletin of the Atomic Scientists, November 1964

A s s h e c a m e t o r e a l i z e the great wrongs committed by Germany, Meitner faced many moral challenges. While she did not completely sever her ties to friends and colleagues from her Berlin days, she distanced herself from the "new" Germany. Immediately after the war, she attempted to use her influence abroad to further the resettlement of German scientists, although she had reservations about several who requested her help. She also sent "care packages" with food to many of her colleagues and their families in the postwar months. Several members of the former Kaiser Wilhelm Institute for Chemistry, which had been severely damaged in the last months of bombing, asked her to vouch for them politically. She filled out the required forms to help them obtain a *Persilschein*, or clearance (taken from the name of a popular German soap powder; the slang term meant "to wash white as snow"), which gave them immunity during the Allied de-Nazification period in Berlin.

During the war Lise Meitner had spent summers in the northern Swedish town of Leksand at a small guest house near her Swedish friends' summer homes. Gudmund Borelius recalled that "she liked very much to walk in the woods, and she used much of her time . . . writing letters. She had a large correspondence with her friend von Laue, who, during the war, managed to let her know about the life of her friends in Germany."[1]

But when the war ended, this correspondence and others were interrupted, as von Laue was arrested by the ALSOS mission and transported to England with other members of the German nuclear research unit. Meitner and her friends in Sweden anxiously awaited news of their former colleagues.

Meitner was in Leksand, at the Borelius' lakeside cottage, on August 6, 1945 when the first atomic bomb was dropped over Hiroshima, Japan. Borelius recalled, "I went to her room to tell her the news from the radio, and will never forget that day."[2] Tears – shock – and then silence.

Lise was stunned. For a time she and her friends could only sit in silent horror; the small city of Hiroshima seemed a world away, but the impact of this huge explosion over a populated region was terrible to contemplate. They were soon roused by a knock at the door — the Borelius cottage did not have a telephone, so a neighbor had run over to tell them that a local newspaper, learning that the physicist Lise Meitner was vacationing in the area, wished to interview her! The reporter arrived, notebook in hand, and began asking the shaken Lise questions about "her work" on the atomic bomb. She tersely informed him she had never worked on any nuclear weapons project; the reporter, determined to get his "scoop," asked her to pose for a picture (with a friendly goat) to show that she was indeed vacationing in their town when the first atomic bomb was deployed.[3]

Lise Meitner was soon sought out again; this time the attention was international. Eleanor Roosevelt, who strongly advocated the formation of an international body of "united nations," had asked about the woman physicist who discovered fission. She informed her staff that she wanted Lise to join her for an NBC trans-Atlantic radio broadcast. The arrangements for the long-distance connection between the Swedish countryside and Washington, D.C. were quickly made: on August 9, three days after the bombing of Hiroshima, Dr. Meitner was interviewed by Mrs. Roosevelt.

In this historic conversation, Mrs. Roosevelt congratulated Meitner on her major contribution in interpreting the process of nuclear fission, and compared her to Marie Curie. She also hoped that Lise might someday visit the United States. Both expressed their belief in the need for world cooperation. Meitner also called for greater involvement of women in the creation of a lasting peace, and for the responsible use of nuclear power: "Women have a great responsibility and they are obliged to try, so far as they can, to prevent another war. I hope that the construction of the atom bomb not only will help to finish this awful war, but that we will be able to also use this great energy that has been released for peaceful work."[4] She had pondered these words with great care.

But the devastation was not over; on August 9 a second atomic bomb was used, this one over Nagasaki. Again Lise Meitner was interviewed on international radio; this broadcast included a tearful reunion with her sisters in New York who lived near the NBC studio.[5] What was important, however, is that now the world knew that Meitner had *not* worked on any military applications related to fission. She and others such as Einstein pledged wholeheartedly to work with other scientists towards a more peaceful world, and did all they could during retirement to alert the world to nuclear dangers.[6]

The war in the Pacific would soon be over. The Japanese surrendered on September 2 after more dreadful battles in the Pacific islands and Philippines in which thousands died. The world was exhausted: World War II had at last ended.

That same month of September, Lise received an invitation from Karl Herzfeld, an Austrian-born physicist who had admired her work for many years, to come to Washington, D.C. and lecture at the Catholic University of America. Herzfeld chaired the physics department and offered her a position as a Visiting Professor for the winter semester 1946.[7]

Surprised and delighted, she promptly telegraphed her acceptance. Here at last was an opportunity to visit the U.S. and have her expenses covered without a permanent move from Europe. The trip would provide a welcome reunion with relatives, and also bring further professional recognition. Meitner's brother-in-law was a professor at the Catholic University, and faculty were honored to have her there. Like the war, her lonely days in Sweden were over.

While arrangements were being made for her January voyage, Meitner was notified in October that she had been elected a Foreign Member of the Swedish Royal Academy of Sciences. The honor of being elected a member of the prestigious academy had been conferred on only two other women in the two hundred years of the its existence: the first on Swedish scientist Eva Ekeblad in 1748, and the second on Madame Marie Curie, elected an Honorary Foreign Member in 1910.[8]

Lise Meitner began to feel that the recognition by her Swedish colleagues was finally being granted, and this validated her decision at the close of the war to make Sweden her permanent home. Yet her citizenship papers were delayed. More debilitated German concentration camp victims arrived, and Meitner wondered if indeed her retirement pension would ever be paid.

But she left all this behind and faced the New Year full of hope. On January 27, 1946 Lise arrived in New York City. She was detained there with the flu for several days, but was met by her sisters' families and caught up with their news. Once on her feet again, the spry, 67-year-old departed for Washington D.C. and the Catholic University, where upon her arrival, the physics department welcomed her warmly. Plans were being made for a lecture tour through many of the Ivy League universities of the East Coast: Lise Meitner was indeed a famous heroine.

She was welcomed on an even grander scale when in February the Women's National Press Club made her the guest of honor at their annual awards ceremony. At this gala event, in the company of Georgia O'Keeffe, Agnes DeMille, and other luminaries, Meitner was presented with the Woman of the Year Award. President Harry Truman and Lise Meitner dined at the same table; Truman was said to have remarked to her, "Ah, so you're the little lady who got us into this mess!"[9] Obviously, he did not know her story of research and escape; and she did not pry him about his decision to drop bombs over Japan.

Two days later, at a Sunday afternoon cocktail party given by Herzfeld in Meitner's honor, Catholic University officials mingled with atomic physicists. Here, ironically, Lise Meitner was introduced to Major General Groves of the Manhattan Project. Mrs. Regina Herzfeld later recalled that the two "did not have much to say to one another."[10] However, Mrs. Herzfeld vividly recalled many of the events of Meitner's reception in the US:

> I remember the dinner given for Meitner by the Women's Press Club which was the occasion for the Truman-Meitner meeting. We were not at the head table, so I can't say how well they got along; Meitner did cause quite a splash and received any number of invitations, many of which she accepted. . . . One invitation she flatly refused was to go to Hollywood – I don't recall to which studio – to supervise a filming of her life. She would not have gone in any case, but she disapproved of the film on Madame Curie on several grounds, one of which was that Hollywood had "prettied" her up too much! I don't know that Meitner was shy, but she was certainly unassuming. There was nothing of the temperamental genius about her. She was always meticulous about her personal appearance, and that reminds me of Hertha Sponer-Franck. The latter told me one day that Lise had told her, when she was just entering the field, that it was very important for a professional woman to be well-groomed, and, for instance, 'never to let her petticoat show!'[11]

Lise's appearance, in her black dresses with lace collars and her hair smoothed into a bun, gave the impression of a proper Viennese woman of a previous era. But her unwavering dedication to science would serve as inspiration for the younger generation of women who began to pursue careers in greater numbers after the war. Younger American women just entering graduate school were particularly grateful for this strong, yet gentle, role model.

The months Lise spent in America in the spring of 1946 were busy ones. She traveled to a number of universities, including Princeton, Harvard, and MIT, where her lectures on her work in Sweden surrounding fission were well received.[12] She also spoke at several women' s colleges – Pembroke (now Brown), Sweet Briar, and Wellesley – on the topic of women in science and higher education. Some of these talks were published in various educational journals and provided a needed forum for postwar discussions of support for women in higher education.[13] During these years, Lise Meitner became increasingly outspoken about the need for greater opportunities for women in academia, and particularly the sciences. In her lectures that winter and spring, she emphasized that women should and must be treated as equals in their chosen fields, and encouraged young women to enter the physics and mathematics professions.

Along with her growing celebrity status came some unwelcome efforts by reporters to delve into her private life, which Lise kept rather closely guarded. In 1946, an article written by a reporter who had interviewed her in Sweden prior to her trip, appeared in the *Saturday Evening Post*. Despite its sensational title, "Is the Atom Terror Exaggerated?", it included the following details of Meitner's personal habits:

> When in the midst of an absorbing experiment, Professor Meitner forgets the clock. She sends out for a sandwich and some apples, and may then keep on until the late hours. She is a frugal eater with decided vegetarian leanings. She loves coffee, and her hobbies are flowers and long walks — if she had time, she does twenty miles on foot every day. 'It keeps me young and alert,' she explains.[14]

The article went on to give some of Meitner's reaction to all the attention she received after the bombings of Hiroshima and Nagasaki, continuing to stress that her contribution to nuclear fission had been her scientific insight, not any of the technological breakthroughs that followed it: "...I do not see why everybody is making such a fuss over me," she adds. "I did not design any atomic bomb. I don't even know what one looks like, nor how it works technically." She continued thoughtfully:

The great strides of atomic science are the results of the combined efforts of all atom[ic] workers. We strove toward a common goal, pooling our knowledge. I must stress that I myself have not in any way worked on the smashing of the atom with the idea of producing death-dealing weapons. You must not blame us scientists for the use to which war technicians have put our discoveries.... We must not be led into drawing too pessimistic conclusions just because the first use to which atomic energy was put happened to be in an engine of destruction. We must look at it as a revolutionizing scientific discovery; but perhaps, even so, only the first step on the road to something greater and still more valuable — mastering the art of using atomic energy for the benefit of mankind.[15]

Meitner continued to advocate for the international pooling of knowledge that had characterized scientific work prior to the war. Her dream of a return to "the spirit of ideal cooperation in which the true scientists of the world have always felt united" was overshadowed by increasing secrecy in nuclear research, and continued suspicion and spying among governments in the postwar world.

Despite the success of her U.S. visit, by mid-1946 Meitner began to tire of the lecture circuit. At spring cap and gown ceremonies in universities from north to south, she was awarded honorary doctorates. Americans loved her. But she was not proficient in English, so after her seven years in Stockholm she now looked forward to returning to Sweden. Her six-month stay had been rewarding both personally and professionally; she had shared in the happiness of her sisters and their families comfortably settled in America, and she had been celebrated and honored for her lifetime achievements, receiving awards, honorary doctorates, and acclaim for her lectures. The experience of teaching at the Catholic University's Physics Department for a semester had brought back happy memories of Berlin University before the war. She could now return to her adopted homeland, to her close friends there, and, she hoped, to a more secure place in Swedish scientific circles as a Member of the Royal Academy. But upon her return in the summer of 1946, she would discover that Swedish recognition for her achievements was not to be taken for granted.

The awarding of Nobel Prizes had been suspended during the last years of World War II. In 1945, the Committee voted not only on the awards for the present year, but for previous years as well. As early as 1940 the tide had changed in the Swedish elite circle of members of the Nobel Committees. Although nuclear fission was an interdisciplinary discovery, the Chemistry Committee Secretary, Anne Westgreen, continued to nominate Hahn *alone* for the Nobel Prize. Both Thé Svedberg and Westgreen marginalized Meitner and Frisch's contributions in reports summarizing the field in 1941 and 1942 — reports which have been critiqued for their *lack* of acknowledgments of the many citations and direct references to Meitner and Frisch's work readily available during the war.[16] The 1944 and 1945 Nobel Prizes in physics and chemistry were decided *after* the war and would be presented at the December 1946 ceremonies.

While Lise Meitner was preparing for her American visit, the nominations for the prizes were publicly announced. Otto Hahn heard the news broadcast in England by the BBC where he was still interned. He, and he alone, was awarded the 1944 Nobel Prize for Chemistry for the "discovery of nuclear fission!" His fellow detainees rejoiced with him, he recalled: "After the announcement by the BBC German service, there was no doubt that it was true, so that evening we were justified in celebrating.... von Laue gave an address, speaking in such moving terms – also about my wife – that I could not restrain my tears."[17] Imagine, however, how Lise Meitner must have felt. A clear bias was shown and the shocking news, learned later by Meitner, was that Otto Hahn, still imprisoned in Farm Hall, England, proposed that Walther Bothe of the University of Heidelberg should receive the 1945 Nobel Prize in Chemistry — Meitner was not even mentioned.[18] Why did Hahn have the honor of writing a formal report from prison internment to the Royal Swedish Academy of Sciences? Because he was formally notified that he had won the 1944 Nobel Prize for Chemistry for "his" discovery of fission.

After six months of confinement by the British, the German scientists were released to an Allied military escort in January 1946. Otto Hahn was returned, escorted by military police, to Germany. The Institute for Chemistry had been completely devastated by a bombing raid; he and his wife would begin a new life in the newly occupied zone of Berlin, still in their lovely house which had escaped harm. Meitner would remain in America for several more months.

The Nobel Committee had decided to award the 1944 Prize in Chemistry to Otto Hahn alone. Nowhere were the names of Fritz Strassmann, Lise Meitner, or Otto Robert Frisch included in this award. The 1944 Prize in Physics went to I.I. Rabi, who had worked closely with J. Robert Oppenheimer at Los Alamos, for prior research; Lise Meitner had also been nominated.[19] Although Meitner and Frisch would later be jointly nominated for the Nobel Prize in Physics, the award for 1945 went to Wolfgang Pauli for his contribution to the understanding of nuclear structure, and for the discovery of the neutrino — which he had first communicated in a letter to Lise Meitner.

The decision to award the Nobel Prize in Chemistry to Otto Hahn alone was not a unanimous one; a number of Lise Meitner's close colleagues sat on the committee, and there was an intense debate about the final choice. The decision, once reached, was reported to her privately by friends. Understandably, Lise Meitner keenly felt the injustice of the situation, as did many of her colleagues. Hahn's experiments, after all, had been interpreted by her, and Frisch had collaborated in the proof.

In his memoirs, Hahn would later characterize his former partner as a "bitter, disappointed woman" who begrudged him his Nobel Prize. He even went so far as to imply that she "blamed him for sending her away" from Berlin in 1938, just as the discovery was about to be made.[20] Of course, Lise's own letter to him, her other personal recollections, and what we know of her character belie these claims. If she was at all unhappy about being passed over for the Prize (for which she and Hahn had been *jointly* nominated over ten times in the past), it was justified. She had faced both gender and racial discrimination, and yet had never ceased to pursue her work, the science that was the passion of her life.

Otto Frisch described his aunt this way:

> She was interested in almost everything; always ready to learn, and ready to admit her ignorance of things outside her own field of study. But within that field, she moved with great assurance and was convinced of the power of the human mind to arrive at correct conclusions from the great laws of nature. When that conviction misled her (as in the belief that B-rays must be monoenergetic, or that certain substances she studied must be transuranic) the recognition that she had been wrong was a shock, as if nature had been unfair to her devoted work. But the advance of knowledge was always her first concern....[21]

This concern for facing the truth whether it coincided with her own theories or not carried over into her personal convictions as well. Her greater moral sense after the war became manifest in the realization of her own responsibility in having stayed in Germany too long; far from regretting that Hahn sent her away "too soon," she had come to believe that she had been wrong not to leave far earlier than she did. Her increasing outspokenness on the future of women in science, and the responsibility of all people to maintain world peace, demonstrate her ethical standards — beliefs which were hard-won and not likely to be shaken by the vagaries of public recognition. Hahn still took it upon himself to speak for her, creating an unjust portrait of Meitner in her later years.

Lise Meitner was there to meet Otto Hahn and his wife when, still escorted by a military guard, they arrived on December 10, 1946 in Stockholm for the Nobel ceremonies.[22] Whatever her initial manner of greeting him was, Meitner never subsequently denied that Otto Hahn deserved to have won the Prize. In fact, she "who could best form an opinion of him," as historian Friedrich Herneck put it, described Hahn as of equal stature to the great chemists of the 20th century who, "through their astonishing intuition and their unsurpassed ability, created the foundations of modern chemistry."[23] Her own contribution in rightly interpreting those chemical results had been published and disseminated worldwide; Bohr, Einstein, and others had acknowledged her work, even if the Swedish Academy of Science's Physics Committee, of which her sponsor Manne Siegbahn was a vocal member, had not.

The day arrived for the formal award of the Nobel Prizes. Arriving in tuxedos, the awardees proceeded to the stage. In his Nobel lecture, Otto Hahn did give credit to his chemistry assistant, Fritz Strassmann (he used the plural "we" throughout), and also to Lise Meitner and Otto Robert Frisch. The attendees at the ceremonies, including the King of Sweden, the members of the Royal Academy of Science, and of course Meitner herself, listened as he recounted his version of the events:

> In the first communication on these tests, which 'were in opposition to all the phenomena observed up to the present in nuclear physics' (*Die Naturwissenschaften*, January 6, 1939), the indicator tests mentioned had not been entirely completed, and we had therefore expressed ourselves cautiously.... Immediately after the first publication on the production of barium from uranium, there appeared as a first communication an article by Lise Meitner and O.R. Frisch in which the possibility of a breakdown of heavy atomic nuclei into two lighter ones, with total charges equal to that of the original nucleus, was explained with the aid of Bohr's model of the original nucleus. Meitner and Frisch also estimated the exceptionally high energy output to be expected from this reaction from the curve of the mass deficiencies of the elements in the Periodic Table. The great repulsive energy of the fragments produced by the splitting was first demonstrated experimentally by Frisch.... Meitner and Frisch soon proved that the active breakdown products, previously considered to be transuraniums, were in fact not transuraniums but fragments produced by splitting.... In quick succession there appeared a whole series of publications from European and American nuclear physics institutes, confirming and expanding the tests described.... The expression 'nuclear fission,' *Kernspaltung*, or *fission nucléaire* is due to Meitner and Frisch.[24]

When Hahn had concluded, the Swedish audience thundered its applause. Perhaps there was some self-congratulation in their reception of his lecture. With the war only a year or so behind them, Swedish officials had demonstrated a degree of forgiveness by awarding the Nobel Prize to a German. What mixed emotions Lise Meitner experienced at this outpouring of enthusiasm can only be imagined. However, Nobel Prizes are often awarded amidst controversy; in Meitner's case, many were shocked.

Otto Hahn gave part of the large financial award that accompanied the Nobel Prize to Lise Meitner. Characteristically, she immediately decided to pass it on in a large contribution to the Emergency Committee of Atomic Scientists in Princeton, New Jersey and sent her check in January 1947. The committee, chaired by Albert Einstein, consisted of scientists who were greatly concerned about the politicization of atomic research. Einstein promptly sent Lise a friendly acknowledgment, which read:

Dear Professor Meitner:

On behalf of my colleagues of the Emergency Committee of Atomic Scientists, I send sincere thanks for your generous help in the great educational task we have undertaken. We value not only the practical support you have sent, but also the good will towards this work and the hope for a reasonable solution of this immense problem which your contribution expresses.

Cordial greetings, Albert Einstein[25]

What Scientists Will Make of
This Newly Found Knowledge: 1947–1968

LISE MEITNER KNEW, as did Einstein, the weight of the moral responsibilities that accompanied nuclear research. They were both willing to work internationally to assist others in relocating and continuing their lives and work in physics in a world torn apart by war.

In postwar Germany, the old institutions were being renamed and replaced by newer entities. The Kaiser Wilhelm Society would soon be known as the Max Planck Society; the Institute for Chemistry had been established in its new location in the city of Mainz. Otto Hahn and Fritz Strassmann were both involved in the rebuilding of various scientific facilities. In 1947 Hahn and Strassmann wrote to Lise Meitner, offering her her old position, which Josef Mattauch had seized when she left in 1938. But she had already decided not to return to Germany to live; she replied to Strassmann, carefully outlining some of the conflicts that such a return would be likely to provoke:

> The question you asked me in your last letter is rather fateful, and I have tried to think it over with all its consequences. Accordingly, it seems to me that I should be better informed before I can reach a decision with a clear conscience; above all I ought to know who are the present members of the staff of the chemical and physical departments. Quite frankly, if the inquiry hadn't been from you, my answer couldn't have been other than 'No,' although the longing for my old sphere of activity has never ceased. But what is left of that circle and what does it look like in the minds of the younger generation?

Add to this the fact that I am neither gifted nor inclined to ignore things that are depressing. I closely followed all the horrible events Hitler's system had brought along with it, and tried to understand them on the basis of their reasons and consequences; and that means that my view of various problems will probably differ from those of the majority of my German friends and colleagues. Would we be able to understand each other? And certainly mutual human understanding is the essential basis of a real collaboration. I have no doubts about you, but that is not enough, you know.[1]

Lise Meitner hence turned down the offer to return to the Institute. And a year later, when Strassmann was hired at the University of Mainz and offered her a position there, she again refused. She wrote to Otto Hahn of this decision, declining the position and stressing that it was not "unfavorable living conditions" but something far deeper that kept her from returning: moral scruples. She carefully wrote to him:

> I do have considerable scruples about the intellectual mentality. If my view about anything which doesn't belong to physics were to differ from that of my colleagues, it would certainly be met by the words: 'Of course she doesn't understand the German situation, because she is an Austrian, or, because she is of Jewish descent.' I also stressed those scruples when I wrote to Strassmann, and he didn't reply except by repeating his assertion of how important I'd be to the institute. So he didn't dare refute my scruples. That means that I couldn't count upon the confidence of the younger colleagues which I formerly had and which, in my opinion, is always – and particularly nowadays – the most important basis of good collaboration. . . . It would become a fight similar to that which I had fought from '33 to '38 with very little success – and today it is very clear to me that it was a grave moral fault not to leave Germany in '33, since in effect by staying there I supported Hitlerism. While those moral scruples don't exist today, in view of the general mentality, my personal situation wouldn't differ much from my situation at that time, and I wouldn't really have the confidence of my colleagues, and therefore I couldn't really be of use. . . .[2]

These contrasts existed on personal levels as well. While Hahn rolled up his sleeves to help rebuild German science after the war, Meitner was committed to asking the hard questions about science's moral and ethical responsibilities. The late 1940s were years of international work in and out of Sweden for Lise Meitner. She participated in the second convention for the peaceful utilization of atomic energy in Geneva as a member of the Swedish Atomic Committee, and in lectures and debates gave voice to her social conscience. She also traveled to Vienna to receive the Lieber Prize for Science and Art in 1947. [3]

In the winter of 1948 she received another invitation to return to Germany — this time, to accept the highest award in German physics, the Max Planck Medal, which would be presented jointly to her and Otto Hahn. In 1949, she made her first trip back to Germany.[4] Despite their many differences, she and Hahn could still enjoy this time together. They were warmly received by the German university community, and Meitner toured Berlin, being rebuilt after the ravages of war. She and Hahn were honored at many dinners and ceremonies, where her long list of academic accomplishments were cited.

Years later, in an address at the 1962 ceremony for the presentation of the prestigious *Orden pour le mérite* award to Meitner and Hahn in Bonn, Werner Heisenberg described the 30-year partnership this way:

> It seems to me that Hahn owed his success primarily to his qualities of character. His untiring energy, his unbendable industry in acquiring new knowledge allowed him to work even more accurately and more conscientiously, to reflect on his own experiments with even more self-criticism, and to carry out even more controls than most of the other scientists who entered the new territory of radioactivity. Lise Meitner's relation to science was somewhat different. She not only asked 'What' but also 'Why.' She wanted to understand...she wanted to trace the laws of nature that were at work in that new field. Consequently, her strength lay in the asking of questions and the interpretation of experiment.[5]

Meitner was happy to leave war-torn Berlin. Her 70th birthday was spent in the company of her friends and colleagues in Sweden. Since coming to Sweden, Meitner had been living on the small salary of an assistant at Siegbahn's Institute. She had been stripped of her pension from the former Kaiser Wilhelm Institute by the Nazi government in 1938. In 1947, her friend and colleague Gudmund Borelius set in motion the creation of a paid position for Meitner at the Royal Institute of Technology (KTH). Even the Swedish Prime Minister, Tage Erlander, had taken an interest in Lise's situation. He offered to arrange a professorship for her in Stockholm, Uppsala, or Lund. But, as she jocularly recalled, Erlander "nearly fell from his chair when I told him that I was 'too old' to get any such position, because of the Swedish laws concerning pensions."[6] Lise did inform the Prime Minister, however, that she had decided upon Swedish citizenship, and had fulfilled all the requirements that would make her eligible for government benefits. The Prime Minister used his influence in concluding arrangements of salary and retirement benefits for Meitner through the Royal Institute of Technology.

With the financial assistance of the Swedish Atomic Energy Committee, a small laboratory was secured for her, and a stipend created for an assistant within the Physics Department. This would give her, as Borelius later wrote, "the possibility of assistance from young students, [to] whom she liked to impart... her knowledge in nuclear physics for the next five years."[7]

Here Meitner worked until she at last retired in 1953 at age 75. Afterwards, she continued to attend lectures and seminars, supervise graduate students, and work closely with doctoral candidates. And she kept up the lifelong habit of daily walking. These years in Sweden were relatively happy ones. In her modest way, she credited her Swedish colleagues with making her continued life in physics a rewarding one. She recalled:

> I was able to watch many interesting new developments in physics. It was mainly Oskar Klein, professor of theoretical physics [at the University of Stockholm] who, in his friendly way, helped me to understand the many new developments in the field of physics. While for the fact that the inner structure of a reactor has not remained entirely a closed book to me, I have to thank Dr. Sigvard Eklund [later Director of the United Nations International Atomic Energy Agency in Vienna], who has always been a very good and helpful friend to me in and outside of physics. Finally, I ought to mention Professor Borelius, whose work has gained greatly in importance, owing to the attention now devoted to semiconductors, a field in which he did much preliminary work. In this way, I can say that in Sweden too physics has brought light and fullness into my life. What still gives ground for anxiety, of course, is what mankind will make of this newly won knowledge, which could come to be used for destruction on a tremendous scale.[8]

The fulfillment that came with sharing knowledge openly was something she truly believed in, and wanted to pass on to the generations of scientists to come. Even while reflecting on the satisfactions of close collaboration, however, she remained clear-sighted about the ethical dilemmas involved in such work.

The years following World War II saw not only the beginnings of the Cold War, but the appearance of many German revisionists. These attempts to rewrite history troubled Lise Meitner. Some of this revisionism affected her directly; in a number of publications, her contributions to the work of the Hahn–Meitner team were being misrepresented. Concerned that her achievements might be erased from the scientific record, in June of 1953 she wrote to Otto Hahn, then the busy President of the Max Planck Society. She chose her words carefully:

> In the report of the Society, the lecture is mentioned that I had given in Berlin (it was purely a physics lecture). In this report, I am called "the long-term coworker of our President." At the same time, I have read an article written by Heisenberg in *Naturwissenschaftliche Rundschau* concerning the relations between physics and chemistry in the last 75 years. In this article, one single mention was made of me.... I quote: "the long-term coworker of Hahn, Fräulein Meitner." In the year 1917, I was officially entrusted by the Governing Board of the Kaiser Wilhelm Institute for Chemistry with the development of the Physics Section, and I led that for 21 years. Now, try to understand this from my perspective! Should it be possible during the last fifteen years – a time

I would not wish on any good friend – that even my scientific past should be taken away? Is this fair? And why is it like this?...What would you say if you were to be characterized as a "long-term coworker" of mine?[9]

Meitner was clearly stung by the injustice of having to defend her stature and substantial body of work to the man who should have known her worth better than anyone. She was as proud of her independent work and research as her joint publications with numerous colleagues, including Geiger, Szilard, Delbrück, and Franck. With over 105 original monographs, books, or articles published in her name, Lise Meitner knew she stood on solid ground as an original "pioneer" of twentieth-century physics.

Late in the 1950s Lise's health began to decline. When she suffered a terrible fall and broke her hip, she reluctantly had to admit that life alone in her Swedish apartment was no longer practical. So in 1960, at the urging of relatives, she packed up her belongings and moved to a small flat in Cambridge, England, close to the home of Otto Frisch and his family. She recuperated there, but adjusted slowly to life in England; she spoke only German with the Frisches and her close friends. However, she was happy, and continued an active correspondence with friends, colleagues, and admirers around the world.

Her travels were not yet over. In 1963, at the age of 84, she was invited to Vienna to speak before a large audience on her fifty years in physics. Her talk was later printed in English in the *Bulletin of the Atomic Scientists* as a retrospective entitled "Looking Back."[10] During her final trip to her home city, she marveled at the changes in Vienna, as well as the horizons and opportunities opening for women at the university where she had received her doctorate in 1906.

In her final long-distance journey, Lise accepted the invitation of her family in the U.S. to spend Christmas holidays during 1964 with them. It was a strenuous trip, but a worthwhile one, as the surviving Meitner siblings were once again united. Somehow these children of the liberal Austrian lawyer and plucky Moravian musician had survived the traumas of war, emigration, and periodic separation, and now could look back on decades of accomplishments. They were all very proud of Lise, and she of them. The trip took its toll on her strength, though: while there, she suffered a heart attack, and returned to England in a weakened condition.

In 1966 she was notified that she was to receive the distinguished Enrico Fermi Award – the first woman ever to be so honored – "for contributions to nuclear chemistry and extensive experimental studies culminating in the discovery of fission." At last, the international scientific community recognized her contributions. Awarded jointly to Hahn, Meitner, and Strassmann, it was wonderful news and confirmation, at long last, of her key role in the discovery and interpretation of fission.[11] Glenn Seaborg came personally to the Frisch residence to present the award. With her $15,000 share of the prize, she was able to establish her library, move all her belongings from Sweden, and participate in the Cambridge community.

Once again in 1967, she fell and broke her hip. This time she was too weak to rally. On October 27, 1968, shortly before her 90th birthday, she passed away. She died without knowing about the deaths of both Otto and Edith Hahn in the summer of that year. It is sad but true that the moral and personal distance between the two former colleagues, Meitner and Hahn, was never bridged in the last ten years of their lives.

During the eighty-nine years of Lise Meitner's life, she witnessed enormous changes in social, political, intellectual, *and* scientific history. Her dramatic life intersected the major events of her times: a mosaic of reshaped nations, scientific breakthroughs, and struggles of women academics to have their work and contributions to science and society valued and *recognized*. But Otto Hahn, after the war, and as we have learned through the history of the Nobel Prize selection process, during the war,[12] did not recognize Meitner as a true equal. His many chances after the war to *rectify* the situation publicly and privately were overshadowed by his own gregarious "championing" of the new rebuilt German scientific establishments. While he rose in post-war Germany as a career scientist and visible administrator (the elected president of the newly "reformed" Max Planck Society), Meitner struggled throughout her lonely retirement years. While some may be angry with her for remaining in Germany from 1933–1938 through a "shallow loyalty" to Germanic ideals in the face of growing atrocities in the Third Reich, we should be aghast at Otto Hahn's behavior, his complacency in the face of Nazi Germany, and the absence of any remorse for his treatment of Meitner and countless others who fled. Blinded by loyalty to what Max Planck also represented, the "German idealism" which kept Meitner out of the university for so long, Otto Hahn had many opportunities after 1945 to recant, recount, and set the record straight by rectifying the many wrongs committed against Meitner. He did not, and Meitner never fully forgave him. She spent her retirement with more trusted colleagues in Sweden and traveled extensively for nearly thirty years.

In 1964 she reflected:

> I believe all young people think about how they would like their lives to develop; when
> I did so, I always arrived at the conclusion that life need not be easy, provided only that

it is not empty. [For the fact] that life has not always been easy – the first and second World Wars and their consequences saw to that – while ... it has indeed been full, I have to thank the wonderful developments of physics during my lifetime, and the great and lovable personalities with whom my work in physics brought me in contact.[13]

Meitner forgave many. Einstein and von Laue as well as new friends such as Sigvard Eklund, one of her doctoral students in Sweden who became the president of the Vienna-based United Nations International Atomic Energy Agency, remained in contact with her. Important questions remain, however, about why she did not receive the Nobel Prize; about German revisionism; about loyalty to a scientific establishment that had lost its respect in the international community; and about lost credit and international recognition for a discovery which literally changed the world and ushered in the "nuclear age." Nazi Germany lost many. Lise Meitner nearly lost her life as well as her credit for fission.

Hence, Meitner will be remembered for who she was and her major accomplishments, but also for what was *not* recognized by the Nobel Committees. She will also be remembered by the international scientific community. Recall that Eleanor Roosevelt and Lise Meitner made a pledge immediately after the bombing of Nagasaki: that women around the world should *actively* work for world peace.

There is now a Hahn–Meitner Institute in Germany for continued research in physics, and even a moon crater that bears her name. Element 109 of the Periodic Table was named "Meitnerium" in 1992, eleven years after this biography was begun. An entire website exists on the Internet devoted to her life and work (http://www.users.bigpond.com/Sinclair/).[14] Lise Meitner's name *should* be preserved in the history of science for her own groundbreaking work in interpreting radioactive processes and elements, as well as the collaborative, interdisciplinary research and discoveries she was part of, including the interpretation of nuclear fission. Many who knew her will also remember her gifts of friendship, love of music and literature, and service to others both inside and outside the science world to which she was so completely dedicated. If this service to science blinded her to political realities until 1938, then we must all learn from the lessons of her life story that *social responsibility* should be the core of scientific enterprise.

The humble gravestone Frisch erected for her in Cambridge, with the equation for nuclear fission carved in stone, captures the spirit of this remarkable person. Her epitaph simply reads, "A physicist who never lost her humanity." But it was not her love of humanity or world peace that cast a shadow over her major scientific achievements. Prejudice, exclusion, and even denial of her major role in scientific discoveries were themes throughout her eighty-nine years — grim reminders of the social and political nature of science, the political realities of the Nobel Prize selection process, and the very clear ties that politics, funding, and choices of technological applications have to the worldwide scientific enterprise itself. When the atomic bomb exploded, the nuclear age was ushered in.[15] Lise Meitner was a pioneer from the very dawn of this nuclear age.

Journals and Periodicals

Am. J. Phys.: American Journal of Physics
Angew. Chem.: Angewandte Chemie
Angew. Chem Int. Ed. Eng.: Angewandte Chemie, International Edition in English
Ann. Phys.: Annalen der Physik
Ark. Mat. Astron. och Fysik: Arkiv för Matematik, Astronomi och Fysik
Ber. dt. chem. Ges.: Berichte der deutschen chemischen Gesellschaft
Biog. Mem. Fel. Roy. Soc.: Biographical Memoirs of the Fellows of the Royal Society
Bull. Atom. Sci.: Bulletin of the Atomic Scientists
Dict. Sci. Biog.: Dictionary of Scientific Biography
Ergebn. Exakt. Naturwiss.: Ergebnisse der Exakten Naturwissenschaften
Hist. Stud. Phys. Sci.: Historical Studies in the Physical Sciences
J. Am. Hist.: Journal of American History
J. Chem. Ed.: Journal of Chemical Education
J. Physique Radium: Le Journal de Physique et le Radium
Kgl. Danske Vid. Selskab, Mat. fys. Med.: Det Kongelige Danske
    Videnskabernes Selskab, Matematisk-fysiske Meddelelser
Naturwiss. Rund.: Naturwissenschaftliche Rundschau
Naturwiss.: Die Naturwissenschaften
New. Per. Phys.: New Perspectives in Physics
Nuc. Phys. Ret.: Nuclear Physics in Retrospect
Phil. Mag.: Philosophical Magazine
Phys. Ab.: Physikalische Abhandlungen
Phys. Bl.: Physikalische Blätter
Phys. Z.: Physikalische Zeitschrift
Phys. Zeit.: Physikalische Zeitung
Phys. Rev.: Physical Review
Phys. Ed.: Physics Education
Phys. Ret.: Physics in Retrospect
Phys. Tod.: Physics Today
Phys. ü Phys.: Physiker über Physiker

Proc. Phil. Soc.: Proceedings of the Philosophical Society
Proc. Roy. Soc.: Proceedings of the Royal Society (London)
Rev. Mod. Phys.: Review of Modern Physics
Ric. Sci.: La Ricera Scientifica
S. Ber. Akad. Wiss. Wien: Sitzungsberichte der Akademie der Wissenschaften
zu Wien
Sci. Am.: Scientific American
Verh. dt. phys. Ges.: Verhandlungen der deutschen physikalischen Gesellschaft
Z. Chem.: Zeitschrift für Chemie
Z. Angew. Chem.: Zeitschrift für Angewandte Chemie
Z. Natur.: Zeitschrift für Naturforschung
Z. Phys.: Zeitschrift für Physik
Z. Phys. Chem.: Zeitschrift für Physikalische Chemie
Z. Ver. Dt. Ing.: Zeitschrift des Vereines Deutscher Ingenieure

End Notes

Notes to Introduction

1. Otto R. Frisch, "Lise Meitner 1878–1968, Elected Foreign Member of the Royal Society 1955," *Bio. Mem. Fell. Roy. Soc.* 16 (1970): pp 405–424.

2. Regina Herzfeld to Patricia Rife, private correspondence, June 10, 1982.

3. Sigvard Eklund, Director of IAEA to Patricia Rife, private correspondence, July 1, 1980; also see Patricia Rife, *Lise Meitner: Ein Leben für die Wissenschaft*, 2nd ed., (Düsseldorf: Claassen Verlag, 1992) and publications list, end of book.

4. Lise Meitner to Otto Hahn, June 27, 1945; reprinted in its original German in Fritz Krafft, *Im Schatten der Sensation: Leben und Wirken von Fritz Strassmann* (Deerfield Beach, FL and Weinheim: Verlag Chemie, 1981), p 181; translated by Patricia Rife.

5. Of the many fine books on this subject, see Spencer Weart, *Nuclear Fear: A History of Images* (Cambridge, MA: Harvard University Press, 1988); Peter Hayes, *Industry and Fear: I.G. Farben in the Nazi Era* (Cambridge: Cambridge University Press, 1987); and current publications by the organization "Scientists for Social Responsibility," Cambridge, MA and Palo Alto, CA, and the Pugwash organization, London, England. Also see Londa Schiebinger, "The History and Philosophy of Women in Science," *Signs* 12, 2 (1987): p 306; and P. Stock, *Better Than Rubies: A History of Women's Education* (Boston: Putnam's Sons, 1978). Women were even blamed, during Lise Meitner's undergraduate years, for their intentions to pursue an education: in 1894, the Prussian Ministry of Education in Berlin justified women's exclusion from the universities on grounds that it was "shameful for the women and unwholesome for the men" to have women attend lectures and that all would suffer from "women's inability to study seriously." For modern efforts to counter such trends, see the Procedings of the annual International GASAT conference (Gender and Science and Technology organization). I wish to thank the National Science Foundation for their support to attend the 1989 conference in Haifa, Israel to lecture on Lise Meitner's life and times to an informed, international audience.

6. Otto R. Frisch, "Lise Meitner 1878–1968, Elected Foreign Member of the Royal Society 1955," *Bio. Mem. Fell. Roy. Soc.* 16 (1970): pp 405–424.

7. Hannah Arendt, *The Human Condition* (Chicago: University of Chicago Press, 1958), p 324.

8. Lise Meitner, "Looking Back," *Bull. Atom. Sci.* 20 (November 1964): p 3.

Notes to Chapter 1

1. Charles Guleck, *Austria from Habsburg to Hitler* (Berkeley, CA: University of California Press, 1948). Also see Meitner collection, Churchill College Archives, Cambridge, England for primary documents related to the Meitner family.

2. Carl Schorske, *Fin-de-Siècle Vienna: Politics and Culture* (New York: Vintage, 1981), p 129.

3. Josef Fränkel, ed., *The Jews of Austria* (London: Vallentine, Mitchell, and Co., 1967). This work provides a clear analysis of the 1867 Constitution's effect upon Jews within the Austro-Hungarian Empire. Also see the chapter by Franz Kobler, "The Contribution of Austrian Jews to Jurisprudence," pp 29–30.

4. Otto R. Frisch, "Lise Meitner," in Charles Gillispie, ed., *Dict. Sci. Biog.* 9 (New York: Scribners, 1974), p 260.

5. Carl Schorske, *Fin-de-Siècle Vienna: Politics and Culture*, p 148.

6. Charlotte Kerner, *Lise, Atomphysikerin: Die Lebensgeschichte der Lise Meitner* (Weinheim and Basel: Beltz & Gelberg, 1987).

7. Lise Meitner, "Looking Back," *Bull. Atom. Sci.* 20 (November 1964): p 2.

I would also like to thank Dr. Emmy Sachs, age 101, for sharing her memories of Vienna and its cultural milieu experienced when working towards her Ph.D. in Latin at the turn of the century. Interview conducted April 10, 1987, Berkeley, CA.

8. Stefan Zweig, *The World of Yesterday* (New York: Viking Press, 1943), p 79.

9. Allan Janik and Stephen Toumlin, *Wittgenstein's Vienna* (New York: Touchstone, 1973), p 47.

10. Otto R. Frisch, "Lise Meitner 1878–1960," *Biog. Mem. Fell. Roy. Soc.* 16 (1970): p 410.

11. See Robert Waissenberger, ed., *Vienna 1890–1920* (New York: Rizzoli International, 1968), essays on the social and political climate of this period. Also see Stephen Bronner and F. Peter Wagner, *Vienna: The World of Yesterday 1889–1914* (New York: Humanities Press, 1996).

12. Charlotte Kerner, *Lise, Atomphysikerin: Die Lebensgeschichte der Lise Meitner*, p 3.

13. O. R. Frisch, "Lise Meitner," p 260.

14. Robert Musil, *The Man Without Qualities* (New York: Perigee Books, 1953); see his views on the psychological "climate" of the Hapsburg Empire.

15. Meitner, "Looking Back," p 2.

16. Ibid.

17. Ibid., p 3.

18. Margrit Twellman, *Die Deutsche Frauenbewegueng im Spiegel Repräsentativer Frauenzeitschriften. Ihre Anfänge und Erste Entwicklung 1843–1889, Volume II* (Frankfurt: Meisenheim am Glan, 1972), p 194.

19. E. Boedeker and M. Meyer Plath, *50 Jahre Habilitation von Frauen in Deutschland* (Göttingen: O. Schwartze, 1974).

20. Fritz Ringer, *Decline of the German Mandarins* (Cambridge, MA: Harvard University Press, 1969). See Chapter 2, "The Mandarin Tradition in Retrospect," for an analysis of the late nineteenth century "German ideal" of the university and its members.

21. Ibid. Also see Georg Simmel, "Der Begriff und die Tragödie der Kultur," *Philosophische Kultur Gesammelte Essais* (Leipzig: Barth Verlag, 1911), p 248.

> Every kind of learning, virtuosity, refinement in a man cannot cause us to attribute true cultivation [*Kultiviertheit*] to him, if these things function ... only as super-additions which come to his personality from a realm external to it and which ultimately remains external to it. In such a case, a man may have cultivated attributes, but he is not cultivated; cultivation comes about only if the contents absorbed out of the suprapersonal realm seem, as through a secret harmony, to unfold only that in the soul which exists within it as its own instinctual tendency and as the inner prefiguration of its subjective perfection.

22. Ringer, *Decline of the German Mandarins*, p 104. Also see Carl Becker, *Vom Wesen der Deutschen Universität* (Leipzig: Steinik Verlag, 1925), pp 1–24.

23. Lise Meitner, "Looking Back," p 3.

24. Andreas Kleinert, *Anton Lampa 1868–1938* (Mannheim: Bionomica-Verlag, 1985).

25. Lise Meitner, "Looking Back," p 3.

26. Ringer, *Decline of the German Mandarins*, p 106.

27. Karl Jaspers, *Die Idee der Universität* (Berlin: Piper Verlag, 1923). In this book, Jaspers admits that absolute truth might never be attained, but he nevertheless insists that it be pursued for its own sake:

> *Wissenschaft* comes into being when first, rational work frees itself from the mere service of life purposes … so that knowledge becomes a goal for its own sake, and when, secondly, the rational does not remain in isolated fragments … when everything rational is systematically to become a whole through being internally related.

It should be noted, Ringer counters, "that even Jasper's position is more extreme than it seems at first. His formulation completely severs the ties between scholarship and all 'life purposes,' and makes truth itself a kind of speculative totality. It thus evokes an image of learning that has little to do with the common sense notion of asking questions and seeking evidence. Indeed, there is very little of the tactile or operational in Jaspers' definition of learning; it does not suggest touching, seeing, or doing something else. In that sense, his argument represents an elitist philosophy of leisure; it is aristocratic, other-worldly, contemplative." Also see Friedrich Paulsen, *Die Deutschen Universitäten und das Universitätsstudium* (Berlin: Steinik Verlag, 1902), for a complete review of the autocratic nineteenth century German university lecture system.

28. Engelbert Broda, *Ludwig Boltzmann* (Woodbridge, CT: Oxbow Press, 1983), and Stephen Brush, "Ludwig Boltzmann," in *Dict. Sci. Biog.*, Volume II, pp 260–268 for further details on Meitner's influential professor of theoretical physics. Also see Yehuda Elkana, "Boltzmann's Scientific Research Program and its Alternatives," in Y. Elkana, ed., *The Interaction between Science and Philosophy* (NJ: Humanities Press, 1972).

29. Lise Meitner, "Looking Back," p 3.

30. C. Bailey, *The Greek Atomists* (London: Oxford University Press, 1928).

31. While British scientists insisted upon mechanical models, most continental scientists labeled this "materialism" and rejected all models or "heuristic representations." See W. Thomson's lectures on "molecular dynamics" in A. Hannequin, *Essai critique sur l'hypothèse des atomes dans la science contemporaine*, 2nd ed. (Paris: F. Alcan, 1899).

32. H.R. Post, "Atomism 1900," *Physics Education* 3 (1968): pp 1–13.

33. One opponent to atomism, Ernst Mach, stated his position as follows: "Intelligible as it is, therefore, that the effort of thinkers have always been bent upon the 'reduction of all physical processes to the motion of atoms,' it must yet be affirmed that this is a chimerical ideal." Ernst Mach, *The Monist* 5 (1894): p 37. Also see Erwin Hiebert, "Mach's Conception of Thought Experiments in the Natural Sciences" in Y. Elkana, ed., *The Interaction between Science and Philosophy*, pp 339–348.

34. D.S.L. Cardwell, ed., *John Dalton and the Progress of Science* (Manchester: Manchester University Press, 1968), for a full discussion of Dalton's work on the atomic theory.

35. Ludwig Boltzmann, *Kinetic Theory*, Volume II [translated by S.G. Brush] (Oxford: Pergamon Press, 1966), pp 119–120.

36. Wilhelm Ostwald, *Lebenslinien*, Volume II (Berlin: Kassing Verlag, 1927).

37. Ibid., p 188.

38. Max Planck, "Die Einheit des Physikalischen Weltbildes, 1908," reprinted in *Vorträge und Reden* (Braunschweig: Vieweg Verlag, 1958), pp 6–29.

39. Ernst Mach, *Populär-Wissenschaftliche Vorträge*, 5th ed. (Leipzig: Barth Verlag), p 237.

40. Arnold Sommerfeld, "Ludwig Boltzmann zum Gedächtinis," *Österreich Chemische Zeitung* 47 (1944): p 25.

41. Wilhelm Ostwald, *Grösse Männer* (Leipzig: Barth Verlag, 1909), p 405.

42. Engelbert Broda, *Ludwig Boltzmann: Mensch, Physiker, Philosoph* (Vienna: Franz Deuticke, 1955). See Broda for further insights into Boltzmann's philosophic influence on the Viennese generation of physicists he encountered and taught. Additionally, see public lectures and non-scientific articles of general interest by Boltzmann collected in *Vorlesungen über Gastheorie*, 2 vols., Leipzig, 1896–1898, translated by S.G. Brush as *Lectures on Gas Theory* (Berkeley, CA: University of

California Press, 1964), for a synthesis of most of Boltzmann's work from these periods.

43. Lise Meitner, "Looking Back," p 3.

44. Lise Meitner to Martin Klein, February 12, 1958, private collection. Also see Martin Klein, *Paul Ehrenfest: The Making of a Theoretical Physicist* (New York: Elsevier, 1970), p 49. I would like to thank Professor Klein for sharing this letter that Lise Meitner sent to him from her Stockholm office in 1958.

45. Lise Meitner, "Looking Back," p 3.

46. Lise Meitner, "Wärmeleitung in inhomogenen Koerpern," *S. Ber. Akad. Wiss. Wien*, IIa, 115 (February 1906): pp 125–137.

47. See D. Flamm, "Life and Personality of Ludwig Boltzmann," *Acta Physica Austriaca*, Suppl. X (Munich: Springer Verlag, 1973), pp 3–16.

48. Lise Meitner, "Über einige Folderungen, die sich den Fresnel'schen Reflexionsformeln ergeben," *S. Ber. Akad. Wiss. Wien*, IIa, Bd. 115 (June 1906): pp 259–286. Meitner always recalled the discovery of radium by Pierre and Marie Curie in 1902 as sparking her interest in the young field of radioactivity; see "Lise Meitner," *Current Biography 1945* (New York: World Press, 1945), p 393.

49. Interview with Lise Meitner by T.S. Kuhn and O.R. Frisch, Cambridge, England, May 12, 1963, Tape 65a (p.16 in German transcript), Archive for the History of Quantum Physics, American Institute of Physics, College Park, MD. After completing her doctoral thesis, Meitner initiated experimental research on alpha and beta rays, publishing two articles during 1906 and 1907: "Über die Absorption der Alpha and Beta-Strahlen," *Phys. Z.* 7 (1906): pp 588–590 and "Über die Zerstreuung von Alpha-Strahlen," *Phys. Z.* 8 (1907): pp 489–491.

50. Lawrence Badash, "Ernest Rutherford," *Dict. Sci. Biog.*, Volume XII, p 28. Also see E. Rutherford, *The Newer Alchemy* (Cambridge: Cambridge University Press, 1937).

51. Ernest Rutherford, *Radioactive Transformations* (London: 1906) and *Radioactive Substances and their Radiations* (Cambridge: Cambridge University Press, 1913).

52. See Max Planck, "Zur Theorie des Gesetzes der Energieverteilung im Normalspektrum," *Verhandlungen der deutschen physikalischen Gesellschaft* 2 (1900). There are numerous sources on Planck's quantum theory: see Max Planck, *The Origin and Development of the Quantum Theory* [Nobel Prize address, June 2, 1920],

(Oxford: Oxford University Press, 1922), for a clear overview in the author's own words.

53. Thomas Kuhn, *Black Body Theory and the Quantum Discontinuity, 1894– 1912* (Oxford: Oxford University Press, 1978).

54. Albert Einstein, *Ann. Phys.* 17 (1905): pp 132–148.

55. Abraham Pais, *Subtle is the Lord: The Science and Life of Albert Einstein* (Oxford: Oxford University Press, 1982).

56. Max von Laue, *Gesammelte Schriften und Vorträge*, Volume III (Braunschweig: Viewig Verlag, 1961).

57. Armin Hermann, *Max Planck in Selbstzeugnissen und Bilddokumenten* (Reinbeck: Rowohlt Verlag, 1973), and John Heilbron, *The Dilemmas of an Upright Man: Max Planck as Spokesman for German Science* (Berkeley, CA: University of California Press, 1986).

Notes to Chapter 2

1. Lise Meitner, "Looking Back," p 2.

2. Hugh Puckett, *Germany's Women Go Forward* (New York: Columbia University Press, 1930), p 188.

3. Arthur Kirchoff, *Die Akademische Frau: Gutachten hervorragender Universitätsprofessoren, Frauenlehrer und Schriftsteller über die Befähigung der Frau zum wissenschaftlichen Studium und Berufe* (Berlin: Hugo Stenik Verlag, 1897).

4. Ibid., Max Planck, pp 256–257.

5. Ibid., Emil Warburg, p 257.

6. Ibid., Wilhelm Ostwald, p 257. Also see Karl Haase, *Die weibliche Typus als Problem der Psychologie und Pädagogik* (Leipzig: Barth Verlag, 1915), p 24.

7. Herbert Spencer, *The Study of Sociology* (London: Appleton & Co, 1900). For an interesting analysis of Herbert Spencer's Social Darwinian views of women, see Elizabeth Fee, "Science and the Woman Problem," in M.S. Teilbaum, ed., *Sex Differences* (New York: Doubleday, 1976), p 185.

8. Charles Darwin, *Origin of Species by Natural Selection or the Preservation of Favored Races in the Struggle for Life*, 1st ed., 1859 (London: John Murray, sixth ed., 1872).

Also see Amaury de Riencourt, *Sex and Power in History* (New York: Delta Books, 1974), p 320; Julie O'Faolain and Laura Martines, eds., *Not in God's Image: Women in History from the Greeks to the Victorians* (Scranton, PA: Harper Torchbooks, 1973); Pnina Abir-Am and Dorinda Outrain, eds., *Uneasy Careers and Intimate Lives: Women in Science 1787–1979* (New Brunswick, NJ: Rutgers University Press, 1987).

For essays on modern "biologically-based" social theories, see Rita Arditti, Pat Brennan, and Steve Covrak, *Science and Liberation* (Boston: South End Press, 1980); Evelyn Fox Keller, *Reflections on Gender and Science* (New Haven: Yale University Press, 1985); Ruth Glater, *Slam the Door Gently* (New York: Fithian Press, 1987); and Mary Morse, *Women Changing Science: Voices from a Field in Transition* (New York: Phenom Press, 1995).

9. Sigmund Freud, *Introductory Lectures on Psychoanalysis: A General Introduction to Psychoanalysis* (New York: Liverhall, 1977).

10. Amy Hackett, "Feminism and Liberalism in Wilhelmine Germany, 1890–1918," in Bernice Carroll, ed., *Liberating Women's History: Theoretical and Critical Essays* (Chicago: University of Illinois Press, 1973), pp 127–136.

Also see Jean Quataert, *Reluctant Feminists in German Social Democracy 1884–1917* (Princeton, NJ: Princeton University Press, 1979), for a vivid account in March 1906 of a German parliamentary debate on removing restrictions to women's right of assembly. The prevalent German attitude seemed to be summed up by a Progressive Party deputy in the Reichstag: women were to be seen as "objects in social life" and "little could be done to encourage their independent thought or action."

Of course, not all German women were subdued by such rhetoric. See Werner Thoennedden, [translated by Joris deBres] *The Emancipation of Women: The Rise and Decline of the Women's Movement in German Social Democracy, 1863–1933* (London: Pluto Press, 1973) for a detailed historical analysis of this time period and women's roles and struggles.

11. George Bernstein and Lottelore Bernstein, "Attitudes Towards Women's Education in Germany, 1870–1914," *International Journal of Women's Studies* 2, 5 (1979) p 475.

12. Lise Meitner, "Max Planck als Mensch," *Naturwiss.* 17 (1958): p 406.

13. Planck's consistent honoring of the "noble truth" in science and scholarship was rooted in the Germanic concern that scientific knowledge and insight should lead to a Weltanschauung, a "world view." See Fritz Ringer and also see John Heilbron, pp 59–60; Max Planck "Die Physik im Kampf um die Weltanschauung," in *Wege zur Physikalischen Erkenntnis* (Leipzig: Verlag von S. Hirzeln, 1944) and Max Planck, *Scientific Autobiography and other Papers* (Westport, CT: Greenwood Press, 1949).

14. Calendar/notebook, 1907, Meitner collection, Churchill College Archives, Cambridge, England. I wish to thank Evelyn Fox Keller for inviting me to lecture on this theme in her 1981 "Women and Science" course when I was a visiting doctoral candidate in the MIT Science, Technology, and Society Program, Cambridge, MA.

15. Meitner, "Looking Back," p 5.

16. Lise Meitner, "Über einige Folgerungen, die sich aus den Fresnel'schen Reflexionsformeln ergeben," *S. Akad. Wiss. Wien* 115 (1906): pp 859–869.

17. Interview with Lise Meitner conducted by O.R. Frisch and T.S. Kuhn, Cambridge, England, May 12, 1963, tape no. 65, side 1(a)-151. (Lise Meitner was 85 years old at the time of this interview, which was conducted primarily in German and transcribed as such). I would like to thank the American Institute of Physics Center for History of Quantum Physics Director Spencer Weart for his assistance. Citation will henceforth be listed as "Meitner Interview, 1963," AIP Archives.

18. Ibid., "Meitner Interview 1963," AIP Archives.

19. Ibid., Hahn and Meitner's autobiographical memoirs differ in many details, but they agree on the place of their meeting.

20. Ernst Berninger, *Otto Hahn* (Munich: List, 1969), p 39.

21. Ernst Kaiser and Eithne Wilkins, translators, *Otto Hahn: My Life* (New York: Herder & Herder, 1970), p 86.

22. Ibid., pp 130–131.

23. Otto Hahn, "Ein neues Zwischenprodukt im Thorium," *Phys. Z.* 9 (1907): p 9. In the biographical article "Otto Hahn," in *Dict. Sci. Biog.*, Volume VI, Charles Gillespie, ed., (New York: Scribners,1974), p 15, Lawrence Badash states:

> Ramsay, famous for his discovery of several 'inert' gases, developed an interest in radioactivity which was furthered by Soddy, who had first worked with Rutherford and then spent a year with Ramsay. The latter was without radiochemical help so, handing his young German visitor (Hahn) a dish containing about 100 grams of barium salt, he asked him to extract the few milligrams of radium in it according to Marie Curie's method. Hahn, an organic chemist whose dissertation had dealt

with bromine derivatives of isoeugenol, was unfamiliar with this subject, but Ramsay observed that he could approach the work 'without preconceived ideas.' Because the sample was small, Ramsay determined the atomic weight of radium preparing it in some organic compounds (thereby greatly increasing the total amount being examined) and calculating the atomic weight from the measured molecular weights.

Chance sometimes favors the unprepared mind and Hahn, who familiarized himself with only the basics of radioactivity, followed the prescribed separation technique and found himself the discoverer of a new radioelement: radiothorium.

24. Alfred Romer, "Henri Becquerel," *Dict. Sci. Biog.*, Volume I, pp 560–561, presents a complete bibliography of Becquerel's published works.

25. P. Curie, M. Curie, and G. Bemont, *Comptes Rendus* 127 (1898): p 1215.

26. E. Kaiser and E. Wilkins, translators, *Otto Hahn: My Life* (New York: Herder & Herder, 1970), p 84.

27. Lawrence Badash, "Ernest Rutherford," *Dict. Sci. Biog.*, Volume XII, p 28. See also Sir James Chadwick, *Collected Works of Lord Rutherford of Nelson*, Volume I (Cambridge: Cambridge University Press, 1962), for Rutherford's published research results from this period. Rutherford had arrived at McGill in Canada in the fall of 1896 – perhaps the finest laboratory in the Western Hemisphere, financed by a tobacco millionaire who considered smoking a "disgusting habit" – with glowing references from his mentor, J.J. Thomson, whom he had studied under at the Cavendish Laboratory of Cambridge University. See Otto Hahn's memories of Montreal with Rutherford's team in Dietrich Hahn, *Otto Hahn: Begründer des Atomzeitalters* (Munich: List Verlag, 1979).

28. Ernest Rutherford, *Radio-activity* (Cambridge: Cambridge University Press, 1904; second edition, 1905).

29. E. Kaiser and E. Wilkins, translators, *Otto Hahn: My Life*, p 86. The *Habilitation* (qualifying examination for university teaching) was the final step in becoming a *Privatdozent* or unsalaried lecturer in the university system; it was one of the lowest rungs on the academic ladder, but a prestigious step nonetheless. For statistics and social analysis on such lecturers and their essential role in the German university system, see Max Baumgart, *Grundsätze zur Erlangung der Doktorwürde bei allen Fakultäten der Universitäten des deutschen Reichs* (Berlin, 1884).

30. Lise Meitner, "Looking Back," p 5.

31. E. Kaiser and E. Wilkins, translators, *Otto Hahn: My Life*, p 86. In his *Scientific Autobiography* (New York: Scribners, 1966), p 52, Hahn recalls that not only eating was dangerous in the radioactive research area:

> It was impossible to carry on chemical research in the woodshop. Luckily Professor Alfred Stock, chief of the inorganic chemistry department of the Chemistry Institute, was willing to give me space in one of his two private laboratories. I can still remember the room very clearly. There was a solid worktable on a raised platform and next to it a large basin filled with mercury, much used by Stock who worked with various gases. At that time the dangers resulting from mercury vapor were not realized, and I paid as little attention to all that mercury next to me as we did later in the Institute of Physics. The countless little spheres of mercury that rested in the cracks of the old wooden flooring were noted but not removed. Later, however, Professor Stock fell ill with mercury poisoning, and then he devoted all his remaining energy to the battle against carelessness in working with mercury and against using mercury amalgam fillings in teeth.

32. Lise Meitner, "Otto Hahn zum 80. Geburtstag," in Dietrich Hahn, *Otto Hahn: Begründer des Atomzeitalters* (Munich: List Verlag, 1979), p 308. Raised in a cultured Viennese home, Lise Meitner was exposed early in life to the many facets of classical music. She played piano (though few of her friends knew this!) and never missed a chance to learn about great composers, enjoying their music and, when in the opera house, following along with a full score in front of her. Her elder sister, Auguste, was a precocious pianist and played by the age of 12 at one of the 'grand soirées' given in the home of a wealthy Viennese friend of the Meitner's. One of Lise's younger brothers, Fritz, later wrote lyrics for Auguste's compositions when she began to study composition and conducting. Auguste was the mother of physicist O.R. Frisch, who would join Meitner in the interpretation of nuclear fission thirty years later. See O.R. Frisch, *What Little I Remember* (Cambridge: Cambridge University Press, 1979).

Hahn also loved music, had belonged to a singing group during his student years in Marburg, and continued his musical evenings throughout his first years in Berlin, meeting socially with groups of university men for lively chorals and public songfests.

33. Lise Meitner, "Looking Back," p 5. One clear vignette of the chauvinism Meitner experienced in her early years at the Chemistry Institute was recalled by Hahn in E. Kaiser and E. Wilkins, translators, *Otto Hahn: My Life*, p 88:

> I remember a time when Lise had great difficulty walking and sitting and was
> obviously in pain. She confided to me that she had a corn on her foot. Some
> months later an acquaintance remarked: 'Lise Meitner had a horrible corn in a
> place that I am sure she did not name to you. That was why she found sitting so
> painful.' It would indeed have been a medical enigma if a corn on her foot had
> been the cause of her difficulty in sitting!

34. Lise Meitner, "Looking Back," p 4.

35. Meitner and von Laue became lifelong friends; she succeeded him as Planck's
assistant in 1912. See Armin Hermann, "Max von Laue" in the *Dict. Sci. Biog.*, Volume VIII, pp 50–53 and A. Hermann, *Max Planck in Selbstzeugnissen und Bilddokumenten* (Reinbeck: Rowohlts Verlag, 1973), for more on von Laue's assistantship
to Planck at the University of Berlin.

36. "Meitner Interview, 1963," AIP Archives, p 14.

37. Lise Meitner, "Über die Zerstreuung von Alpha-Strahlen," *Phys. Z.* 8 (1907):
pp 489–491.

38. Otto Hahn and Lise Meitner, "Über die Absorption der Beta-Strahlen einiger
Radioelemente," *Phys. Z.* 9 (1908): pp 321–333.

39. O.R. Frisch, "Lise Meitner 1878-1968," pp 406–407.

40. Otto Hahn and Lise Meitner, "Actinium C, ein neues kurzlebiges Produkt des
Actinium," *Phys. Z.* 9 (1908): pp 697–702.

41. O.R. Frisch, "Lise Meitner 1878–1968," pp 406–407.

42. E. Kaiser and E. Wilkins, translators, *Otto Hahn: My Life*, p.90.

43. O.R. Frisch, "Lise Meitner 1878–1968," p 407.

44. Willy Ley, ed. and trans., *Otto Hahn: A Scientific Autobiography* (New York:
Scribners, 1966), p 61.

45. In "Otto Hahn 1879–1968," *Biog. Mem. Fel. Roy. Soc.* (1970): p 285, R.
Spence explains:

> Following up on earlier work by Meyer & von Schweidler in Vienna, Hahn and
> Meitner showed that the radioactive deposit derived from actinium contained
> actinium X. It was found that deposition of actinium X from a radioactinium
> source increased when a negative potential was applied to the collection surface
> and decrease when it was positive. Evidently actinium X ions were expelled from
> the source by the recoil accompanying alpha particle emission from radioactinium.
> By using the very thin active deposits as sources, new species could be separated
> from them by this recoil method. Thorium D (later called thorium G″), actinium G

(later called actinium C″) and radium C2 (later radium C″) were all characterized in this way.

46. Willy Ley, ed. and trans., *Otto Hahn: A Scientific Autobiography* (New York: Scribners, 1966), p 61.

47. Ibid., p 89.

48. E. Kaiser and E. Wilkins, translators, *Otto Hahn: My Life*, p 89.

49. Otto Hahn, *Mein Leben* (München: Piper Verlag, 1969), p 88.

50. See E. Kaiser and E. Wilkins, translators, *Otto Hahn: My Life*, for details of Hahn's salaried position and the Max-Planck-Gesellschaft Archives, Berlin–Dahlem for documents relating to Hahn's professional career. I wish to thank the archivists at the MPG during my research trip to what was then West Berlin in 1988, and the Leo Baeck Institute, NY, and Deutsches Akademische Austausdienst for their grant support before the Berlin Wall came down.

51. Meitner was actually awarded the title "Professor" by the Ministry of Science, Art and Education on July 31, 1919, but did not actually teach at the university until several years later when she was awarded the *Habilitation*. See documents, Meitner Collection, Churchill College Archives, Cambridge, England and Ruth Sime, *Lise Meitner: A Life in Physics* (Los Angeles & Berkeley, CA: University of California Press, 1996).

52. Lise Meitner, "Looking Back," p 5. Also see Thomas Kuhn's questions in "Meitner Interview, 1963," AIP Archives, p 8.

53. E. Kaiser and E. Wilkins, translators, *Otto Hahn: My Life*, p 9.

54. Lise Meitner, "Looking Back," p 5.

55. Lise Meitner, "Einige Erinnerungen an das Kaiser-Wilhelm-Institut für Chemie in Berlin–Dahlem," *Naturwiss.* 41, 97 (1954): p 9.

On this same theme, within his article "On Academic Scientists in Wilhelmian Germany," *Daedalus* (Summer 1974), p 157, Russell McCormmach notes:

> ...in the Wilhelmian years, Germany was deeply divided by its powerful, quickening industrialization, and scientists and the public faced similar questions about the place of science in a society in change and conflict. Wilhelmian physical scientists found their work identified with a scientifically intensive technology that underlay much of Wilhelmian industrialization and that nourished a one-sided practical view of the world; they also found it identified with a one-sided rational view of the world that offended a segment of educated Germans who formed an

avant-garde of an antimodernist culture imbued with artistic, mystical, and nature-worship values. By and large, Wilhelmian scientists steered a middle course. They upheld the values of nonutilitarian research and rational and empirical modes of inquiry; at the same time, they promoted industrial technology in limited contexts, as in the creation of applied scientific disciplines, and they recognized nonscientific components in the collective culture and in the worldview of the individual.

Notes to Chapter 3

1. Lise Meitner, "The Nature of the Atom," *Fortune Magazine* (February 1946): pp 137–188.

2. J.S. Ames, *Discovery of Induced Electric Currents* (New York: American Book Company, 1900) gives a full explanation of Faraday's works; also see L.P. Williams, "Michael Faraday and the Physics of 100 Years Ago," *Science* 156 (1967): pp 1335–1342.

3. J.J. Thomson, "The Cathode Rays," *Phil. Mag.* 44 (October 1897): p 312.

4. J.J. Thomson, *Recollections and Reflections* (London: Bell Publishers, 1936) gives a thorough description of Thomson's diverse research paths leading to the discovery of the electron.

5. Sanford P. Bordeau, *Volta to Hertz: The Rise of Electricity* (Minneapolis, MN: Burgess Publishing, 1982).

6. Lawrence Badash, "Ernest Rutherford," *Dict. Sci. Biog.,* Volume XII, p 29. Also see E.N. Andrade, *Rutherford and the Nature of the Atom* (London: Peter Smith, 1990).

7. Barbara Cline, *The Questioners: Physicists and Quantum Theory* (New York: Thomas Crowell, 1965) gives a vivid description of Marsden's work with Rutherford; also see A.S. Eve, *Rutherford* (Cambridge: Cambridge University Press, 1939), a well-documented biography of the charismatic "pioneer of radioactivity," and Lawrence Badash, ed., *Rutherford and Boltwood: Letters on Radioactivity* (New Haven: Yale Studies in the History of Science, No. 4, Books Demand).

8. Meitner related this tale to her Swedish physicist friend Gudmund Borelius, many years later. Interview with G. Borelius, April 8, 1981, KTH (Royal Institute of Technology), Stockholm, Sweden.

9. See Bibliography for full citations of all published articles.

10. Herman Minkowski, "The Principle of Relativity and the Inferences to be Drawn from It," *Jahrbuch der Radioaktivität und Elektronik* (1908). This famous lecture by Minkowski was delivered to the Society of German Scientists and Physicians in Cologne, September 1908. His opening words were: "Gentlemen! The ideas on space and time which I wish to develop before you grew from the soil of experimental physics. Therein lies their strength. Their tendency is radical. From now on, space by itself and time by itself must sink into the shadows, while only a union of the two preserves independence."

In *Einstein: The Life and Times* (New York: Vintage, 1978), p 160, Ronald Clark states:

> For Minkowski, relativity had become a central fact of life. After he and David Hilbert had visited an art exhibition at Kassel, Hilbert's wife asked what they thought of the pictures. 'I do not know,' was the reply; 'we were so busy discussing relativity that we never really saw the art.' Minkowski was among the most austere and dedicated of mathematicians. He was the last man to 'popularize,' to play to the gallery. Yet he had sounded the trumpet for relativity in no uncertain fashion. He was still only forty-four and in the early winter of 1908, it would not have been too outrageous to speculate on the prospects of future long-term collaboration between Minkowski in Göttingen and Einstein in Berne. Then, towards the end of the year, he fell ill. He was taken to the hospital and died of peritonitis on January 12, 1909, regretting on his deathbed, according to a legend which has more than a touch of plausibility: 'What a pity that I have to die in the age of relativity's development.

For a further account of this period, see Leopold Infeld, *Albert Einstein: His Work and its Influence on our World* (New York: Dover, 1950).

11. From the numerous biographies of Albert Einstein, I suggest: Max Born, *The Born-Einstein Letters* (London: 1971); Louis de Broglie, Louis Armand et. al., *Einstein* (New York: Peebles Press, Farrar-Strauss, 1976); Arthur Eddington, *The Theory of Relativity and its Influence on Scientific Thought* (Oxford: Oxford University Press, 1922); Gerald Holton, *Thematic Origins of Scientific Thought: From Kepler to Einstein* (Cambridge, MA: Harvard University Press, 1973); Banesh Hoffmann, *Albert Einstein: Creator and Rebel* (New York: New American Library, 1972); and Abraham Pais, *Subtle is the Lord: The Science and the Life of Albert Einstein* (Oxford: Oxford University Press, 1978).

12. Leopold Infeld, *Albert Einstein: His Work and its Influence on our World*, p 119.

13. Max Born, *My Life: Recollections of a Nobel Laureate* (London: Taylor & Francis, 1978).

14. Later published as Albert Einstein, "Über die Entwicklung unserer Anschauungen über das Wesen und die Konstitution der Strahlung," *Phys. Z.* 10 (1909): pp 817–825. Einstein's paper on the photoelectric effect ("On a Heuristic Viewpoint Concerning the Production and Transformation of Light") published four years earlier was later described by Louis de Broglie as "falling like a bolt from the blue." See L. de Broglie, *New Perspectives in Physics* (New York: Basic Books, 1962), p 134.

15. Lise Meitner, "Looking Back," p 4. Also Christopher Ray, *The Evolution of Relativity* (Bristol: Adam Hilger, 1987).

16. Ibid. Also see Lise Meitner and Otto Hahn, "Über eine typische Beta-Strahlung des eigentlichen Radiums," *Phys. Z.* 10 (1909): pp 741–745.

17. Lise Meitner, "Über die Beta-Strahlen der radioaktiven Substanzen," *Naturwiss. Rund.* 25 (1910): pp 337–340. Also see Bibliography at end of text.

18. Lise Meitner, "Status of Women in the Professions," *Phys. Tod.* (August 1960): p 20.

19. "Meitner Interview, 1963," AIP Archives.

20. John T. Blackmore, *Ernst Mach: His Life, Work & Influence* (Los Angeles & Berkeley, CA: University of California Press, 1972). The controversies between the two physicists are summarized in Max Planck's, "Naturwissenschaften und reale Aussenwelt," *Naturwiss.* 28 (1940): pp 778–799 and J. Heilbron, *Dilemmas of an Upright Man*, Chapter 2 "Defending the World Picture."

21. "Meitner Interview, 1963," AIP Archives.

22. See Armin Hermann, "Max von Laue," *Dict. Sci. Biog.*, Volume VIII, pp 33–50.

23. "Meitner Interview, 1963," AIP Archives, p 13.

24. Ibid.

25. Planck to von Laue, March 22, 1934, in Armin Hermann, *Max Planck*, p 86.

26. James Franck and Lise Meitner, "Über radioaktive Ionen," *Verhandlungen der deutschen physikalischen Gesellschaft* 13 (1911): p 671.

27. J. Heilbron, *Dilemmas of an Upright Man*, pp 64–66.

28. Geheimes Staatsarchiv, Preussischer Kulturbesitz, Rep. 90, Nr. 452a, "Althoffs Pläne für Dahlem," Abschrift zu U. IV, Nr. 1840. U.I. and Adolf von Harnack, *Die Institute und Unternehmunger der Kaiser-Wilhelm-Gessellschaft zur Förderung der Wissenschaften* (Berlin: Reichsdruckerei, 1917).

29. See *Dokuments zur Gründung der KWG und der MPG zur Förderung der Wissenschaften* (Munich: MPG, 1981), pp 41–48.

30. Russell McCormmach, "On Academic Scientists in Wilhelmian Germany," *Daedalus* (Summer 1974), p 160.

31. Walther Nernst, "Antrittsreden," *Sitzungsberichte der Königlichen Akademie der Wissenschaften zu Berlin* (1906): p 551.

32. Russell McCormmach, "On Academic Scientists in Wilhelmian Germany," p 162.

33. "Adolf Harnacks Denkschrift von Nov. 1909," in Lothar Burchart, *Wissenschaftspolitik im Wilhelmischen Deutschland* (Göttingen: Vandenhoeck & Ruprecht, 1975), p 31.

34. L. Burchart, *Wissenschaftspolitik im Wilhelmischen Deutschland*, Chapter 4, "Die Aufbringung der privaten Finanzierungskomponente;" also see Adolf von Harnack, *Die Institute und Unternehmungen der Kaiser-Wilhelm-Gesellschaft zur Förderung der Wissenschaften*.

35. Adolf von Harnack, in Max-Planck-Society, eds., *50 Jahre Kaiser-Wilhelm-Gesellschaft und Max-Planck-Gesellschaft zur Förderung der Wissenschaften* (Göttingen: MPG, 1961); Christa Kirsten and Hans-Günther Köber, *Physiker über Physiker Band 1: Wahlvorschläge zur Aufnahme von Physikern in die Berliner Akademie 1870 bis 1930* (Berlin: Akademie–Verlag, 1975–1979); and Günter Wendel, *Die Kaiser-Wilhelm-Gesellschaft, 1911–1914: zur Anatomie einer Imperialistischen Forschungsgesellschaft* (Berlin: Akademie–Verlag, 1975). Einstein was nominated to head the Institute for Physics, which did not receive state funding in the early formation of the Society for a variety of financial reasons. Hence, Einstein remained at the University, and never took the position.

36. Lise Meitner, "Einige Erinnerungen...," *Naturwiss.* 41, 97 (1954): p 10.

37. W. Ley, ed. and trans., *Otto Hahn: A Scientific Autobiography*, pp 70–71.

38. Jost Lemmerich, *Geschichte der Entdeckung der Kernspaltung* (Berlin: Universitätsbibliothek, 1988), pp 66–67.

39. W. Ley, ed. and trans., *Otto Hahn: A Scientific Autobiography*, pp 70–71.

40. Daniel Kelves, *The Physicists: The History of a Scientific Community in America* (New York: Vintage, 1971), p 91.

41. See E.N. Andrade, *Rutherford and the Nature of the Atom.*

42. Ibid., p 111.

43. Ernest Rutherford, "The Scattering of Alpha and Beta Rays and the Structure of the Atom," *Proc. Phil. Soc.* 55 (1911): pp 18–20.

44. Niels Bohr, *Collected Works*, Volume I, L. Rosenfeld et. al., eds., (Amsterdam: North Holland Press, 1972–1980). The first and last papers in the series published by Bohr during 1913–1915 are: "On the Constitution of Atoms and Molecules," *Phil. Mag.* 26 (1913): pp 1–25, and "On the Quantum Theory of Radiation and the Structure of the Atom," *Phil. Mag.* 30 (1915) pp 394–415.

45. George von Hevesy to Ernest Rutherford, October 14, 1913, Rutherford Papers, Cambridge University, Cambridge, England. Also see Max Jammer, *The Conceptual Development of Quantum Mechanics* (New York: McGraw–Hill, 1966), Chapter 2, for a clear and thorough explanation of Bohr's theories and their consequences for 20th century quantum physics. Jammer quotes George von Hevesy recalling: "Speaking with Einstein on different topics, we came to speak about Bohr's theorie [*sic*]. He told me that he had once similar ideas but he did not dare publish them. 'Should Bohr's theories be right, so it is from [*sic*] the greatest importance.' When I told him about the Fowler Spectrum, the big eyes of Einstein looked still bigger and he said: 'Then it is one of the greatest discoveries.' "

46. See Albert Einstein, "On a Heuristic Point of View about the Creation and Conversion of Light," in D. ter Haar, *The Old Quantum Theory* (New York: Pergamon Press, 1967), pp 91–107.

47. "Meitner Interview, 1963," p 3.

48. Ibid., pp 8–10.

49. Ibid., p 10.

50. Planck to Wien, August 25, 1925, Archives for History of Quantum Physics, Office for History of Science and Technology, University of California, Berkeley.

51. Interview with Lise Meitner and her sister Frieda, August 11, 1945, NBC Radio Archives, New York, NY.

52. Meitner to Elisabeth Schiemann, December 22, 1915, in Jost Lemmerich, *Geschichte der Entdeckung der Kernspaltung*, p 75.

Notes to Chapter 4

1. Historians and sociologists of science are still debating these issues; for diverse viewpoints, see: Joseph Ben-David, "German Scientific Hegemony and the Emergence of Organized Science" in *The Scientist's Role in Society: A Comparative Study* (Englewood Cliffs, NJ: Prentice Hall, 1971); Paul Forman, "Scientific Internationalism and the Weimar Physicists: The Ideology and Its Manipulation in Germany after World War I," *Isis* 64 (1973) pp 158–180; David Stevenson, *Armaments and the Coming War: Europe 1904–1914* (Oxford: Oxford University Press); Russell McCormmach, "On Academic Scientists in Wilhelmian Germany," *Daedalus* (Summer 1974), in which he reflects (p 167): "...The unprecedented importance of science in World War I assured that science would be implicated in the moral outrage provoked by the war."

2. William Carr, *A History of Germany: 1815–1945* (New York: St. Martin's Press, 2nd ed., 1979), Chapter 8, "The Foreign Policy of Imperial Germany, 1890–1914" provides a good explanation of the complex factors involved in Germany's support of Austria throughout this period. Also see Paul Christopher, *The Ethics of War and Peace: An Introduction to Legal and Moral Issues* (Englewood Cliffs, NJ: Prentice Hall, 1994).

3. R. Spence, "Otto Hahn," *Biog. Mem. Fel. Roy. Soc.* 16 (1970): p 286.

4. Lise Meitner, "Einige Erinnerungen...," *Naturwiss.* 98 (1954).

5. X-rays were originally known as "Röntgenstrahlen" after their discoverer Wilhelm Röntgen. See Henry Boorse and Lloyd Motz, eds., *The World of the Atom* (New York: Basic Books, 1966) and Alex Keller, *The Infancy of Atomic Physics: Hercules in his Cradle* (Oxford: Oxford University Press, 1983).

6. Lise Meitner, "The Status of Women in the Professions," *Phys. Tod.* 20 (August 1960).

7. Dietmar Grieser, "Im Schatten der Bombe: Lise Meitner 1878–1978" in *Köpfe* (Vienna: Österreichischer Bundesverlag, 1991), p 117.

8. For Planck's role in these negotiations, see correspondence in Planck material, University of California, Bancroft Library collection, Berkeley, CA: Planck to Fischer, May 1914 and Meitner to Fischer, August 2, 1914; and J. Heilbron, p 39. For a graphic illustration of the weeks leading to Germany's decision to declare war that summer, see Barbara Tuchman, *The Guns of August* (New York: Bantam, 1962).

9. Hahn's contract of June 14, 1912 was for 5000 marks: see J. Lemmerich, *Geschichte der Entdeckung der Kernspaltung*, pp 66–67. Fischer's first offer to Meitner came May 14, 1914: see Jeffrey A. Johnson, *The Kaiser's Chemists: Science and Modernization in Imperial Germany* (Chapel Hill, NC: University of North Carolina Press, 1990), pp 173–175, for details about Fischer's decision. Her 1916 pay increase would be funded by I.G. Farben, the major industrial chemical firm; see J.A. Johnson, pp 194–196.

In E. Kaiser and E. Wilkins, translators, *Otto Hahn: My Life*, p 126, Hahn states:

> . . . in the period shortly before the First World War, all the radiothorium produced by Knöfler and the Auer Company and all the available radium was sold to medical and other institutes. I remember a telephone conversation I had with Knöfler early in 1914. They asked 'into which account' the sum of 66,000 marks, the commission due to me on the sales of the mesothorium, was to be paid! It was an extraordinarily large sum in those days. I gave a tenth of it to my colleague Lise Meitner, whose work on fractional crystallization was of very great help to me. A year later, I had to pay a fairly large sum in 'victory' tax, and I bought war-bonds with the rest. Thus this money, and also the 40,000 marks I received in 1915, went the way of all flesh. During the war, manufacture ceased entirely because it was no longer possible to import thorium salts.

10. E. Kaiser and E. Wilkins, translators, *Otto Hahn: My Life*, pp 112–115. For an overview of the military policy and philosophies behind Germany's alliance with Austria-Hungary, see General Graf Joseph Stürgkh (Austrian representative at OHL) *Im deutschen grossen Hauptquartier* (Leipzig: List Verlag, 1921). Also see Lothar Burchart, *Wissenschaftspolitik im Wilhelminischen Deutschland* (Göttingen: Vandenhoeck & Ruprecht, 1975), Chapter 7, "Staat-Wirtschaft-Wissenschaft," pp 131–144.

11. The "Manifesto to the Civilized World" was written by Hermann Sudermann and promoted by Matthias Erzberger, a leader in the Catholic Centrist Party, who later became a Chief of Propaganda in Hitler's Third Reich. The Manifesto has been published in its entirety in G.F. Nicolai, *Die Biologie des Kriegs* (Zürich: Orell Fussli, 1916). A section reads:

Were it not for German militarism, German culture would have been wiped off the face of the earth. That culture, for its own protection, led to militarism since Germany, like no other country, was ravaged by invasion for centuries. The German army and the German people today stand shoulder to shoulder, without regard to education, social position, or partisan allegiance. We cannot wrest from our enemies' hand the venomous weapon of the lie. We can only cry out to the whole world that they bear false witness against us. To you who know us, who have hitherto stood with us in safeguarding mankind's most precious heritage — to you we cry out: Have faith in us! Have faith in us when we say that we shall wage this fight to the very end as a civilized nation, a nation that holds the legacy of Goethe, Beethoven, and Kant no less sacred than hearth and home.

Within days of the Manifesto's publication, G.F. Nicolai issued a pacifist challenge to such sentiments, entitled "Manifesto to Europeans," which was also circulated among the faculty of the University of Berlin. Only three men signed this petition: one was Albert Einstein. For a clear account of the entire scenario, see Otto Nathan and Heinz Norden, eds., *Einstein on Peace* (New York: Simon & Schuster, 1960), pp 1–7.

12. Lise Meitner, "Max Planck als Mensch," *Naturwiss.* 45 (1948): pp 406–408. Reflecting upon Planck's reaction to the "Manifesto of 93," Lise later wrote: "Don't (his) words establish Planck's courage and, in the middle of the worst national strife, his composed moral truthfulness and need for justice?"

Also see Stephen Kern, *The Culture of Time and Space, 1860–1918* (London: George Weidenfeld & Nicholas Ltd., 1983) concerning the "August 1914" crisis and the German response to it.

13. Planck to Lorentz, August 2, 1915, in Otto Nathan and Heinz Norden, eds., *Einstein on Peace*. On p 11, they too mention Planck's ignorance of the Manifesto's text.

14. See Sabine Ernst, ed., *Lise Meitner an Otto Hahn: Briefe aus den Jahren 1912 bis 1924 — Edition und Kommentierung* (Stuttgart: Wissenschaftliche Verlagsgesellschaft, 1992), pp 130–136.

15. Lise Meitner to Otto Hahn, October 14, 1915, literary estate of Otto Hahn, Max-Planck-Gesellschaft Archives, Berlin–Dahlem; translated from P. Rife, *Lise Meitner: Ein Leben für die Wissenschaft*, 2nd ed., (Düsseldorf: Claassen Verlag, 1992), p 110. Also see *Handbuch der Aerztlichen Erfahrungen im Weltkriege 1914–1918*, Volume X, *Röntgenologie* (Leipzig: Artze Verlag, 1922) for a detailed illustrated history of World War I X-ray ambulance trucks and equipment.

16. Richard Willstätter, *From My Life: The Memoirs of Richard Willstätter* (New York: W.A. Benjamin, 1963).

17. O.R. Frisch, "Lise Meitner," p 408.

18. Lise Meitner to Otto Hahn, January 9, 1916, literary estate of Otto Hahn, Max-Planck-Gesellschaft Archives, Berlin–Dahlem; original German letter cited in Sabine Ernst, ed., *Lise Meitner an Otto Hahn*, p 135; translated by P. Rife.

19. Lise Meitner to Otto Hahn, November 16, 1916, literary estate of Otto Hahn, Max-Planck-Gesellschaft Archives, Berlin–Dahlem; translated from P. Rife, *Lise Meitner: Ein Leben für die Wissenschaft*, 2nd ed., (Düsseldorf: Claassen Verlag, 1992), p 116.

20. Lise Meitner, "Einige Erinnerungen ...," p 98.

21. See Dietrich Stoltzenberg, *Fritz Haber: Chemiker, Nobelpreisträger, Deutscher, Jude* (Weinheim: VCH Verlagsgesellschaft, 1994); Morris Goran, *The Story of Fritz Haber* (Norman, OK: University of Oklahoma Press, 1967); and David Nachmansohn, *German-Jewish Pioneers in Science 1900–1933* (New York: Springer Verlag, 1979).

22. Lise Meitner, "Einige Erinnerungen ...," p 98.

23. For more insight into her research ideas, see Meitner to Hahn, October 25, 1916, literary estate of Otto Hahn Max-Planck-Gesellschaft Archives.

24. Dr. Prof. Hatjidakis to Lise Meitner, 1915, Meitner collection, Churchill College Archives; quoted in part in R. Sime, *Lise Meitner: A Life in Physics*, p 61 as:

> I would like to have the honor of marrying you. I admire you and the other Germans and your wonderful country. I hope you take my offer of marriage seriously. Also I would like your photograph. Please answer me.
>
> P.S. Greece is now all for the Germans.

Meitner kept the note, with its "flowery writing, purple ink and all."

25. Lise Meitner, "Einige Erinnerungen ...," p 98.

26. Meitner had repeatedly published that she returned in 1917, but her correspondence proves otherwise. See Fritz Krafft, "Lise Meitner 7/11/1878 – 27/10/1968" in Willi Schmidt and Christoph J. Scriba, eds., *Frauen in den exakten Naturwissenschaften* (Stuttgart: Franz Steiner Verlag, 1990), pp 33–70.

27. Otto Hahn and Lise Meitner, "Über die Alpha-Strahlung des Bismuts aus Pitchblende," *Phys. Z.* 16 (1915): pp 4–6.

28. Before being transferred to Haber's regiment, Hahn was given the command of a machine-gun section armed with captured weapons during the opening months of the war, and was awarded the Iron Cross Class II for his part in the autumn offensive. By Christmas, Hahn was in a front-line position in the neighborhood of Messines, and he took part in the fraternization that occurred between German and British troops on Christmas Eve and Christmas Day through his good command of English, which he had acquired during his days in Rutherford's Canadian labs and Ramsay's London labs. See R. Spence, "Otto Hahn," *Biog. Mem. Fell. Roy. Soc.*, p 286. Also see Lawrence Badash, "Otto Hahn, Science, and Social Responsibility," in Hans Grätzer, ed., *Otto Hahn and the Rise of Nuclear Physics* (Amsterdam: North Holland Press, 1983); and Samuel Dumas and K.D. Vedel-Petersen, *Losses of Life Caused by War* (Oxford: Oxford University Press, 1923).

29. Adolf Hitler, *Mein Kampf*, Volume I (Munich: 1930).

30. Lise Meitner to Otto Hahn, February 22, 1917, literary estate of Otto Hahn, Max-Planck-Gesellschaft Archives, Berlin; translated from P. Rife, *Lise Meitner: Ein Leben für die Wissenschaft*, 2nd ed., (Düsseldorf: Claassen Verlag, 1992), 111.

31. For a detailed account of their discovery process, see Ruth Sime, "The Discovery of Protactinium," *J. Chem. Ed.* 63 (August 1986): p 653.

32. See Hahn, O. and Meitner, L. "Die Muttersubstanz des Actiniums, ein neues radioaktives Element von langer Lebensdauer," *Phys. Z.* 19 (1918): pp 208–212.

33. E. Kaiser and E. Wilkins, translators, *Otto Hahn: My Life*, p 132. Also see Mario Satori, *The War Gases* (New York: Van Nostrand, 1940), and Joseph Ben-David, *The Scientists' Role in Society: A Comparative Study*, p 134 and n 46. Ben-David identifies the values of scientific universalism and altruism, and discusses the historical interactions between the developments of science, and social and moral value systems.

34. Raymond Fredette, *The Sky on Fire* (New York: Harcourt Brace Jovanovich, 1976).

35. Lise Meitner, "Einige Erinnerungen . . . ," p 98.

36. Lise Meitner to Otto Hahn, May 7, 1917, literary estate of Otto Hahn, Max-Planck-Gesellschaft Archives, Berlin–Dahlem; original German letter cited in Sabine Ernst, ed., *Lise Meitner an Otto Hahn*; translated by P. Rife.

37. Lise Meitner to Otto Hahn, July 27, 1917, literary estate of Otto Hahn, Max-Planck-Gesellschaft Archives, Berlin–Dahlem; translated from P. Rife, *Lise Meitner: Ein Leben für die Wissenschaft*, 2nd ed., (Düsseldorf: Claassen Verlag, 1992), pp 124–125.

38. Lise Meitner to Otto Hahn, August 6, 1917, literary estate of Otto Hahn, Max-Planck-Gesellschaft Archives, Berlin–Dahlem; original German letter cited in Sabine Ernst, ed., *Lise Meitner an Otto Hahn*; translated by P. Rife.

39. Stefan Meyer to Lise Meitner, June 5, 1918, Meitner collection, Churchill College Archives, Cambridge, England, in R. Sime, *Lise Meitner: A Life in Physics*, p 70.

40. Hahn, O. and Meitner, L. "Die Muttersubstanz des Actiniums, ein neues radioaktives Element von langer Lebensdauer," *Phys. Z.* 19 (1918): pp 208–212. For a detailed account of their discovery process, see Ruth Sime, "The Discovery of Protactinium," *J. Chem. Ed.* 63 (August 1986): p 653.

41. Willy Ley, ed. and trans., *Otto Hahn: A Scientific Autobiography* (New York: Scribners, 1966); Lise Meitner, "Einige Erinnerungen...," p 99; Lise Meitner, "Looking Back," p 6.

42. For the financial ties among I.G. Farben, Bayer Dye Works (with over 10,000 employees by 1914), and other industrial firms and the Kaiser Wilhelm Institute for Chemistry and Physical Chemistry, see Morris Goran, *The Story of Fritz Haber*.

43. Brigitte Schröder-Gudehus, *Deutsche Wissenschaft 1914–1928* (Geneva: Dumaret & Golay, 1966).

44. Max Planck, *Acht Vorlesungen über theoretische Physik gehalten an der Columbia University in the City of New York im Frühjahr 1909* (Leipzig: Hirzel Verlag, 1910), pp 126–127.

45. Lise Meitner, "Max Planck als Mensch," *Naturwiss.* 45 (1948) pp 406–408.

46. A. Einstein to L. Meitner, Berlin to Berlin–Dahlem, September 14, 1918, Collected Papers of Albert Einstein, Boston University, Boston and Hebrew University, Jerusalem.

47. A. Einstein to L. Meitner, October 29, 1919, Collected Papers of Albert Einstein, Boston University, Boston and Hebrew University, Jerusalem.

48. Max Born, "Max Planck," *Obituary Notices of Fellows of the Royal Society* 6 (1948): p 175 and John Heilbron, *Dilemmas of an Upright Man: Max Planck* as *Spokesperson for German Science*, p 83. Heilbron quotes a letter Planck wrote during this time to his Dutch colleague Lorentz: "Now I mourn both my dearly loved children in bitter sorrow and feel robbed and impoverished. There have been times when I doubted the value of life itself."

Planck to Lorentz, December 21, 1919, Lorentz Papers, Archives for the History of Quantum Physics, Office of the History of Science and Technology, University of California, Berkeley.

49. Reginald Pound, *The Lost Generation of 1914* (London: Coward–McCann, 1964).

50. Lise Meitner, "Einige Erinnerungen...," p 99.

51. Peter Gay, *Weimar Culture: The Outsider as Insider* (New York: Harper & Row, 1968).

52. Paul Forman, "Scientific Internationalism and Weimar Physicists...," *Isis* 64 (1973) pp 158–180; see *The Hague Declaration (IV, 2) of 1899 concerning asphyxiating gases*, Washington, DC: Carnegie Endowment for International Peace, 1915, for the illegality of this type of warfare; Lawrence Badash, "British and American Views of the German Menace during World War I," *Notes Rec. Roy. Soc.* 34 (1979): pp 91–121 for an overview of the nationalistic fervor which also swept Allied scientists; and Willibalt Apelt, *Geschichte der Weimarer Verfassung* (Munich: Biederstein, 1946) for the German response.

53. Soddy, F. and Cranston, J., *Proc. Roy. Soc.* A94 (1918): p 384.

54. Willy Ley, ed. and trans., *Otto Hahn: A Scientific Autobiography* (New York: Scribners, 1966), p 93. For a more general analysis of such disputes, see Robert Merton, "Priorities in Scientific Discovery: A Chapter in the Sociology of Science," *Am. Soc. Rev.* 22 (1957): pp 635–659.

55. Erich Bagge, *Die Nobelpreisträger der Physik* (Munich: Heinz Moos Verlag, 1964), pp 26–29 and Thomas Kuhn, *Black-Body Theory and the Quantum Discontinuity 1894–1912* (Oxford: Oxford University Press, 1978).

56. A. Hermann, "Max von Laue," *Dict. Sci. Biog.*, Volume VIII, p 51.

57. Arthur Stanley Eddington, *Report on the Relativity Theory of Gravitation* (London: Fleetway Press, 1918).

58. For Planck's Nobel Prize lecture, see M. Planck, "Nobel Prize Address," *A Survey of Physical Theory* (New York: Dover, 1960), p 102, and M. Planck,

Physikalische Abhandlungen und Vorträge (Braunschweig: Vieweg Verlag, 1958), Volume III, p 121. For detailed analysis of Stark's "Aryan Physics" movement, see Alan Beyerchen, *Scientists under Hitler: Politics and the Physics Community under the Third Reich* (New Haven, CT: Yale University Press, 1977). Also see Nobel Prize Archives Internet site at

http://www.almaz.com/nobel/nobel.html

59. Paul Forman, "Scientific Internationalism...," *Isis* 64 (1973) p 152. Also see Peter Gay, *Freud, Jews and other Germans: Masters and Victims in Modernist Culture* (Oxford: Oxford University Press, 1978).

60. John Heilbron, *Dilemmas of an Upright Man: Max Planck as Spokesperson for German Science.* To sense the rabid nationalist views Planck encountered during this time, see Philipp Lenard, *England und Deutschland zur Zeit des grossen Krieges* (Heidelberg, 1914) for the bitter distinctions drawn between "German physics" and "English physics" at the outbreak of war. Also see Akademie der Wissenschaften, Berlin, *Die Entwicklung Berlins als Wissenschaftszentrum 1870–1930* 1 (1981).

61. Bridgitte Schröder-Gudehus, "The Argument for Self-Government and Public Support for Science in Weimar Germany," *Minerva* (1972): pp 537–570.

62. Lise Meitner to Otto Hahn, literary estate of Otto Hahn, literary estate of Otto Hahn, Max-Planck-Gesellschaft Archives, Berlin–Dahlem; original German letter cited in Sabine Ernst, ed., *Lise Meitner an Otto Hahn*; translated by P. Rife.

63. Max Horkheimer, *Critical Theory: Selected Essays* (Boston: Seabury Press, 1972), p 3.

64. E. Kaiser and E. Wilkins, translators, *Otto Hahn: My Life*, p 122.

Notes to Chapter 5

1. Lise Meitner, "Looking Back," p 6.

2. In his biographical volume *Physics in the 20th Century* (Cambridge, MA: MIT Press, 1978), pp 52–65, Victor Weisskopf comments that by the late 1920s, young physicists such as Kramers, Pauli, Heisenberg, Gamow, Bloch, Casimir, Landau, and himself — all German or Russian-speaking — were chosen by Bohr for collaboration. "It was his great strength to assemble around him the most active, the most gifted, the most perceptive physicists of the world," regardless of nationality.

See essays by L. Rosenfeld and Erik Ruedinger, "The Decisive Years 1911–1918" and O. Klein "Glimpses of Niels Bohr as Scientist and Thinker" in Stefan

Rozenthal, ed., *Niels Bohr: His Life and Work as seen by his friends and colleagues* (New York: Wiley & Sons, 1967) and Paul Forman, "Scientific Internationalism and the Weimar Physicists: The Ideology and Its Manipulation in Germany after World War I, *Isis* 64 (1973).

3. Lise Meitner, "Looking Back," p 6. In Ruth Moore's biography of Bohr, *Niels Bohr*, p 111, she states:

> The early visits of Franck and Meitner confirmed the ties of the institute with the German groups in Berlin and Göttingen. But more than this, the friendships with Bohr and his wife Margrethe assured that, despite national differences, intellectual and cultural affinities transcended one's profession and provided a bond for countless physicists who enjoyed evenings of classical music and stimulating conversation at the Copenhagen Institute.

4. Private interviews with Erik Bohr, London, August 3, 1984 and Copenhagen, September 6, 1988.

5. Lise Meitner to Niels Bohr, microfilm reel 5, Bohr Scientific Correspondence, Archives for the History of Quantum Physics, American Institute of Physics, College Park, MD.

6. Cited in Hans Hartmann, *Max Planck als Mensch und Denker* (Basel: Ott Verlag, 1953), pp 32–33.

7. See "Emmy Noether," *Dict. Sci. Biog.*, Volume X, pp 137–139.

8. Max Planck, *The Philosophy of Physics* (New York: Norton, 1936) and Adolf von Harnack, 1911, as quoted in Carl Seelig, *Albert Einstein und die Schweiz* (Zürich: Europa–Verlag, 1952), p 45. Also see Alexander Deubner, "Die Physik an der Berliner Universität von 1910 bis 1960," *Wissenschaftliche Zeitschrift der Humboldt Universität zu Berlin* 14 (1964): p 87 for the role of the colloquium in the Berlin community of scientists and its philosophic as well as physical science debates.

9. Alan Beyerchen, *Scientists under Hitler*, p 4. Also see A.J. Nicholls, *Weimar and the Rise of Hitler* (New York: St. Martin's, 1979).

10. Ibid., Beyerchen. Beyerchen also states that Hitler declared that his movement "was above politics." National Socialism, Beyerchen wryly comments, promised above all a "national uplifting" instead of a "national leveling" of society.

11. Otto Hahn, "Wandlungen einer Forschungsstaette," 1919, literary estate of Otto Hahn, Max-Planck-Gesellschaft Archives, Berlin–Dahlem.

12. Paul Forman, "The Financial Support and Political Alignment of Physicists in Weimar Germany," *Minerva* 13 (January 1974): pp 62–63.

13. Stefan Richter, *Forschungsförderung in Deutschland 1920–1936: Dargestellt am Beispiel der Notgemeinschaft der detuschen Wissenschaft und ihrem Wirken für das Fach Physik. Technikgeschichte in Einzeldarstellung,* Nr. 23, (Düsseldorf: VDI-Verlag, 1972), p 74.

14. Quoted in Phillip Frank, *Einstein: sein Leben und seine Zeit,* (Munich, 1949), p 193.

15. See Philosophische Fakultät der Friedrich-Wilhelms-Universität to Lise Meitner, August 7, 1922, Meitner collection, Churchill College Archives, Cambridge, England.

16. Lecture schedule, "Prof. Lise Meitner als Dozentin an der Philosophischen Fakultät der Friedrich-Wilhelms-Universität Berlin, 1923–1933," Akademie der Wissenschaften central Archiv, Humboldt Universität, Berlin. I wish to thank Dr. Horst Melcher, Potsdam, for bringing these documents to my attention.

17. Akademie der Wissenschaften central Archive, Humboldt Universität, Berlin; Habilitation Lett. H., No.1, Volume 41, pp 173–191; No.1, Volume 32, pp 28–34 (letter thanking Planck).

18. Inaugural lecture August 31, 1922; see Lise Meitner "The Status of Women in the Professions," *Phys. Tod.* 13 (August 8, 1960): pp 16–21.

19. Laurie Brown and Donald Moyer, "Lady or Tiger? The Meitner–Hupfeld Effect and Heisenberg's Neutron Theory," *Am. J. Phys.* 52 (February 2, 1984): pp 130–135.

20. Lise Meitner, "Radioaktivität und Atomkonstitution," Festschrift der Kaiser-Wilhelm-Gesellschaft, 1921, pp 154–161; also see Lise Meitner, "Über eine notwendige Folgerung aus dem Comptoneffekt und ihre Bestätigung," *Z. Phys.* 22 (1924): pp 334–342; Lise Meitner, "Radioaktivität und Atomkonstitution," *Naturwiss.* 9 (1921): pp 423–427; and Lise Meitner, "Über die verschiedenen Arten des radioaktiven Zerfalls und die Möglichkeit ihrer Deutung aus der Kernstruktur," *Z. Phys.* 4 (1921): pp 146–156.

21. Lise Meitner to Otto Hahn, June 23, 1918, literary estate of Otto Hahn, Max-Planck-Gesellschaft Archives, Berlin–Dahlem; translated from Patricia Rife, *Lise Meitner: Ein Leben für die Wissenschaft,* 2nd ed., (Düsseldorf: Claassen Verlag, 1992).

22. Richard Evans, *The Feminist Movement in Germany, 1894–1933* (London: Sage, 1976), pp 20–21. He notes that during this same time, girls' elementary school education "concentrated on domestic science, elementary pedagogy and the care of children, with perhaps religious instruction and 'aesthetic' subjects thrown in. The Prussian government did little to encourage girls to prepare for university or professional life."

23. Hugh Trevor Llyod, *Suffragettes International: The Worldwide Campaign for Women's Rights* (New York: American Heritage Press, 1971).

24. Franco Rasetti to Helen Garbutt, February 1979, author's private collection.

25. Elisabeth Schiemann, "Freundschaft mit Lise Meitner," *Neue Evangelische Frauenzeitung* (January–February 1959).

26. See "Wahlvorschlag für Otto Hahn," in Christa Kirsten and Hans-Jürgen Treder, *Physiker über Physiker* (Berlin: Akademie–Verlag, 1975) and "anti-Einstein" articles written by Philipp Lenard, *Völkischer Beobachter*, issues throughout 1933.

27. Werner Heisenberg, *Physics and Beyond: Encounters and Conversations* (New York: Harper & Row, 1971), pp 43–44. Heisenberg participated in the typical "lecture schedule" of many undergraduates in Germany, traveling among Göttingen, Munich, and other cities to attend lectures by the top physicists of this era.

28. Heisenberg was representative of the new generation of German students Meitner and other physicists of her generation were to confront during the 1920s: intellectually pioneering and yet politically nationalistic to the point of xenophobia. This nationalism would hold ominous consequences under the Third Reich. See David Cassidy, *Uncertainty: The Life and Science of Werner Heisenberg* (New York: WH Freeman, 1993).

29. See Nernst and other physicists quoted in Paul Forman, "Weimar Culture, Causality, and Quantum Theory . . ." Part III: Dispensing with Causality – Adaptation to the Intellectual Environment, in McCormmach, ed., *Hist. Stud. Phys. Sci.* (1971) pp 63–108.

30. See Max Planck "Ansprache des vorsitzenden Sekretars, gehalten in der öffentlichen Sitzung zur Feier des Leibnizschen Jahrestages, 29 June 1922," Preussische Akademie der Wissenschaften, Sitzungsberichte; reprinted in *Max Planck in seinen Akademie-Ansprachen: Erinnerungsschrift der deutschen Akademie der Wissenschaften zu Berlin* (Berlin, 1948), pp 46–48.

31. Ilse Rosenthal-Schneider, *Reality and Scientific Truth* (Detroit: Wayne State University Press, 1980), p 113, ft. 22.

32. Koppel Pinson, *Modern Germany*, 2nd ed. (New York: Macmillan, 1966), Ch 18, "Economic Life in Weimar Germany."

33. Paul Forman, "Weimar Culture, Causality, and Quantum Theory, 1918–1927," *Hist. Stud. Phys. Sci.* 3 (1971).

34. Lise Meitner, "Einige Erinnerungen . . . ," p 98.

35. William Carr, *A History of Germany: 1815–1945*, 2nd ed. (New York: St. Martins, 1979), p 287.

36. See documents housed in Meitner collection, Churchill College Archives, Cambridge, England.

37. Lise Meitner, "Looking Back," *Bull. Atom. Sci.* 20 (November 1964): p 8. The Executive Directors of the Kaiser Wilhelm Institute for Chemistry played a major role in Hahn and Meitner's careers. Ernst Beckmann led the Institute from its formation in 1913 to 1921, when he was succeeded by Alfred Stock, Director from 1922 to 1927. In 1928, Otto Hahn became the Executive Director of the entire Institute. Department chairs throughout the 1920s were: Inorganic Chemistry, Ernst Beckmann; Organic Chemistry, Richard Willstätter; Chemistry of Radioactive Compounds, Otto Hahn. Hahn's department was officially divided into a section for "chemical radioactive research" and for "physical radioactive research," of which Meitner was made head in 1917. However, both sections were budgeted as a single unit (the Hahn/Meitner department) until Hahn became Executive Director of the Institute for Chemistry in 1928.

38. Margaret Rossiter, *Women Scientists in America: Struggles and Strategies to 1940* (Baltimore: Johns Hopkins University Press, 1982), p 306. Rossiter cites the example of physicist Robert Millikan's attitude towards women scientists in American universities. He had rejected German physicist Hertha Sponer, one of Meitner's close friends in later years, as a postdoctoral fellow in his Pasadena laboratories. Millikan later reflected upon his decision in a cryptic letter which Rossiter analyzes on p 191: he apologized that his letter had meant "no reflection on the individual concerned, but it is a mere expression of opinion as to the *policy* of bringing women into a university department of physics." He then went on to spell out and justify what might be called the "bright young man" theory of academic advancement.

Millikan sincerely believed that the future of physics in the United States depended on extending every opportunity to the "bright young men" (his emphasis) in the field, and allowing the talents of the rest to go undeveloped. He seemed to start with the assumption that although some men were better than others, all were superior to the women, and to adjust all subsequent "reasoning" to support that view. Although Madame Curie and Lise Meitner had made worthwhile contributions to physics, he did not think that this indicated that other women would also. In fact, rather than "opening the doors" to other women, these two examples had the opposite effect, and justified continued *exclusion* – for, to Millikan at least, they showed how *unlikely* it was that any other women physicists would ever attain their high level. It is important to note in this letter that Millikan saw Meitner not only as an intelligent "exception" but that her worthwhile contributions made it "*more* difficult" to attain her "level!" Such chauvinism was rampant at this time, and this convoluted logic of scientific achievement remains a social question – and continuing enigma – for women in science.

39. February 2, 1924, Akademie der Wissenschaften central Archiv, Humboldt Universität, Berlin. Also see W. Schlicker and L. Stern, *Die Berliner Akademie der Wissenschaften in der Zeit des Imperialismus*, Part II: 1917–1933 (Berlin: Akademie–Verlag, 1975), p 277 for a description of the Leibniz Medal criteria.

40. O.R. Frisch, "Lise Meitner," *Biog. Mem. Fel. Roy. Soc.* 16, 1970, p 415.

41. Swedish Royal Academy of Science, Nobel Archives, Stockholm, Sweden. The Academy accepts only those nominations from previous Nobel laureates and members of the Swedish Academy of Science. I would like to thank Bengt Nagel, Secretary of the Physics Committee, for his archival assistance in procuring names and dates in reference to nominations for both Otto Hahn and Lise Meitner from 1924–1937. Also see E. Crawford, J.L. Heilbron, and R. Ullrich, *The Nobel Population, 1901–1937: A Census of the Nominators and Nominees for the Prizes in Physics and Chemistry*, Office of History of Science and Technology, University of California, Berkeley and Office for History of Science, Uppsala University, Uppsala, 1987.

42. Robert Gerwin, *The Max-Planck-Gesellschaft and Institutes* (Munich: MPG, 1977), p 112.

43. Otto Hahn, "Das Kaiser-Wilhelm-Institut für Chemie," in *Jahrbuch der Max-Planck-Gesellschaft* (Göttingen: MPG, 1951), pp 175–190.

44. O.R. Frisch, "Lise Meitner," *Biog. Mem. Fel. Roy. Soc.* 16, 1970, p 415.

45. Fritz Krafft, "Lise Meitner: Her Life and Times," *Angew. Chem. Int. Ed. Eng.* 17 (1978): p 829. In 1924, Meitner published an article with Kurt Freitag, "Photographischer Nachweis von Alpha-Strahlen langer Reichweite nach der Wilsonschen Nebelmethode," *Naturwiss.* 12 (1924): pp 634–635. Meitner introduced many innovations in the use of the cloud chamber during this productive decade, and together with her assistant Kurt Philipp, published work on the study of slow neutrons at greatly reduced speeds by analyzing cloud chamber results in the early 1930s. See Lise Meitner and Kurt Philipp, "Die Anregung positiver Elektronen durch gamma Strahlen von ThC," *Naturwiss.*, 21 (1933): p 468.

46. Otto Hahn and Lise Meitner, "Die Beta-Strahlspektren von Radioactinium und seinem Zerfallsprodukten," *Z. Phys.* 34 (1925): pp 795–806; and "Bemerkungen zu einer Arbeit über die künstliche Umwandlung von Uran in Uran X," *Naturwiss.* 13 (1925): p 907.

47. Max von Laue and Lise Meitner, "Die Berechnung der Reichweitstreuung aus Wilsonaufnahmen," *Z. Phys.* 41 (1927): pp 397–406. Also see Horst Kant, "Einige Betrachtungen zur Geschichte der sogenannten 'Geiger-Zaehler'," *Physik in der Schule* 16 (July 8, 1978), pp 299–303.

48. Paul Forman, "Financial Support and Political Alignment of Physicists," *Minerva* 12 (January 1974): p 48. Also see Albert Vögler "Wissenschaft, Technik, und Wirtschaft...," *Naturwiss.* 14 (1926): pp 44–99 for a detailed position on the industrialists' "search for allies" against state interference during this period.

49. Beyerchen, *Scientists under Hitler*, pp 3–4; pp 85–111.

50. Fritz Strassmann, "Zur Erforschung der Radioaktivitat, Lise Meitner zum 75. Geburtstag," *Angew. Chem.* 66 (1954): p 93; also reprinted in Krafft, "Lise Meitner: Her Life and Times," *Angew. Chem. Int. Ed. Eng.* 17 (1978): pp 826–827.

51. Thomas Mann, *Essays of Three Decades* (New York: Knopf, 1947).

52. See Diane Crane, *Invisible Colleges: Diffusion of Knowledge in Scientific Communities* (Chicago: University of Chicago Press, 1972) for discussion of this social phenomenon.

53. Stefan Rozenthal, ed., *Niels Bohr*, pp 38–109. Bohr's Institute was founded immediately after World War I. During this same period, he was offered a position in Manchester by Rutherford. His decision was characteristically phrased in the following reply to his former mentor (see Ruth Moore, *Niels Bohr*, p 95):

Dec. 15, 1918

...I don't know how to thank you for your letter of November 17, which I have just received and which besides being the source of the greatest pleasure for me has at the same time been the object of sorrowful consideration. You know that it has always been my ardent wish to be able to work in your neighborhood and to take a part in the enthusiasm and imagination which you impart to all your surroundings and of which I have so much benefitted myself. At the same time I am not in a position to accept the splendid offer, for which I feel more thankful than I can express because of the undeserved confidence in me that it contains. The fact is that I feel I have morally pledged myself to do what I can to help in the development of the scientific physical research here in Denmark, and in which the small laboratory will play a part. I should like so much to settle down in Manchester again and I know it would be of the greatest importance to my scientific work, but I feel I cannot accept the post you write about because the university [in Copenhagen] has done all that they could do to place all external means necessary for my work. Of course the pecuniary means, personal allowance as well as that to the running of the laboratory, will be far below English standards. I feel it is my duty here to do my best, though I feel very strongly the result will never be the same as if I could work with you.

54. See English translations of Schrödinger's papers from this period in Gunter Ludwig, *Wave Mechanics* (Pergamon Press, 1968), particularly E. Schrödinger, "On the Relationship of the Heisenberg–Born–Jordan Quantum Mechanics to Mine," reprinted with permission from *Ann. Phys.* 79 (1926): pp 734–756.

55. Interview with Max Delbrück by Carolyn Harding, California Institute of Technology Archives, Pasadena, CA, July 20, 1978, tape 3, side 1, p 39 in transcript.

56. Interview with Max Delbrück, tape 3, side 1, page 40. Wolfgang Pauli, always the skeptic, wrote to Kramers during this time that he did not "want to be mistaken for one of the true believers" of Bohr's theories: both were in their twenties, and eager to challenge causal boundaries. See W. Pauli to H.A. Kramers, July 27, 1925, Archives for the History of Quantum Physics, American Institute of Physics, microfilm nr 8, sec 9. Also see E. Schrödinger, "Bohrs neue Strahlunghypothese und Energiesatz," *Naturwiss.* 12 (September 5, 1924): pp 720–724. Peculiarly, Schrödinger seems to have found in the Bohr–Kramers–Slater proposal an attempt to rid the quantum theory of discontinuities, a goal he then pursued within his wave mechanics which he began to develop late in 1925. In this same interview (ibid., pp 40–41), Delbrück recalled:

Bohr very vigorously asked the question whether this new dialectic wouldn't be important also in other aspects of science. He talked about that a lot, especially in relation to biology, in discussing the relation between life on the one hand, and physics and chemistry on the other — whether there wasn't an experimental mutual exclusion, so that you could look at a living organism either as a living organism or as a jumble of molecules; you could do either, you could make observations that tell you where the molecules are, or you could make observations that tell you how the animal behaves, but there might well exist a mutually exclusive feature, analogous to the one found in atomic physics. He talked about that in biology and in psychology, in moral philosophy, in anthropology, in political science, and so on, in various degrees of vagueness, which I found both fascinating and very disturbing, because it was always so vague.

I wish to thank both Carolyn Harding and Max Delbrück, with whom I carried on a lively correspondence before his death.

57. Paul Forman, "Weimar Culture, Causality, and Quantum Theory," *Hist. Stud. Phys. Sci.* (1971): p 97. Forman also points out that Max Planck himself was "by no means silent" in these debates. (p 101, n. 229). See Planck's lecture "Physikalische Gesetzlichkeit im Lichte neuerer Forschuung," *Vorträge und Erinnerungen* (Stuttgart, 1949), pp 183–205, especially pages 184 and 194–196, also reprinted in Planck, *Phys. Ab.*, Braunschweig, 1958, Volume III, pp 159–171. This lecture was delivered on February 14, 1926 in Düsseldorf and again on February 17 in the Auditorium Maximus of the University of Berlin. Also see Werner Heisenberg, "Über den anschaulichen Inhalt der quanten theoretischen Kinematik und Mechanik," *Z. Phys.* 43 (1927): pp 172–198. For later views of Heisenberg's concerning his contributions to physics and the philosophic questions/implications they posed, see his *Philosophic Problems of Nuclear Science* (London: Faber & Faber, 1952); "Kausalgesetz und Quantenmechanik," *Annalen der Philosophie und philosophischen Kritik* 2 (1931): pp 172–82; and "Die Entwicklung der Quantentheorie, 1918–1928," *Naturwiss.* 17 (1929).

58. Lise Meitner to Otto Hahn, March 10, 1927, literary estate of Otto Hahn, Max-Planck-Gesellschaft Archives, Berlin–Dahlem.

59. See Delbrück interview by Carolyn Harding, California Institute for Technology Archives, July 20, 1978. In a letter to the author, January 9, 1980, Delbrück stated: I was Meitner's assistant for five years, 1932–1937. In 1937 I moved to the U.S.A. and into biology." In Carolyn Harding's interview with Delbrück, pages 42, 51, and 55, he states:

When the question came up of what job I would take after a year with Bohr and Pauli (and another half year in Bristol), and I had the choice of either going to Berlin to become an assistant of Lise Meitner or to Zurich to be an assistant of Pauli, I chose to go to Berlin because of the vicinity of the Kaiser Wilhelm Institutes for biology to the Institute (for Chemistry) I was going to work in. When I did go to Berlin (1932), my job was to be a theoretical physicist, as it were, a consultant for Lise Meitner. Lise Meitner was an experimental physicist working on radioactive substances, a very good experimental physicist, and there were quite a few new developments all the time. I was supposed to keep up with the theoretical literature and watch out for what happened, and also presumably be productive as a theoretical physicist, and write theoretical physics papers. This sort of black-market research was going on, I mean it was moonlighting; I was supposed to be the theoretical physics advisor to Lise Meitner, but actually took all this time out to work in biophysics.

60. Paul Forman, "Scientific Internationalism and Weimar Physicists," *Isis* 64 (1973): pp 176–177.

61. See Ernst Peter Fischer & Carol Lipson, *Thinking about Science: Max Delbrück and the Origins of Molecular Biology* (New York: Norton, 1988), and interview with Max Delbrück by Carolyn Harding, California Institute for Technology Archives, July 20, 1978. He stated: "... This little club which started out as theoretical physics, and then brought in genetics, also brought in biochemists and photosynthesis physiologists... an important piece of what could be called molecular biology came out of these discussions."

62. In Gertrud Szilard and Spencer Weart, eds., *Leo Szilard: His Version of the Facts* (Cambridge, MA: MIT Press, 1978), pp 13–44, Szilard states:

In 1932, my interest shifted to nuclear physics and I moved to the Harnack House in Berlin–Dahlem with the thought of taking up some experimental work in one of the Kaiser Wilhelm Institutes there. I discussed the possibility of doing experiments in nuclear physics with Miss Lise Meitner in the Kaiser-Wilhelm-Institut für Chemie, but before this led to a final discussion one way or the other, the political situation in Germany became tense and it seemed advisable to delay a final decision. I reached the conclusion something would go wrong in Germany very early.

63. O.R. Frisch, *What Little I Remember* (Cambridge: Cambridge University Press, 1979), p 30.

64. Hendrik Casimir, *Haphazard Reality* (New York: Harper & Row, 1983).

65. Cited by Emilio Segrè in Roger Stuewer, ed., *Nuclear Physics in Retrospect: Proceedings of a Symposium on the 1930s* (Minneapolis, MN: University of Minnesota Press, 1972), p 79.

66. In a fascinating article by Anthony Michelis entitled "How Nuclear Energy Was Foretold," *New Scientist* 276 (March 1, 1962): pp 507–509, there is an analysis of how H.G. Well's book *The World Set Free* (London: Free Press, 1914), together with four other 'fiction' stories, which identify and create applications for the release and production of atomic energy, led many to speculate on "atom power" as early as World War I. Also see Hans Dominik, *Atomgewicht 500* (Berlin: Verlag Scherl, 1935) which influenced a new generation of scientists as well as a curious general public.

67. *Pittsburgh Post Gazette*, December 29, 1934. Such newspaper reports did not go unnoticed by the younger generation of scientists in Europe during this time. Leo Szilard recalled: "H.G Wells had described the discovery of artificial radioactivity and puts it in the year of 1933, the year in which it actually occurred." See Ion Hobana, "Nuclear War Fiction in Eastern Europe," in *Nuclear Texts and Contexts Newsletter* (January 1990): pp 7–8.

68. Quoted by Eugene Hecht, *Physics in Perspective* (New York: Addison–Wesley, 1962).

69. Lise Meitner and Wilhelm Orthmann, "Über eine absolute Bestimmung der Energie der primären Beta-Strahlen von Radium E," *Z. Phys.* 60 (1930): pp 143–155.

70. O.R. Frisch, "Experimental Work with Nuclei," in R. Stuewer, ed., *Nuclear Physics in Retrospect*, p 75.

71. See "Zur älteren und neueren Geschichte des Neutrinos," in Wolfgang Pauli, *Collected Scientific Papers*, R. Kronig and V. Weisskopf, eds., (New York: Wiley, 1964), Volume II, pp 1313–1337, especially pp 1316–1317.

72. Lise Meitner, "Einige Erinnerungen . . . ," p 98.

73. S. Watkins, "Lise Meitner and the Beta-Ray Energy Controversy: An Historical Perspective," *Am. J. Phys.* 51 (June 6, 1983).

74. Lise Meitner and K. Philipp, "Die Anregung positiver Elektronen durch Gamma-Strahlen von Th C," *Naturwiss.* 21 (1933): p 428.

75. Lise Meitner, *Z. Phys.* 9 (1922): pp 145–152; Lise Meitner, *Z. Phys.* 17 (1923): p 54 and R. Sietmann, "False Attribution: A Female Physicist's Fate," *Physics Bulletin* 39, 8 (1988): pp 316–317.

76. Peter Gay, *Wiemar Culture: The Outsider as Insider*, p 87.

77. Ibid.

78. Erwin Schrödinger, *Über Indeterminismus in der Physik. Zwei Vorträge zur Kritik der Erkenntnis von Naturwissenschaften* (Leipzig: Barth, 1932), pp 38–39. Also see Walter Moore, *Schrödinger: Life and Thought* (Cambridge: Cambridge University Press, 1989).

79. See E. Crawford, J.L. Heilbron, and R. Ullrich, *The Nobel Population, 1901–1937: A Census of the Nominators and Nominees for the Prizes in Physics and Chemistry*, pp 276 – 280. Organizational developments and budgets within the Kaiser Wilhelm Institute for Chemistry were directly influenced by the socio-political developments of the Weimar Republic. Meitner's and Hahn's memories of the period give us some insight into the obstacles and hardships with which scientists were confronted in their efforts to maintain their research and to train a new generation of physicists and chemists during the severe worldwide Depression years.

Notes to Chapter 6

1. Quoted in Hannah Vogt, *The Burden of Guilt* (New York: Oxford University Press, 1964), p 106. Also see Winston Churchill, *Great Contemporaries* (New York: Putnams, 1937), for a vivid portrait of von Hindenburg.

2. A brief list of references concerning this period which focuses on the consolidation of the National Socialist Party and its racist policies include: K.D. Bracher, *The German Dictatorship: The Origins, Structure and Consequences of National Socialism* (New York: Free Press, 1970); Alan Bullock, *Hitler: A Study in Tyranny* (London Penguin, 1952); J. Ball-Kaduri, *Leben der Juden in Deutschland im Jahre 1933* (Munich: List Verlag, 1980); G. Schulz, *Aufstieg des Nationalsozialismus: Krise und Revolution in Deutschland* (Frankfurt: Viewig Verlag, 1973); Lucy Dawidowicz, *The War Against the Jews, 1933–1945* (New York: Bantam, 1975); and Y. Bauer, *A History of the Holocaust* (New York: Watts Press, 1982).

3. Viktor Reimann, *Goebbels*, [translated by S. Wendt] (New York: Doubleday, 1976). Also see Walter Lacquer, *Weimar: A Cultural History* (New York: Putnams, 1974), Ch. 8 "An End With Horror," pp 270–277.

4. See William Shirer, *The Rise and Fall of the Third Reich*, Volume I, "The Rise," (New York: Simon & Schuster, 1959), Chapter 6 "The Last Days of the Republic," pp 180–187. There have been numerous studies on the poltical seizure of power by Hitler in 1932 to 1933. In Joachim Fest's, *Hitler* (New York: Harcourt Brace Jovanovich, 1974), pp 387–390, the author reflects upon the mood of the German people in the midst of the political turmoil of 1933:

> Skeptical predictions were legion. Hitler would run afoul of the power of his conservative partners in the coalition of Hindenburg and the army, of his resistance of the masses, of the multiplicity and difficulty of the country's economic problems. Or else there would be foreign intervention. Or his amateurishness would be exposed at last. But Hitler gave all these prophecies the lie in an almost unprecedented process of conquering power. Granted, every detail of his action was not so minutely calculated in advance as may sometimes appear in historical hindsight. But he never forgot for a moment what he was after: namely, to gather all the threads of power into his own hands by the time the 85-year-old President died. And he knew how to go about it: namely, to continue to use those tactics of 'legality.'

5. In Dietrich Hahn, ed., *Otto Hahn: Begründer des Atomzeitalters* (List Verlag, 1979), a section from the Hahn family guestbook is reproduced with the following poem written by Lise Meitner, February14, 1933, in commemoration of Otto's trip to America:

> That Otto Hahn, you all must know
> is to America, next to go
> and says to Edith: 'Your worry belay
> a USA trip is just child's play.
> I went there when I was a young Doktor
> and now I am going as a pro-fessor
> naturally I won't begrudge myself First Class
> and will run in races, flirting over bar glass
> on deck reside as a gentleman grand
> and dance over parquet to the wonderful band.
> To eat and drink, I want haute cuisine
> and for refreshment,Vasano daily (per diem).'
> That good spirits ne'er these things him deny,
> Our dear Otto, whereever he may lie!

6. Lise Meitner to Otto Hahn, March 21, 1933, literary estate of Otto Hahn, Max-Planck-Gesellschaft Archives, Berlin–Dahlem. Also see Morris Goran, *The Story of Fritz Haber* (Norman, OK: University of Oklahoma Press, 1967).

7. See Charles Maier, Stanley Hoffmann, and Andrew Gould, eds., *The Rise of the Nazi Regime: Historical Reassessments* (Boulder, CO: Westview Press, 1986), and Susanne Everett, *Lost Berlin* (New York: Hamlyn, 1979).

8. Interview with Theodore von Laue by Patricia Rife, Worcester, MA, January 1981.

9. Lise Meitner to Otto Hahn, April 2, 1933, literary estate of Otto Hahn, Max-Planck-Gesellschaft Archives, Berlin–Dahlem. Of course, others such as Einstein swore even after the war that "the conduct of German intellectuals – as a group – was no better than the rabble." See his 1949 letter to Otto Hahn, and other documents from the MPG Archives, in K.E. Boeters and Jost Lemmerich, eds., *Gedächtnisausstellung zum 100. Geburtstag von Albert Einstein, Otto Hahn, Max von Laue, Lise Meitner* (Bad Honnef: Physik Kongress-Ausstellungs-und Verwaltungs, 1979).

10. Gertrud Szilard and Spencer Weart, *Leo Szilard: His Version of the Facts*, p 14.

11. This ordinance and others relating to the April 7 dismissal of 'non-Aryan' and other university staff and faculty can be found translated in English in E.Y. Hartshorne, *The German Universities and National Socialism* (London: George Allen & Unwin, 1937), pp 175–177.

12. Franck accepted a guest lectureship at Johns Hopkins University by June 1933, after a storm of controversy was raised over his public resignation from the University of Göttingen faculty in April 1933. Franck had not wanted to leave Germany, but in his letter of resignation, he stated: "This decision is an inner necessity for me because of the attitude of the government toward German Jewry." Quoted in Alan Beyerchen, *Scientists under Hitler*, pp 15–17. Four Nobel Prize laureates were officially dismissed under the April laws: Otto Meyerhof, James Franck, Fritz Haber, and Albert Einstein. See the *New York Times*, March 5, 1934, p 10.

13. See accounts of this event of May 10, 1933 in J.C. Jackman and Carla Borden, eds., *The Muses Flee Hitler: Cultural Transfer and Adaptation* (Washington, DC: Smithsonian Institution Press, 1983) and Alan Beyerchen, David Cassidy, Mark Walker, *Dimensions* 10, 2 (1996), special issue on science and the Third Reich.

14. Otto Hahn interview, *Toronto Star Weekly*, April 8, 1933. Also see the *Cornell Daily Sun*, March 8, 1933.

15. For a thorough account of Einstein's dismissal from the University in Berlin, see Banesh Hoffmann, *Albert Einstein: Creator and Rebel* (New York: Viking, 1972), pp 157–172.

16. Max Planck reported this incident to Max Born when they met in 1946 at the Newton celebrations in London. See Max Born, *Mein Leben: Erinnerungen des Nobelpreisträgers* (Munich: Nymphenburger, 1975), p 353.

17. Morris Goran, *The Story of Fritz Haber*, p 160.

18. Ibid., p 16.

19. Fritz Krafft, "Lise Meitner und ihre Zeit: Zum hundertsten Geburtstag der bedeutenden Naturwissenschaftlerin," *Angew. Chem.* 90 (1978): pp 876–892; and *Angew. Chem. Int. Ed. Eng.* 17 (1978): p 832.

20. Meitner to Born, January 11, 1957; see Max Born, *My Life*, p 263.

21. O.R. Frisch, "Experimental Work with Nuclei: Hamburg, London, Copenhagen," in Roger Stuewer, ed., *Nuclear Physics in Retrospect: Proceedings of a Symposium on the 1930s* (Minneapolis, MN: University of Minnesota Press, 1979), p 67.

22. For Bohr's untiring assistance to refugees, including Frisch, who would leave London and go to Bohr's Copenhagen Physics Institute until 1939, see Finn Aaserud, *Redirecting Science: Niels Bohr, Philanthropy, and the Rise of Nuclear Physics* (New York: Cambridge University Press, 1990). In Ruth Sime, *Lise Meitner: A Life in Physics*, a letter from James Franck to Lise Meitner, June 27, 1933 is cited in which Franck reported that Dirk Coster was also interested in having Frisch work with him in Groningen, Holland. See Franck Papers, Regenstein Library, University of Chicago.

23. See Sir Rudolf Peierls, "Otto Robert Frisch, 1 October 1904 – 22 September 1979" *Biog. Mem. Fel. Roy. Soc.* 27 (1981): pp 283–306.

24. O.R. Frisch in *Niels Bohr*, S. Rozenthal, ed., p 137.

25. See Lucy Dawidowicz, *The War Against the Jews, 1933–1945* in the work cited; Tom Bower, *The Paperclip Conspiracy: The Hunt for the Nazi Scientists* (Boston: Little Brown & Co, 1987); Hannah Arendt, *Eichmann in Jerusalem* (New York: Harcourt Brace Jovanovich, 1958); H. Arendt, *The Origins of Totalitarianism*

(New York: Harcourt Brace Jovanovich, 1973; and Heinz Boberach, *Meldungen aus dem Reich. Auswahl aus den Geheimen Lagerberichten des Sicherheitsdienstes der SS 1933–1944* (Berlin: Neuwied & Berlin, 1965).

26. See Charles Weiner, "A New Site for the Seminar: The Refugees and American Physics in the Thirties," in Bernard Bailyn and Donald Fleming, eds, *The Intellectual Migration: Europe and America, 1930–1960* (Cambridge, MA: Harvard Universtiy Press,1969), p 204. Also see *The Manchester Guardian Weekly*, "Nazi 'Purge' of the Universities: A Long List of Dismissals," May 19, 1933, p 399.

27. Selig Hecht to Alfred Cohn, June 20, 1933, L.C. Dunn Papers, American Philosophical Society Library, Philadelphia, PA.

28. Selig Hecht to Mrs. G. Hecht, June 20, 1933, L.C. Dunn Papers, American Philosophical Society Library, Philadelphia, PA.

29. E.Y. Hartshorne, *The German Universities and National Socialism*, p 133.

30. Lise Meitner to Otto Hahn, June 1933, literary estate of Otto Hahn, Max-Planck-Gesellschaft Archives, Berlin–Dahlem.

31. Herbert Steiner, "Lise Meitners Entlassung," *Österreich in Geschichte und Literatur* 9 (1965): pp 462–466.

32. Einstein to Planck, July 17, 1931; reprinted in Ronald Clark, *Einstein: The Life and Times* (New York: Avon Books, 1971), p 510.

33. Interview by Evelyn Seeley, *New York World Telegram*, March 10, 1933.

34. Reprinted in Ronald Clark, *Einstein: The Life and Times*, p 561.

35. Einstein to Planck, March 9, 1933, Leiden, Holland; see Einstein Collection, Boston University, Boston, MA.

36. Planck to Einstein, March, 1933; see Einstein Collection, Boston University, Boston, MA.

37. Reprinted in Albert Einstein, "The World As I See It," in *Living Philosophies* (New York: Simon and Schuster, 1931), p 85.

38. Ibid., Einstein to Prussian Academy of Sciences, March 28, 1933. Also see Fritz Stern, "Einstein's Germany" in Holton G. and Elkana Y., *Albert Einstein: Historical and Cultural Perspectives* (Princeton, NJ: Princeton University Press, 1982).

39. Edith Hahn to James Franck, in Jost Lemmerich, ed., *Max Born, James Franck: Physiker in ihrer Zeit, der Luxus des Gewissens* (Berlin: Staatsbibliothek Preussischer Kulturbesitz, 1982), p 115.

40. For details concerning the Schrödinger family and their emigration from Berlin, see Walter Moore, *Schrödinger: Life and Thought* (Cambridge: Cambridge University Press, 1989); Auguste Dick, G. Kerber, and W. Kerber, eds., *Dokumente, Materialien und Bilder zur 100. Wiederkehr der Geburtstages von Erwin Schrödinger* (Vienna: Fassbänder, 1987).

41. See Otto Hahn, *Vom Radiothor zur Uranspaltung: Eine Wissenschaftliche Selbstbiographie* (Braunschweig: Viewig Verlag, 1962), p 91 for Hahn's memories of his resignation from the University.

42. See Ronald Clark, *Einstein: The Life and Times*, p 572.

43. See Phillip Frank, *Einstein: His Life and Times* (London: Jonathan Cape, 1948). Also see Otto Nathan and Heinz Norden, eds., *Einstein on Peace* (New York: Avenel Books, 1981).

44. See Alan Beyerchen, *Scientists Under Hitler*, Chapter 5, "The Aryan Physicists" for a thorough account of Lenard's development of his anti-Semitic stance in relation to the sciences, and of his citations of speeches published in the popular German newspapers of the time. Also see Beyerchen, "Anti-Intellectuallism and the Cutural Decapitation of Germany under the Nazis," in J.C. Jackman and Carla Bohrden, eds, *The Muses Flee Hitler: Cultural Transfer and Adaptation 1930–1945* (Washington, DC: Smithsonian Institution Press, 1983), pp 29–44.

45. Adolf Hitler, *Mein Kampf*, org. 1927 edition (New York: Houghton Mifflin, 1971).

46. Archives of the Swedish Academy of Sciences, Stockholm; nominations for the Nobel Prize in Chemistry, 1937. I wish to thank Bengt Nagel, Secretary of the Nobel Committees for Physics and Chemistry, and the Stockholm based Swedish Institute for their support during my research at the Nobel Archives in 1981 and 1988. In a contemporary study of those women scientists nominated for the Nobel Prize, Harriet Zuckermann observes that most nominees have studied or worked with a Prize recipient. See J. Cole, *Fair Science: Women in the Scientific Community* (New York: Free Press, 1979), p 32. It is important to note, however, that Planck's nominations of both Hahn and Meitner for the Nobel Prize in Chemistry was received no later than January 31, 1933, and hence, his nomination was most likely *not* for political reasons.

47. Revocation of Lise Meitner's lectureship at the University of Berlin, September 9, 1933, Prussian Minister für Wissenschaft, Kunst und Volksbildung: "Auf Grund von no. 3 des Gesetzes zur Wiederherstellung des Berufsbeamtentums vom 7.4.1933," reprinted in Herbert Steiner, "Lise Meitners Entlassung," *Österreich in Geschichte und Literatur* 9 (1965): p 463.

48. E. Kaiser and E. Wilkins, translators, *Otto Hahn: My Life*, p 146.

49. Hahn to Ministerialrat Achelis, July 27, 1933; published in German in Herbert Steiner, "Lise Meitners Entlassung," *Österreich in Geschichte und Literatur*, p 463.

50. Planck to Minister für Wissenschaft, Kunst und Volksbildung, August 30, 1933, in Steiner, *Österreich in Geschichte und Literatur*, p 463.

51. Hahn to Planck, October 16, 1933, literary estate of Otto Hahn, Max-Planck-Gesellschaft Archives, Berlin–Dahlem; German letter cited in E. Berninger, *Otto Hahn in Selbstzeugnissen und Bilddokumenten* (Reinbeck: Rowohlt, 1974), p 58; translated by P. Rife.

52. E. Kaiser and E. Wilkins, translators, *Otto Hahn: My Life*, p 146.

53. Fritz Krafft, "Lise Meitner: Eine Biographe," speech delivered January 1988, Hahn-Meitner Institute, Glienicker Str. 100, Berlin.

54. See Otto Hahn's memories of this period in Dietrich Hahn, *Otto Hahn: Erlebnisse und Erkenntnisse* (Düsseldorf: Econ Verlag, 1975), p 54.

55. Ibid., p 54. Also see Fritz Krafft, *Im Schatten der Sensation: Leben und Wirken von Fritz Strassmann* (Weinheim: Verlag Chemie, 1981), p 43.

56. See Ronald Clark, *Einstein: The Life and Times*, p 571.

57. See D. Hahn, *Otto Hahn: Erlebnisse und Erkenntnisse*, p 54.

58. Max Delbrück to Patricia Rife, February 5, 1980. In this letter, he stated: "I do not recall when I first met Lise Meitner; possibly in Copenhagen in 1931. My job with her started in the fall of 1932." Delbrück remained at the Institute for Chemistry as Lise Meitner's paid physics section assistant until 1937.

59. Interview with Max Delbrück by Carolyn Harding, California Institute of Technology Archives, Pasadena, CA, July 20, 1978, session 3.

60. See Fritz Krafft, *Im Schatten der Sensation: Leben und Wirken von Fritz Strassmann* (Weinheim: Verlag Chemie, 1981). One of Hahn's biographers, R. Spence, wrote in "Otto Hahn," *Biog. Mem. Fell. Roy. Soc.* 16 (1970): p 293:

> [Strassmann] originally joined the Institute in 1929 and was by 1935 an experienced radiochemist. As he wished to enter academic life, Hahn had urged him to apply to the University for admission to the academic staff but when he was informed that he must first join one of the Nazi organizations, he refused to proceed with the application.

61. E.Y. Hartshorne, *The German Universities and National Socialism*, p 112.

62. "The Academic Situation in Germany and the American Association of University Professors," *School and Society* 37 (June 10, 1933): p 742.

63. "A Call to Scientists," *The New Republic* 80 (August 29, 1934): pp 76–77.

64. See Charles Weiner, "A New Site for the Seminar: The Refugees and American Physics in the Thirties," in B. Bailyn and D. Fleming, eds., *The Intellectual Migration: Europe and America, 1930–1960*, p 209 and "Science and the State in Germany," *Nature* 131 (August 5, 1933): pp 198–199.

65. See original letter, Wolfgang Pauli to Lise Meitner and Hans Geiger, December 4, 1930, Meitner collection, Churchill College Archives, Cambridge, England; reprinted in Laurie Brown "The Idea of the Neutrino," *Phys. Tod.* 31 (September 9, 1979): pp 23–28 and in R. Kronig and Victor Weisskopf, eds, *Wolfgang Pauli Collected Scientific Papers* (New York: Interscience Press, 1964), pp 1316–1317.

66. E. Kaiser and E. Wilkins, translators, *Otto Hahn: My Life*, p 146.

67. See Dietrich Hahn, *Otto Hahn: Begründer des Atomzeitalters*, p 138. In a postcard to his wife Edith, Hahn wrote on September 14, 1934: "We have just come from a banquet in the castle of Katharina II and the late Czar, with fireworks, splendid food, and quantities of alcohol. The magnificence of the Czar's castles is frightening. Herr Biitz will take this card to the airport on his way to Germany. I will arrive Sunday evening. Please telephone Lise's Marie [her housemaid] that she will also arrive then.

68. E. Kaiser and E. Wilkins, translators, *Otto Hahn: My Life*, p 147.

69. Enrico Fermi, "Possible Production of Elements of Atomic Number Higher than 92," *Nature* 133 (1934): pp 898–899. Also see Edoardo Amaldi, "Neutron Work in Rome 1934–1936 and the Discovery of Uranium Fission," *Riv. Stor. Sci.* 1 (1984): pp 1–24.

70. Ibid., Fermi, p 898. A compendium of the original articles relating to the history of nuclear fission can be found in Horst Wohlfarth, ed., 40 *Jahre Kernspaltung: Eine Einführung in die Originalliteratur* (Darmstadt: Wissenschaftlich Buchgesellschaft, 1979). Also see Esther Sparberg, "A Study of the Discovery of Fission," *Am. J. Phys.* 32 (1964): p 3. In this article, Sparberg states: "The published reports of Fermi's work in *Nuovo Cimento* and *Nature* were so fascinating to Meitner that immediately she "persuaded Otto Hahn to renew our direct collaboration...with a view to investigating these problems." So it was that in 1934, after an interval of more than twelve years, we started working together again, with the especially valuable collaboration, after a short time, of Fritz Strassmann."

71. Ibid., Fermi, p 899.

72. E. Kaiser and E. Wilkins, translators, *Otto Hahn: My Life*, p 147.

73. See E. Berninger, *Otto Hahn in Selbstzeugnissen und Bilddokumenten*, p 16. Hahn also participated extensively in two important international commissions: the International Radium Standards Commission, set up in 1910 under the instigation of Mme. Curie, E. Rutherford, F. Soddy, and others; and the Atomic Weights Commission, which had, since the end of the 19th century, held the task of providing a critical assessment of work on setting atomic weights and compiling tables with standards of reliable atomic weights. Hahn was on this commission until 1938, except for a break during World War I, when Germany created its own "national" Atomic Weights Commission for a time.

74. Alan Bullock, *Hitler: A Study in Tyranny*, p 373.

75. Speech by Adolf Hitler, delivered in Munich, November 8, 1938. See Norman Baynes, *The Speeches of Adolf Hitler*, Volume II (Oxford: Oxford University Press, 1942), p 1551.

76. R. Sime, *Lise Meitner: A Life in Physics*, p 432, nr 74; original in Meitner to Max von Laue, July 13, 1947, Lise Meitner collection, Churchill College Archives, Cambridge, England.

77. Lise Meitner, "Max Planck als Mensch," *Naturwiss.* (1948): p 407; translated by P. Rife. Planck's speech was reprinted in its entirety as "Die Physik im Kampf um die Weltanschauung," in Max Planck, *Wege zur Physikalischen Erkenntnis* (Leipzig: Verlag von S. Hirzelin, 1944).

78. Ibid., p 408. Also see Fritz Krafft, *Im Schatten der Sensation*, pp 44–45 for issues surrounding Planck's authority in hiring – or not hiring – younger scientists within the Kaiser Wilhelm Society.

79. Fritz Krafft, "Lise Meitner: Eine Biographe," speech delivered January 1988, Hahn–Meitner Institute, Glienicker Str. 100, Berlin.

80. See Claudia Koonz, "Women between God and Führer," in C.S. Maier, S. Hoffmann, A. Gould, eds., *The Rise of the Nazi Regime: Historical Reassessments* (Boulder, CO: Westview Press, 1986), pp 69–79. Also see National Committee to Aid Victims of German Fascism, "Women Under Hitler Fascism," (New York: Worker's Library, 1934), pp 2–3.

81. E.Y. Hartshorne, *The German Universities and National Socialism*, pp 135–136.

82. See Helmuth Albrecht, ed., *Naturwissenschaft und Technik in der Geschichte. 25 Jahre Lehrstuhl für Geschichte der Naturwissenschaft und Technik am Historischen Institut der Universität Stuttgart* (Stuttgart: Verlag für Geschichte der Naturwissenschaften und der Technik, 1993); J.L. Heilbron, *The Dilemmas of an Upright Man: Max Planck as Spokesman for German Science*; Leon Poliakov and Josef Wulf, eds., *Das Dritte Reich und seine Denker* (Berlin: Arani Verlag, 1959); Hans Hartmann, *Max Planck als Mensch und Denker* (Basel: Ott Verlag, 1953).

83. Max Planck, "Mein Besuch bei Adolf Hitler," *Phys. Bl.* 3 (1947): p 143. Fritz Krafft explains that Planck's "attitude of mind" in his attempt to petition Hitler was "rather a traditional one of absolute loyalty to public authority." (See Krafft, "Lise Meitner: Her Life and Times," *Angew. Chem. Int. Ed. Eng.* 17 (1978): p 832.) That the 75-year-old Planck "found the issue most perplexing" also demonstrates his political naiveté in the face of Hitler's beliefs and tactics.

84. Ibid., Planck, p 144.

85. E.Y. Hartshorne, *The German Universities and National Socialism*, p 133. Also see Anne Becker, Kurt A. Becker, and Jochen Block, and others, "Fritz Haber," in Wilhelm Treue and Gerhard Hildebrandt, eds., *Berlinische Lebensbilder: Naturwissenschaftler* (Berlin: Colloquium Verlag, 1987); David Nachmansohn, *German-Jewish Pioneers in Science, 1900–1933*, pp 168–195.

86. Willy Ley, ed. and trans., *Otto Hahn: A Scientific Autobiography* (New York: Scribners, 1966), p 112, in which Hahn recalled: "Haber's death was not mentioned in the German press; only Max von Laue had the courage to write an obituary in

Die Naturwissenschaften which almost got him in serious trouble." Correspondence between Weizmann and Haber demonstrates that Haber had been offered a position in Cambridge, which was then denied, before he decided to relocate to Palestine. See Weizmann correspondence, Hebrew University, Jerusalem. I thank Prof. Dr. J.H. Block, Director of the Fritz Haber Institute, Berlin–Dahlem, for bringing this correspondence to my attention.

87. See in the work cited, Interview with Max Delbrück by Carolyn Harding, California Institute of Technology Archives, Pasadena, CA.

88. Otto Hahn, *Vom Radiothor zur Uranspaltung*, p 92.

89. E. Kaiser and E. Wilkins, translators, *Otto Hahn: My Life*, p 146.

90. See the description of the audience and participants in Otto Hahn, *Vom Radiothor zur Uranspaltung*, p 92.

91. William Shirer, *The Rise and Fall of the Third Reich, Volume I*, "The Rise," p 231.

92. Editorial, "Nazi Socialism and International Science," *Nature* 136 (December 14, 1935): pp 927–928.

93. William Shirer, *The Rise and Fall of the Third Reich, Volume I*, "The Rise," pp 231–232.

Notes to Chapter 7

1. Lise Meitner, "Wege und irrwege zur Kernenergie," *Naturwiss. Rund.* 5 (May 1963): p 167.

2. Enrico Fermi, *Nature* 133 (1934): pp 898–899. Also see Fermi, Amaldi, d'Agostino, Rasetti and Segrè, *Proc. Roy. Soc.* 146 (1934): p 483. Otto Hahn also noted in an article for *Sci. Am.* 198 (February 1958): p 79:

> There was one product which, unlike Fermi's extremely short-lived "uranium isotopes," had a half-life of 23 minutes. Its life was sufficiently long for us to establish chemically that it in fact was an isotope of uranium. Since it emitted a beta particle, it was evident that this isotope must become an isotope of element 93, which we called eka-rhenium. We looked for the new element, but were unable to detect it. If we had not been convinced that we had already identified two other isotopes of element 93 — an erroneous assumption, as it turned out — we would have prepared stronger samples of the material and made a determined effort to find the disintegration product of our 23-minute uranium isotope. We should then have had the pleasure of discovering element 93. Later, Edwin M. McMillan and Philip

H. Abelson in the U.S. identified an isotope element neptunium. Subsequently, when we ourselves obtained a stronger neutron beam for irradiating uranium, we had no trouble in detecting neptunium.

3. Lise Meitner, "Wege und irrwege zur Kernenergie," p 167.

4. Ibid.

5. See Fritz Krafft, *Im Schatten der Sensation: Leben und Wirken von Fritz Strassmann* (Weinheim: Verlag Chemie, 1981), pp 40–46 and the literary estate of Fritz Strassmann, Chemistry Department, Mainz Universität, Mainz, Germany.

6. E. Kaiser and E. Wilkins, translators, *Otto Hahn: My Life*, p 148. Also see Laura Fermi, *Atoms in the Family* (Chicago: University of Chicago Press, 1954), p 157, who quotes her husband Enrico:

> We did not have enough imagination to think that a different process of disintegration might occur in uranium... and we tried to identify the radioactive products with elements close to uranium on the periodic table... we did not know enough chemistry to separate the products... and we believed that we had four, while actually their number was close to fifty.

7. Ida Noddack, "Über das Element 93," *Angew. Chem.* 47 (1934): pp 653–655.

8. Related in an interview by the author with Fritz Krafft, July 17, 1981, Mainz, Germany. Also see R. Sime, *Lise Meitner: A Life in Physics*, p 273.

9. See Thomas Kuhn, *Structure of Scientific Revolutions*, 2nd ed. (Chicago: University of Chicago Press, 1970), pp 82–83.

10. Video taped interview with Emilio Segrè by Patricia Rife, Orinda, CA, September 8, 1984, deposited at the American Institute of Physics Emilio Segrè Visual Archives, College Park, MD.

11. See Aristide von Grosse and H. Agruss, "The Chemistry of Element 93 and Fermi's Discovery," *Phys. Rev.* 46 (1934): p 241. In this short paper, Grosse questioned whether the Rome team had really found transuranic elements after all. He reported experiments suggesting that Fermi might just as easily have been observing a new radioactive isotope of protactinium, which Hahn and Meitner had discovered in 1917 and trained many of their students to recognize. Hahn and Meitner's reply was published first as "Die Transurane und ihre chemischen Verhaltensweisen" (with Fritz Strassmann), *Ber. dt. chem. Ges.* 67 (1934) and later refined in "Über die künstliche Umwandlung des Urans durch Neutronen," *Naturwiss.* 23 (1935): pp 37–38 (submitted December 22, 1934).

12. Carl von Weizsäcker to Patricia Rife, August/September 1988.

13. Lise Meitner and Max Delbrück, *Der Aufbau der Atomkerne: Natürliche und künstliche Kernumwandlungen* (Berlin: Verlag Julius Springer, 1935).

14. See Emilio Segrè, *Enrico Fermi: Physicist* (Chicago: University of Chicago Press, 1970), pp 70–71.

15. Otto Hahn, "The Discovery of Fission," *Sci. Am.* 198, 2 (1958): p 77.

16. See Spencer Weart, "The Discovery of Fission and a Nuclear Physics Paradigm," pp 91–133 in William Shea, ed., *Otto Hahn and the Rise of Nuclear Physics* (Dordrecht and Boston: Reidel, 1983).

17. Ibid., p 133.

18. Lise Meitner and Otto Hahn, "Über die künstliche Umwandlung des Uran durch Neutronen," *Naturwiss.* 23 (1935): p 38.

19. Video taped interview with Emilio Segrè by Patricia Rife, September 1984, Orinda, CA; also see Edoardo Amaldi, "Neutron Work in Rome in 1934–1936 and the Discovery of Uranium Fission," *Ric. Sci.* 1, 1 (1984): pp 1–24.

20. Otto Hahn to Pummerer, cited in Krafft, "Lise Meitner: Eine Biographe," speech delivered January 1988, Hahn–Meitner Institute, Glienicker Str. 100, Berlin.

21. See Elisabeth Crawford, John Heilbron, Rebecca Ullrich, eds., *The Nobel Population 1901–1937*, Office for History of Science & Technology, University of California, Berkeley and Office for History of Science, Uppsala University, Uppsala, Sweden, 1987.

22. In T.W. MacCallum and S. Taylor, eds., *Nobel Prize Winners 1907–1937* (Zürich: Central Publishing Times, 1938); the following vignette of Ossietzky, author and philosopher who was imprisoned by the S.S. [under accusation of treason for exposing the Lufthansa bribery scandal] is given:

> In the question of the international peace movement, Ossietzky was one of those who visualized a solution of the peace problem in the promises of the Peace of Versailles, demanding that the victorious powers should reduce their armaments to the level attained in the disarmament of Germany. His pacifist convictions made it impossible for him to approve either of German rearmament or of the armaments in other countries in a corresponding proportion. His passionate literary advocacy of disarmament and international peace brought him into conflict with the German State authorities. Ossietzky was imprisoned in the Papenburg concentration camp the same year he was awarded the Nobel Prize for Peace.

23. F. Glum, *Zwischen Wissenschaft, Wirtschaft und Politik* (Bonn: Bouvier, 1964), p 378.

24. Planck to von Laue, December 22, 1936; reprinted in Armin Hermann, *Max Planck in Selbstzeugnissen und Bilddokumenten* (Hamburg: Rowohlts Monograph, 1973); translated by P. Rife.

25. See Nobel Archives, Swedish Royal Academy of Sciences Archives. I wish to personally thank Bengt Nagel, Secretary of the Physics Committee, for his kind assistance during my research in Stockholm during 1981 and 1988.

26. See Robert Weber, *Pioneers of Science: Nobel Prize Winners in Physics* (London: Institute of Physics, 1980).

27. See Spencer Weart, "The Discovery of Fission and a Nuclear Physics Paradigm," in William Shea, ed., *Otto Hahn and the Rise of Nuclear Physics* (Dordrecht and Boston: Reidel, 1983), p 102; for details see G. Gamow in *International Conference on Physics, London, 1934* (Cambridge: Cambridge University Press, 1935), Volume I, pp 60–66; and Lise Meitner, "Künstliche Umwandlungsprozesse beim Uran," pp 24–42 in E. Bretscher, ed., *Kernphysik: Vorträge gehalten am Physikalischen Institut der E.T.H. Zürich in Sommer 1936* (Berlin: Springer Verlag, 1936).

28. See Otto Hahn, "Die 'falschen' Transurane. Zur Geschichte eines wissenschaftlichen irrtums," *Natur. Rund.* 15 (1962): pp 43–47 for a detailed account of the team's scientific work during this period. Also see Kurt Starke, "The Detours Leading to the Discovery of Nuclear Fission," *J. Chem. Ed.* 56 (December 1979): p 772 and Hermann Günther, "Five Decades Ago: From the Transuranics to Nuclear Fission," *Angew. Chem. Int. Ed. Eng.* 29 (1990): pp 481–503.

29. Hahn to Rutherford, November 12, 1935. Also see Meitner and Hahn to Rutherford, February 19, 1937. Rutherford correspondence, American Institute for Physics, Archive for the History of Quantum Physics microfilm collection, College Park, MD.

30. O.R. Frisch and E.T. Sorensen, *Nature* 136 (August 17, 1935): p 258.

31. Lise Meitner and Otto Hahn, "Künstliche radioaktive Atomarten aus Uran und Thor," *Z. Angew. Chem.* 49 (1936): pp 127–128 and Lise Meitner, "Radioaktive und chemische Untersuchungen an den Transuranelementen," *Z. Angew. Chem.* 49 (1939): p 692. Later that year, Meitner would publish a longer "overview" of this work which had originally been delivered as a lecture in Switzerland. See Lise Meitner, "Kunstliche Umwandlungsprozesse beim Uran," pp 24–42 in Egon

Bretscher, ed. *Kernphysik: Vorträge gehalten am Physikalischen Institut der E.T.H. Zürich in Sommer 1936* (Berlin: Springer Verlag, 1936), p 41.

32. Leo Szilard and T.A. Chalmers, *Nature* 135 (1936): p 98.

33. C. von Weizsäcker, *Naturwiss.* 24 (1936): pp 813–814. Also see von Weizsäcker, *Die Atomkerne: Grundlagen und Anwendungen ihrer Theorie* (Leipzig: Akademische Verlagsgesellschaft, 1937).

34. Niels Bohr and F. Kalckar, *Kgl. Danske Vid. Selskab., Mat. fys. Medd.* 14 (1937): p 10.

35. Rutherford to Chadwick, February 17, 1936, quoted in Charles Weiner and Elsperth Hart, eds., *Exploring the History of Nuclear Physics* American Institute of Physics Conference Proceedings, 7 (New York: A.I.P., 1972), p 189.

36. Spencer Weart, "The Discovery of Fission...," p 103.

37. See Thomas Kuhn, *Structure of Scientific Revolutions*, Chapter 7, "Crisis and the Emergence of Scientific Theories," pp 66–76.

38. I. Curie, H. von Halban and P. Preiswerk, *J. Phys.* (1935): p 367. See Diebner and Grassmann, *Phys. Z.* 37 (1936): p 378, who skirted the priority controversy by giving joint credit to the Berlin *and* Paris teams for this "discovery!"

39. See Walther Gerlach in Dietrich Hahn, *Otto Hahn: Begründer des Atomzeitalters* (Munich: List Verlag, 1979), p 12.

40. See G. von Droste, *Z. Phys.* 110 (1938): pp 84–94.

41. I. Curie, P. Savitch, "Sur les radioelements formés dans l'uranium irradié par les neutrons, (I)," *J. Physique Radium* 8 (1937): pp 385–387.

42. See Strassmann's role in Krafft, *Im Schatten der Sensation: Leben und Wirken von Fritz Strassmann*, pp 204–207.

43. N. Bohr, *Nature* 137 (February 29, 1936): pp 344–348.

44. E. Amaldi, *Riv. Stor. Sci.* 1,1 (1984): pp 1–24.

45. See J.R. Tillman, B.P. Moon, "Selective Absorption of Slow Neutrons," *Nature* 36 (June 27, 1934): pp 66–67 and E. Amaldi, E. Fermi, "Sull'assorbimento dei neutroni lenti-I," *Ric. Sci.* 6 (November 15, 1935): p 2.

46. G. Breit and E. Wigner, "Capture of Slow Neutrons," *Phys. Rev.* 49 (April 1, 1935): pp 344–348.

47. See E. Amaldi, *Riv. Stor. Sci.* 1, 1 (1984): p 21 and "From the Discovery of the Neutron to the Discovery of Fission," *Phys. Rep.* 111 (1984): pp 276–295.

48. Hahn and Meitner to Curie, January 1938, Joliot–Curie Papers, Radium Institute, Paris.

49. See M.L. Pool, J.M. Cork, and R.L. Thornton, *Phys. Rev.* 52 (1937): pp 239–240 and Elisabeth Rona and Elisabeth Neuninger, *Naturwiss.* 24 (1936): p 491 for reviews of the literature on this topic.

50. For the Vienna group's work, see Elisabeth Rona and Elisabeth Neuninger, *Naturwiss.* 24 (1936): p 491. The Ann Arbor group published their results as M.L. Pool, J.M. Cork, and R.L. Thornton, *Phys. Rev.* 52 (1937): pp 239–240. Also see Cornelius Keller, "Wie Stellt man Transurane her?" *Bild der Wissenschaft* 2 (1980): pp 112–115.

51. P. Abelson, "A Graduate Student with Ernest O. Lawrence," pp 24–34 in Jane Wilson, ed., *All in our Time: The Reminiscences of Twelve Nuclear Pioneers* (Chicago: Bulletin of the Atomic Scientists, 1975).

52. See interview by John Bennett with Bretscher's widow, Hedy, videotaped, Cambridge, England, 1984, American Institute of Physics Emilio Segrè Visual Archives, College Park, MD, and P. Abelson, *Phys. Rev.* 53 (1938): p 211.

53. See Fritz Krafft, *Im Schatten der Sensation*, "Kernspaltung: Das Team Hahn–Meitner–Strassmann," pp 204–207.

54. I. Curie and P. Savitch, *J. Physique Radium* 9 (1938): p 355.

55. Paul Forman, "Scientific Internationalism and Weimar Physicists: The Ideology and Its Manipulation in Germany After World War I," *Isis* 64 (1973): p 156.

56. Dr. Berliner suffered great hardship under the Nazi regime and even after the war. See Ruth Sime, *Lise Meitner: A Life in Physics*, pp 151–152; 226; 270–271.

57. *New York Times*, March 9, 1936. Also see Willi Menzel, "Deutsche Physik und jüdische Physik," *Völkischer Beobachter* (January 29, 1936): p 5.

58. F. Joilot-Curie, *Textes Choisis* (Paris: Editions Sociales, 1959), p 35.

59. Spencer Weart, "The Discovery of Fission...," p 26.

60. O. Hahn and F. Strassmann, *Naturwiss.* 26 (1938): p 755; also see R. Spence, "Otto Hahn," *Biog. Mem. Fel. Roy. Soc.*, pp 294–295 and Fritz Strassmann, *Kernspaltung: Berlin, Dezember, 1938* (Mainz: Hans Krach, 1978).

Notes to Chapter 8

1. Jürgen Gehl, *Austria, Germany and the Anschluss* (New York: Oxford University Press, 1963), p 167.

2. Brice Pauly, *Hitler and the Forgotten Nazis: The History of the Austrian National Socialists* (Chapel Hill, NC: University of North Carolina Press, 1981).

3. Charles Gulick, *Austria from Habsburg to Hitler*, Volume II (Berkeley: University of California Press, 1948).

4. See O.R. Frisch, *Nuclear Physics in Retrospect: Proceedings from a Symposium on the 1930's*, R. Stuewer, ed., (Duluth, MN: University of Minnesota Press, 1979), pp 68–69.

5. Johannes Stark, "The Pragmatic and the Dogmatic Spirit in Physics," *Nature* 141 (April 30, 1938): p 722.

6. Beyerchen, *Scientists under Hitler*, and Richard Rhodes, *The Making of the Atom Bomb*, Chapter 7, "Exodus," (New York: Simon & Schuster, 1986).

7. Otto Hahn, *Mein Leben: Die Erinnerungen des grossen Atomforschers und Humanisten*, p 149.

8. David Nachmansohn, *German-Jewish Pioneers in Science* (New York: Springer–Verlag, 1979).

9. Reprinted in F. Krafft, "Lise Meitner: Her Life and Times," *Angew. Chem. Int. Ed. Eng.* 17 (1978): p 835.

10. See Meitner calendar, March 20–21, 1938, Meitner collection, Churchill College Archives, Cambridge, England. I would like to thank Mrs. Ulla Frisch for permission to use the Archives for research in 1984.

11. Reprinted in F. Krafft, "Lise Meitner: Her Life and Times," *Angew. Chem. Int. Ed. Eng.* 17 (1978): p 835.

12. For further analysis of this period see Meitner's calendar and diary entries from April 23 to May 7, in R. Sime, *Lise Meitner: A Life in Physics*, pp 188–189. On April 23, Planck's 80th birthday, Meitner met physicist A.D. Fokker from the Netherlands, and sat with him at the celebration dinner. He would become an important ally in her emigration.

13. Reprinted in Fritz Krafft, *Im Schatten der Sensation*, p 173; translated by P. Rife. Wilhelm Frick was one of Hitler's visible government officials, of high rank after the April 1933 civil service laws which he carried out ruthlessly, and cosigner of the March 13, 1938 degree of the "annexation" of Austria. See Edwin Hartshorne, *The German Universities and National Socialism*, p 33, in which he cites statistics about dismissals of "politically unreliables," "nominal Protestants, nominal Catholics," etc. Only half of the dismissed scholars throughout Germany between 1933–1943 were Jewish, and a tenth were dismissed on the pretext of "simplifying the administration."

14. Meitner diary, June 6, 1938; see R. Sime, "Lise Meitner's Escape from Germany," *Am. J. Phys.* 58, 3 (1990): pp 262–267. Debye, who left Germany in 1940, worked to assist his friends during the 1930's, but his son was an active Nazi Party member and often conversations would be held in his presence, making it all the more difficult for Meitner to assess what her colleague really meant.

15. See Lise Meitner to Dirk Coster, June–July, 1938, private collection of Dr. Ada Klokke-Coster, the Netherlands.

16. Dirk Coster to A.D. Fokker, June 11, 1938, Fokker papers, Museum Boerhaave, Leiden, Netherlands; see R. Sime, *Am. J. Phys.* 58, 3 (1990): p 263.

17. Reprinted in F. Krafft, "Lise Meitner: Her Life and Times," *Angew. Chem. Int. Ed. Eng.* 17 (1978): pp 835–836.

18. Scherrer to Meitner, June 17, 1938; Meitner collection, Churchill College Archives; see R. Sime, *Am. J. Phys.* 58, 3 (1990): p 263.

19. P. Debye to N. Bohr, June 16, 1938; Fokker papers, Museum Boerhaave, Leiden, Netherlands; see R. Sime, *Lise Meitner: A Life in Physics*, p 196.

20. Meitner diary, Churchill College Archives, Cambridge; see R. Sime, *Lise Meitner: A Life in Physics*, p 195.

21. See Bohr correspondence in Fokker papers, Museum Boerhaave, Leiden, Netherlands.

22. Bohr to Coster, June 21, 1938; see R. Sime, *Am. J. Phys.* 58, 3 (1990): p 264.

23. Fokker to Coster, June 17, 1938; in the same place.

24. D. Cohen to Fokker, June 20, 1938; in the same place.

25. Coster/Fokker to Minister of Justice, June 28, 1938, Fokker papers, Museum Boerhaave; see R. Sime, *Lise Meitner: A Life in Physics*, p 197.

26. P.F.S. Otten to Fokker, June 28, 1938; see R. Sime, *Am. J. Phys.* 58, 3 (1990): p 264.

27. Fokker to Coster, June 27, 1938; in the same place, p 265.

28. Coster to Debye, recorded in a letter from Lise Meitner to Otto Hahn, May 13, 1966; Max-Planck-Gesellschaft Archives.

29. See Hugo Atterling, "Karl Manne Georg Siegbahn 1886–1978," *Biog. Mem. Fell. Roy. Soc.* 37 (1991) and Svante Lindqvist, ed., *Center on the Periphery: Historical Aspects of 20th Century Swedish Physics* (Canton, MA: Science History Publications, 1993). Siegbahn had been watching with envy as California became growing center of cyclotron research, and he wanted to make Europe – and particularly his own new Institute – a well-known physics research center as well. For more on cyclotron developments in America, see J.L. Heilbron and Robert Seidel, *Lawrence and his Laboratory: Nuclear Science at 1931–1961* (Los Angeles: Universtiy of California Press, 1981). Lawrence began to cultivate an international blend of young physicists as soon as his laboratories opened, inviting many refugees to his growing circle throughout the 1930's.

30. To analyze Meitner's personal thoughts during this time, see her brief diary entries, June 1938, Meitner Collection, Churchill College Archives.

31. Debye to Fokker (in German), July 11, 1938, Fokker papers, Museum Boerhaave, Leiden; see R. Sime, *Am. J. Phys.* 58, 3 (1990): p 265.

32. See extensive Bohr correspondence, Meitner file, 1933/1934, Rockefeller Archives Center, Tarrytown, New York, and Meitner file, Nobel Research Institute for Physics, Stockholm.

33. Debye to Coster, July 6, 1938; Dirk Coster papers, University of Groningen, Netherlands; see R. Sime, *Lise Meitner: A Life in Physics*, p 202.

34. Coster to Debye, July 9, 1938; see R. Sime, *Am. J. Phys.* 58, 3 (1990): p 265.

35. Debye to Fokker (in German), July 11, 1938, Fokker papers, Museum Boerhaave, Leiden; see R. Sime, *Lise Meitner: A Life in Physics*, p 203.

36. E. Kaiser and E. Wilkins, translators, *Otto Hahn: My Life*, p 149.

37. Arnold Kramish, *The Griffin: The Greatest Untold Story of World War II* (New York: Houghton–Mifflin, 1986), p 49.

38. E. Kaiser and E. Wilkins, translators, *Otto Hahn: My Life*, p 149.

39. Kramish, *The Griffin: The Greatest Untold Story of World War II*, p 49. Coster's widow later recounted that she thought the German border guards may have overlooked "Frau Professor" because they assumed that Lise Meitner was Coster's wife!

40. Hahn to Coster family, July 15, 1938; reprinted in F. Krafft, "Lise Meitner: Her Life and Times," *Angew. Chem. Int. Ed. Eng.* 17 (1978): p 836.

41. Pauli wrote to Coster immediately, and so did others; from Zürich, Paul Scherrer sent Meitner his first uncoded letter in months lamenting his isolation and the cryptic messages from colleagues in Berlin.

42. Mentzel to Telschow, August 11, 1938; German letter reprinted in F. Krafft, *Im Schatten der Sensation*, p 78; translated by P. Rife. Rudolf Mentzel was the Committee Chair for Sciences and Technology in the Nazi Ministry of Education, and was also the new Nazi Party member/President of the German *Forschuungsgemeinschaft* in 1938, succeeding Johannes Stark, well known for his "Aryan physics" diatribes.

43. Telschow to Hahn, August 18, 1938; Meitner personnel file, Max Planck Institute for Chemistry, Mainz, Germany; German letter reprinted in F. Krafft, *Im Schatten der Sensation*, p 78; translated by P. Rife.

44. O.R. Frisch, "Lise Meitner," *Biog. Mem. Fel. Roy. Soc.*, p 411.

45. Meitner to Hahn, August 24, 1938; Reprinted in F. Krafft, "Lise Meitner: Her Life and Times," *Angew. Chem. Int. Ed. Eng.* 17 (1978): p 837.

Notes to Chapter 9

1. Hahn to Meitner, September 12, 1938, Lise Meitner collection, Churchill College Archives, Cambridge, England; translated and reprinted in Fritz Krafft, "Lise Meitner: Her Life and Times," *Angew. Chem. Int. Ed. Eng.* 17 (1978): p 842.

2. Hahn to Meitner, October 1, 1938, Lise Meitner collection, Churchill College Archives; translated and reprinted in Fritz Krafft, "Lise Meitner: Her Life and Times," *Angew. Chem. Int. Ed. Eng.* 17 (1978): p 841–842.

3. Meitner to Hahn, September 25, 1938; October 15, 1938; November 6, 1938; translated and reprinted in Krafft, "Lise Meitner: Her Life and Times," *Angew. Chem. Int. Ed. Eng.* 17 (1978): p 837.

4. Meitner to Hahn, December 5, 1938; translated and reprinted in Krafft, "Lise Meitner: Her Life and Times," *Angew. Chem. Int. Ed. Eng.* 17 (1978): p 837. Also see Meitner personal notebooks, Studsvik Reactor Archives, Nykoeping, Sweden.

5. E. Kaiser and E. Wilkins, translators, *Otto Hahn: My Life*, p 150.

6. Meitner stayed in Copenhagen from November 10–17, 1938; see guestbook, Niels Bohr Archives, Institute for Theoretical Physics. I wish to thank Erik Bohr for his support in researching documents relating to Meitner's trip to Copenhagen in 1938.

7. On October 1, 1938, Meitner was at last granted emeritus status in Berlin but her pension was put into a frozen account under the Third Reich's laws. Also see author's interview with Gudmund Borelius, Physics Dept., KTH, Stockholm.

8. Hahn to Meitner, November 19, 1938; reprinted in Jost Lemmerich, ed., *Die Geschichte der Entdeckung der Kernspaltung* (Berlin: Technische Universität Berlin, Universitätsbibliothek, 1988), p 184; translated by P. Rife. Also see O.R. Frisch, "A Walk in the Snow," *New Scientist* 60 (1973): p 833. For Baron von Weizsäcker's involvement in Meitner's case, see Raul Hilberg, *The Destruction of the European Jews* (New York: Octagon Books, 1978), p 93. von Weizsäcker's directives in July 1938 may have even endangered Meitner further by calling attention to the fact that a German passport had been denied to her after the annexation of Austria to the Third Reich.

9. See Meitner collection, Churchill College Archives, Cambridge, England. Also see Anthony Read and David Fisher, *Kristallnacht: The Nazi Night of Terror* (New York: Random House, 1989).

10. Hahn to Meitner, December 19, 1938, Meitner collection, Churchill College Archives, Cambridge, England; reprinted in Patricia Rife, *Lise Meitner: Ein Leben für die Wissenschaft*, 2nd ed., (Düsseldorf: Claassen Verlag, 1992), pp 259–260; translated by P. Rife. I would like to thank Dr. Hofacher of the University of Michigan German Department for assistance in translations into English.

11. Spencer Weart, "The Discovery of Fission and a Nuclear Physics Paradigm," p 28. Also see Mark Rollins, *Mental Imagery: On the Limits of Cognitive Science* (New Haven, CT: Yale University Press, 1989).

12. Hahn collection, 1938 laboratory notebook, Deutsches Museum, Munich. Also see Kurt Starke, "Detours Leading to the Discovery of Nuclear Fission," *J. Chem. Ed.* 56 (December 1979): p 774.

13. Otto Hahn, "The Discovery of Fission," *Sci. Am.* 198 (February 1958): p 82.

14. Otto Hahn, *Vom Radiothor zur Uranspaltung: Eine Wissenschaftliche Selbstbiographie* (Braunschweig: Viewig Verlag, 1962), p 21.

15. Interviews and personal records of Fritz Strassmann, reprinted in Fritz Krafft, "Lise Meitner: Her Life and Times," p 834. Also see Fritz Strassmann, *Kernspaltung: Berlin Dezember 1938* (Mainz: Privatdruck, 1978).

16. E. Kaiser and E. Wilkins, translators, *Otto Hahn: My Life*, p 150.

17. Gottfried von Droste had been an assistant of Meitner's when their search for alpha emissions from the transuranics was a daily task. Interview with Max Delbrück by Patricia Rife, 1980.

18. Meitner to Hahn, December 21, 1938; reprinted in Patricia Rife, *Lise Meitner: Ein Leben für die Wissenschaft*, 2nd ed., (Düsseldorf: Claassen Verlag, 1992), pp 254–256; translated by P. Rife.

19. See E. Kaiser and E. Wilkins, translators, *Otto Hahn: My Life*, p 152 and the crucial analysis of Strassmann's exclusion by Hahn in Krafft, *Im Schatten der Sensation*.

20. Otto Hahn and Fritz Strassmann, "Über den Nachweis und das Verhalten der bei der Bestrahlung des Urans mittels Neutronen entstehenden Erdalkalimetalle," *Naturwiss.* 27 (1939): p 11. A complete volume of all the scientific texts preceding and related to the discovery of nuclear fission in the original languages they were published in has been edited by Horst Wohlfarth, *40 Jahre Kernspaltung: Eine Einführung in die Origniallitertatur* (Darmstadt: Wissenschaftliche Buchgesellschaft,1979).

21. See Hahn's telephoned additions to the page proofs, December 27, 1939, in Krafft, *Im Schatten der Sensation*, pp 266–267 and Patricia Rife, *Lise Meitner: Ein Leben für die Wissenschaft*, 2nd ed., (Düsseldorf: Claassen Verlag, 1992), p 250.

22. This paragraph is part of the final article, Hahn and Strassmann, "Über den Nachweis und das Verhalten der bei der Bestrahlung des Urans mittels Neutronen entstehenden Erdalkalimetalle," *Naturwiss.* 27 (1939): p 11. It is fascinating to note that Hahn was leaving an "out" for himself in this speculation. Since Meitner had not been available to interpret their results, Hahn was delimiting his expertise to that of his own field – nuclear chemistry – and if this paragraph is read closely, while his four pages of chemical results were published, he would NOT put forward with confidence his speculations on the nucleus "bursting" or renaming all their prior four years of "transuranics;" hence, this soon-to-be famous article is a publication of chemistry results and concludes with a speculation, not an announcement, concerning the "possibility" of nuclear "bursting."

23. Hahn to Meitner, December 21, 1938; reprinted in Dietrich Hahn, ed., *Otto Hahn: Erlebnisse und Erkenntnisse*, p 80; translated by P. Rife.

24. O.R. Frisch, *What Little I Remember*, p 114 and O.R. Frisch, "The Interest is Focusing on the Atomic Nucleus," in Rozenthal, ed., *Niels Bohr*, p 115.

25. O.R. Frisch, *What Little I Remember*, p 115. For the liquid drop model see N. Bohr, *Nature* 137 (1936): pp 354–355. Also see Roger Stuewer, "The Origin of the Liquid-Drop Model and the Interpretation of Nuclear Fission," *Perspectives on Science* 2 (1994): pp 76–129.

26. See Augustine Brannigan, *The Social Basis of Scientific Discoveries* (Cambridge: Cambridge University Press, 1981); Karl Popper, *The Logic of Scientific Discovery* (New York: Basic Books, 1959); Gerald Holton, *Thematic Origins of Scientific Thought* (Cambridge, MA: Harvard University Press, 1973); Paul Feyerabend, *Against Method* (London: New Left Books, 1975); and others mentioned throughout this chapter.

27. N. Bohr and F. Kalckar, "On the Transmutation of Atomic Nuclei by Impact of Material Particles. I. General Theoretical Remarks," *Kgl. Danske Vid. Selskab, Mat. fys. Med.* 14, 10 (1937): pp 1–40. Also see N. Bohr, *Nature* 137 (1936): pp 354–355.

28. See article by George Gamow within *International Conference on Physics, London 1934* (Cambridge: Cambridge University Press, 1935), pp 60–61. Gamow had even speculated upon such a nuclear model as early as 1931; see George Gamow, *Constitution of Atomic Nuclei and Radioactivity* (Oxford: Clarendon Press, 1931).

29. O.R. Frisch, in Rozenthal, ed., *Niels Bohr*, pp 144–145.

30. O.R. Frisch, *What Little I Remember*, p 116.

31. O.R. Frisch, in Rozenthal, ed., *Niels Bohr*, p 145.

32. Lise Meitner to Otto Hahn, December 29, 1938, literary estate of Otto Hahn, Max-Planck-Society Archives, Berlin–Dahlem; reprinted in Dietrich Hahn, ed., *Otto Hahn: Erlebnisse und Erkenntnisse*, p 59; translated by P. Rife.

33. Frisch later recalled that they had several phone conversations about fission before his aunt returned to Copenhagen. Also, see Roger Stuewer, "Bringing the News of Fission to America," *Phys. Tod.* (October 1985) concerning Bohr's work during these weeks as well as Niels Blaedel, *Hamoni Og Enked: Niels Bohr en Biograf* (Copenhagen: Rhodos, 1985).

34. Reprinted in Esther Sparberg, "Study of the Discovery of Fission," *Am. J. Phys.* 32 (1964): p 4.

35. Otto Hahn to Lise Meitner, December 28, 1938, reprinted in part in D. Hahn, *Otto Hahn: Erlebnisse und Erkenntnisse*, p 82; translated by P. Rife.

36. Dietrich Hahn has edited his grandfather's letters. Historians Fritz Krafft and Spencer Weart have noted the serious omission of the sentence "Strassmann ist verreist" [Strassmann is away] from this letter published in D. Hahn, *Otto Hahn: Erlebnisse und Erkenntnisse*, p 82. This crucial sentence indicates that Otto Hahn was alone with his thoughts – and doubts – when composing their "joint" article for *Die Naturwissenschaften*. See Krafft, "Lise Meitner: Her Life and Times," pp 826–842.

37. See Bernard Barber, "Resistance by Scientists to Scientific Discovery," *Science* 134 (August 25, 1961) for a fascinating social/historical study of this complex subject of scientists' own resistance to their data.

38. See Thomas Kuhn, *The Structure of Scientific Revolutions* 2nd ed., (Chicago: University of Chicago Press,1970). Also see Edwin Boring, "Psychological Factors in Scientific Progress," in *Psychologist at Large: An Autobiography and Selected Essays* (New York: Basic Books, 1961), pp 314–337. Boring discusses his views in this essay of the role of the "zeitgeist" (the spirit of the times) in scientific progress. In his article, "The Discovery of Fission and a Nuclear Physics Paradigm," in Shea, ed., *Otto Hahn and the Rise of Nuclear Physics*, pp 121–122, Spencer Weart discusses the role of scientific models in the work leading to the discovery of fission:

> In sum, Hahn, Meitner and many of their colleagues worked with a paradigm, which had several components. There was a concept of transmutation, possibly connected with some sort of mental picture of the nuclei. There was a choice of notations by which transmutations could be represented. And there were cases showing how to use a notation.... The paradigm was not used rigidly but was modified to agree with new observations. Almost from the start, Meitner and Hahn were willing to postulate reactions slightly different from any reaction seen before. In their early 1936 paper, as we have seen, they listed the three known types of neutron reaction but showed that not all their results could fit into this pattern. So they postulated a fourth type of reaction, not so different from the three already known, wherein two neutrons come flying out together. This reaction turned out to be a valid one, another case in the collection of cases that was part of the paradigm. When double neutron reaction nevertheless failed to explain their results, the Berlin group turned to the concept of isomers. Meitner took one of the accepted types of reaction and extended it, used it as a template so

to speak, to accommodate not one nucleus but two. Since isomerism does exist, the expansion of the paradigm was again valid, a permanent addition to knowledge of the nucleus.... In the end the paradigm was not altogether overthrown. Even the notations, so inadequate to portray fission, remains in use to this day for other types of nuclear processes. Superficially, fission simply expanded the paradigm to include another case, another nuclear process in the list.

39. Lise Meitner to Otto Hahn, January 1, 1939, reprinted in D. Hahn, *Otto Hahn: Erlebnisse und Erkenntnisse*, p 84; translated by P. Rife.

40. Lise Meitner to Otto Hahn, December 29, 1938, literary estate of Otto Hahn, Max-Planck-Gesellschaft Archives, Berlin–Dahlem; reprinted in Sime, *Lise Meitner: A Life in Physics*, p 239 and Christoph Wolff, "Sie was dabei, als das Atomzeitalters begann," *Die Welt* (November 8, 1963): p 79; translated by P. Rife.

41. Lise Meitner to Otto Hahn, December 21, 1938, in E. Kaiser and E. Wilkins, translators, *Otto Hahn: My Life*, p 151.

Notes to Chapter 10

1. See Roger Stuewer, "Bringing the News of Fission to America," *Phys. Tod.* (October 1985): pp 49–56.

2. O.R. Frisch to Lise Meitner, January 3, 1939, Lise Meitner collection, Churchill College Archives, Cambridge, England; reprinted in Krafft, *Im Schatten der Sensation*, p 271.

3. See O.R. Frisch, *Phys. Tod.* (November 1967): p 47.

4. O.R. Frisch to Lise Meitner, January 8, 1939, Lise Meitner collection, Churchill College Archives, Cambridge, England; reprinted in Jost Lemmerich, *Geschichte der Entdeckung* (Berlin: Universitätsbibliothek, 1988), pp 179–181. In this letter, he continues:

> When Hevesy saw the paper, he immediately said that [Irène] Curie had told him already in the fall that she found very light elements from uranium, but she obviously did not trust herself to publish this. But she can be satisfied: she already has the Nobel Prize.

5. O.R. Frisch, "Physical Evidence for the Division of Heavy Nuclei Under Neutron Bombardment," *Nature* 143 (February 18, 1939): p 276.

6. See Frisch's confirmation of this mailing in O.R. Frisch to Lise Meitner, January 17, 1939, Lise Meitner collection, Churchill College Archives, Cambridge, England.

7. O.R. Frisch and Lise Meitner, "Disintegration of Uranium by Neutrons: A New Type of Nuclear Reaction," *Nature* 143 (February 11, 1939): pp 239–240. A complete volume of all the scientific texts preceding and related to the discovery of nuclear fission in the original languages they were published in has been edited by Horst Wohlfarth, *40 Jahre Kernspaltung: Eine Einführung in die Originialliteratur* (Darmstadt: Wissenschaftliche Buchgesellschaft, 1979). It is interesting to note that the English language publication of this article may have assisted in the rapid spread of the experimental verification of fission, and yet, also retained the priority of Frisch and Meitner with various other labs soon publishing their own findings.

8. O.R. Frisch and L. Meitner, *Nature* 143 (February 11, 1939): p 240.

9. Cited by Frisch and Meitner as footnote 6: N. Bohr, *Nature* 137 (1936): pp 344, 355.

10. Cited by Frisch and Meitner as footnote 7: N. Bohr and F. Kalckar, *Kgl. Danske Vid. Selskab, Mat. fys. Med.* 14 (1937): p 10.

11. O.R. Frisch, *What Little I Remember*, p 117. Frisch recalled in an article "How It All Began," *Phys. Tod.* (November 1967): p 47:

> I told the whole story to George Placzek, who was in Copenhagen, before it even occurred to me to do an experiment. At first Placzek did not believe the story that these heavy nucleii already known to suffer from alpha instability, should also be suffering from this extra affliction. 'It sounds a bit' he said, 'like the man who is run over by a motor car and whose autopsy shows that he had a fatal tumor and would have died within a few days anyway.'

12. Lise Meitner, "New Products of the Fission of the Thorium Nucleus," *Nature* 143 (1939): p 637.

13. See Léon Rosenfeld, ed., in *Cosmology, Fusion and other Matters: George Gamow Memorial Volume* (Boulder, CO: Colorado Associated University Press, 1972); also see A.P. French and P.J. Kennedy, eds., *Niels Bohr: A Centenary Volume* (Cambridge, MA: Harvard University Press, 1985).

14. O.R. Frisch, in Rozenthal, ed. *Niels Bohr*, p 146. Also see Frisch's chapter, "Experimental Work with Nuclei: Hamburg, London, Copenhagen," in Stuewer, ed., *Nuclear Physics in Retrospect*, p 72.

15. Lise Meitner to Otto Hahn, January 14, 1939, reprinted in D. Hahn, *Otto Hahn: Erlebnisse und Erkenntnisse*, p 91; translated by P. Rife.

16. Meitner may have been referring to a paper she had published in *Scientia* 63 (1938): p 13 before her emigration, which Spencer Weart describes as "the elaborate, largely mistaken description of neutron reactions with uranium developed in Berlin."

17. Laura Fermi, *Atoms in the Family* (Chicago: University of Chicago Press, 1954). Also see Laura Fermi, *Illustrious Immigrants: The Intellectual Migration from Europe, 1930–1941* (Chicago: University of Chicago Press, 1968).

18. See Allan Debus, ed., *World Who's Who In Science* (New York: Marquis Who's Who, 1968).

19. See John A. Wheeler, "Some Men and Moments in the History of Nuclear Physics: The Interplay of Colleagues and Motivations," in R. Stuewer, ed., *Nuclear Physics in Retrospect*, p 78.

20. Interview with John A. Wheeler by Patricia Rife, Princeton University, September 1984.

21. Ibid. Wheeler recalled that Bohr was so concerned for Lise Meitner's safety and publication credit that he did not want to publish his own interpretation with Wheeler until he knew that Frisch and Meitner had received credit.

22. Interview with Erik Bohr by Patricia Rife, London, England, June 1984; videotape deposited with The American Institute for Physics, Emilio Segrè Visual Archives, College Park, MD.

23. O.R. Frisch, "The Interest is Focusing...," in Rozenthal, ed., *Niels Bohr*, p 146.

24. Niels Bohr to O.R. Frisch, January 20, 1939, Bohr Correspondence microfilm, Office of the History of Science, University of California, Berkeley.

25. See Roger Stuewer, "Bringing the News of Fission to America," *Phys. Tod.* (October 1985).

26. See O.R. Frisch to Niels Bohr, January 22, 1939, Bohr Correspondence microfilm, University of California, Berkeley, and O.R. Frisch, *What Little I Remember*.

27. See Tuve Papers, Library of Congress, Washington, DC for documents initiated on November 30, 1938 when Tuve, George Gamow, and Edward Teller drew up a proposal that the Carnegie Institute and George Washington University would sponsor the Fifth Conference on Theoretical Physics in Washington, DC between January 21 and 30, 1939.

28. John A. Wheeler, *Phys. Tod.* (November 1967): p 50.

29. E. Fermi, *Phys. Tod.* (November 1955): p 12. Also see Esther Sparberg, "A Study of the Discovery of Fission," *Am. J. Phys.* 32 (1964): pp 2–8.

30. See L. Badash, E. Hodes, A. Tiddens, "Nuclear Fission: Reaction to the Discovery in 1939," IGCC Research Paper 1, San Diego: Institute on Global Conflict and Cooperation, University of California.

31. See H.L. Anderson, E.T. Booth, J.R. Dunning, E. Fermi, G.N. Glasoe, G.F. Slack, *Phys. Rev.* 55 (1939): p 511,

32. See *Science Service*, January 30, 1939.

33. See documentary film "Atomic Power," 1958, Defense Nuclear Agency, Washington, DC and Patricia Rife, "Chain Reaction: Dawn of the Nuclear Age," documentary video, 1988.

34. O.R. Frisch, "The Interest is Focusing . . . ," in Rozenthal, ed., *Niels Bohr*, p 147.

35. R.D. Fowler, R.W. Dodson, *Phys. Rev.* 55 (1939): p 417.

36. R.B. Roberts, R.C. Meyer, L.R. Hafstad, *Phys. Rev.* 55 (1939): p 416. This was the group which invited Fermi and Bohr in the historic "midnight experimental conference" to view fission fragments as they were measured flying off. See W. Davis and R.D. Potter, *Science News Letter*, February 11, 1939, p 87, and *Science Supplement*, February 10, 1939, p 5.

37. See Philip Abelson, "A Graduate Student with Ernest O. Lawrence," in Jane Wilson, ed., *All in our Time: The Reminiscences of Twelve Nuclear Pioneers*, and G.K. Green, L.W. Alvarez, *Phys. Rev.* 55 (1939): p 417.

38. Niels Bohr to O.R. Frisch, February 3, 1939, Bohr Correspondence microfilm, University of California, Berkeley.

39. L. Meitner and O.R. Frisch, *Nature* 143 (1939): pp 239–240.

40. O.R. Frisch, *Nature* 143 (1939): p 276.

41. Niels Bohr, *Nature* 143 (1939): p 330.

42. *Time* 33 (1939): p 21.

Notes to Chapter 11

1. Quoted in Ruth Moore, *Niels Bohr: The Man, His Science, and the World They Changed* (New York: Knopf, 1966), p 231.

2. Quoted by O.R. Frisch in S. Rozenthal, ed. *Niels Bohr*, p 147.

3. Meitner to Hahn, February and March, 1939; reprinted in Krafft, "Lise Meitner: Her Life and Times," *Angew. Chem. Int. Eng. Ed.* 17 (1978): p 838.

4. O.R. Frisch interview by Peter Walton, in "Scientists Under Hitler," BBC Open University documentary, 1978. I wish to thank BBC producer Peter Walton for his insights during my research on Lise Meitner in London and Cambridge, England.

5. Hahn to Meitner, February 7, 1939; see R. Sime, *Lise Meitner: A Life in Physics*, p 256. Nazi party members Otto Erbacher and Kurt Philipp (who was promoted to the rank of professor soon after Meitner's departure) made it clear that the Institute's own physicists, Gottfried von Droste and Siegfried Fluegge, might have been given "priority" credit for their March 1939 results if published earlier. Yet in that same period, Wilfrid Wefelmeier and Carl Friedrich von Weizsäcker were still publishing articles in which they were drawing incorrect theoretical conclusions about the "transuranics." In Elizabeth Rona, *How It Came About: Radioactivity, Nuclear Physics, Atomic Energy* (Oak Ridge, TN: Oak Ridge Associated Universities, 1978), p 45, she also recalls that Hahn felt his colleagues held much against him, and he "complained bitterly."

6. O.R. Frisch and L. Meitner, "Products of the Fission of the Uranium Nucleus," *Nature* 143 (March 18, 1939): p 470. Hahn to Meitner, February 7, 1939: reprinted in R. Sime, *Lise Meitner: A Life in Physics*, p 262.

7. Meitner to Bragg, undated; reprinted in Krafft, *Im Schatten der Sensation*, p 326.

8. Meitner to Hahn, March 10, 1939; literary estate of Otto Hahn, Max-Planck-Gesellschaft Archives, Berlin–Dahlem; see R. Sime, *Lise Meitner: A Life in Physics*, p 267.

9. O.R. Frisch and L. Meitner, "Products of the Fission of the Uranium Nucleus," *Nature* 143 (1939): p 471.

10. L. Meitner and O.R. Frisch, *Nature* 143 (March 18, 1939): p 239 and N. Bohr, *Nature* 143 (1939): p 330.

11. O.R. Frisch, *Nature* 143 (1939): p 276.

12. R.D. Fowler and R.W. Dodson, *Nature* 143 (1939): p 233 and W. Jentschke and F. Prankl, *Naturwiss.* 27 (1939): p 134.

13. O.R. Frisch, *Nature* 143 (1939): p 276.

14. F. Joliot, *Comptes Rendus* 208 (1939): p 341.

15. L. Meitner, *Nature* 143 (1939): p 637.

16. O.R. Frisch, "The Interest is Focusing...." in Rozenthal, ed., *Niels Bohr*, p 147. Most important to note that, in April 1939, Frisch's colleague Moeller suggested that fission *fragments* might contain enough energy to send out a neutron.

17. Henrick Kramers, *The Atom and the Bohr Theory of its Structure* (Cambridge: Cambridge University Press, 1923).

18. Hans Dominik, *Atomgewicht 500* (Berlin: Wissenschaft Verlag, 1935). Also see Anthony Michaelis, "How Nuclear Energy was Foretold," *New Scientist* (March 1962): p 276, and John Tierney, "If This is the Future...," *Science* (January 1984): p 84.

19. See "Atom Energy Hope is Spiked by Einstein," *Pittsburgh Post-Gazette*, December 29, 1934.

20. Gertrud Szilard and Spencer Weart, eds., *Leo Szilard: His Version of the Facts*, p 54.

21. See "Leo Szilard Home Page," World Wide Web at
`http://www.peak.org/~danneng/szilard.html`
for documents, photographs, and interviews. Szilard's unpublished papers and correspondence are held by the Mandeville Department of Special Collections, Central Library, University of California, San Diego.

22. Ibid.

23. See Leo Szilard, in Bernard Bailyn, ed., *The Intellectual Migration* (Cambridge, MA: Harvard University Press, 1969).

24. Szilard to Joliot–Curie, February 2, 1939; reprinted in G. Szilard and S. Weart, eds., *Leo Szilard: His Version of the Facts*, p 69.

25. O.R. Frisch, in Rozenthal, ed., *Niels Bohr*, p 148.

26. Niels Bohr and John A. Wheeler, *Phys. Rev.* 56 (1939): pp 426–450.

27. O.R. Frisch, in Rozenthal, ed., *Niels Bohr*, p 148.

28. Gordon Craig, *Germany: 1866–1945* (New York: Oxford University Press, 1978).

29. Hans von Halban, Jr., Frederic Joliot, and Lew Kowarski, "Liberation of Neutrons in the Nuclear Explosion of Uranium," *Nature* 143 (1939): pp 470–471. In Otto Hahn to Lise Meitner, March 20, 1939, Hahn wrote: "In France, Joliot is regarded as the discoverer of fission. From this one sees how systematic suppression and false citations have their intended effect." Reprinted in Sime, *Lise Meitner: A Life in Physics*, p 267

30. Lise Meitner to Otto Hahn, March 28, 1939; reprinted in Krafft, "Lise Meitner: Her Life and Times," *Angew. Chem. Int. Eng. Ed.* 17 (1978): p 838.

31. See a summary of such articles in L.A. Turner, "Nuclear Fission," *Rev. Mod. Phys.* 12 (January 1940): p 29.

32. Interview with Gudmund Borelius by P. Rife, Royal Institute for Technology (KTH), Stockholm, May, 1980.

33. See John Heilbron, "Creativity and Big Science," *Phys. Tod.* 45, 11 (November 1992): pp 42–47.

34. See J.L. Heilbron, *Lawrence and his Laboratory: A History of the Lawrence Berkeley Laboratory, Volume I* (Los Angeles and Berkeley: University of California Press, 1989).

35. Nomination letter for Nobel Prize in Chemistry, January 31, 1939, *Protokoll vid Kung Vetenskapsakakademiens*, Samankoster foer Behandling af Aeenden Roerande Nobelstiftelsen Aer, 1939. On March 31, 1939, Thé Svedberg submitted a four-page review summarizing the proliferation of publications and scientific proofs for nuclear fission in support of his nomination of Hahn and Meitner for the 1939 Nobel Prize in Chemistry.

36. Lise Meitner to Otto Hahn, October 27, 1939; reprinted in F. Krafft, "Lise Meitner: Her Life and Times," *Angew. Chem. Int. Ed. Eng.* 17 (1978): p 838.

Notes to Chapter 12

1. Mark Walker, *German National Socialism and the Quest for Nuclear Power, 1939–1945* (Cambridge: Cambridge University Press, 1989); and David Irving, *The German Atomic Bomb* (New York: Simon & Schuster, 1965). While Irving has made several mistakes about Allied efforts during the war, his conclusion is interesting: that the German nuclear scientists failed to fire Speer's imagination with the possibilities of atomic fission was their greatest shortcoming.

2. Max von Laue, "Otto Hahn zum 60. Geburtstag," *Naturwiss.* 27 (1939): p 153.

3. Elisabeth Rona, *How It Came About: Radioactivity, Nuclear Physics, Atomic Energy*, p 45.

4. R. Sime, *Lise Meitner: A Life in Physics*, p 278.

5. Esther Caulkin Brunauer to Florence Sabin, April 22, 1938; Florence Sabin's reply, April 29, 1938; cited in Margaret Rossiter, *Women Scientists in America: Struggles and Strategies to 1940.* Lise's younger sister Lola married Dr. Rudolph Allers and moved to Washington, DC, where he taught at the Catholic University. Her sister Frieda, who earned her Ph.D. in Chemistry at the University of Vienna during World War I, emigrated with her husband, a lawyer, to New York. Both of her younger sisters wanted Lise to retire in America soon after the Anschluss. A third sister, Lotte Meitner Graf, resided in London, where her exquisite photographic portraits of famous scientists and artists became well known.

6. O.R. Frisch, *What Little I Remember*, p 108.

7. See videotaped interview with Sir Rudolf Peierls by Patricia Rife, Oxford, England, 1984. Niels Bohr Library, Center for History of Physics, American Institute of Physics, College Park, MD.

8. Quoted in Spencer Weart, "Scientists with a Secret," *Phys. Tod.* (February 1976): p 24.

9. O.R. Frisch, "Investigating Fission," in Jane Wilson, ed., *All in our Time: The Reminiscences of Twelve Nuclear Pioneers*, p 54.

10. Richard Rhodes, *The Making of the Atomic Bomb*, and tape recorded interview with Victor Weisskopf by Patricia Rife, MIT, Cambridge, MA, 1981.

11. H. von Halban, F. Joliot, L. Kowarski, *Nature* 143 (1939): p 680.

12. See Spencer Weart, *Scientists in Power* (Cambridge, MA: Harvard University Press, 1979) for a thorough overview of the Joliot–Curie team's work during this time. Also see Mark Walker, *German National Socialism and the Quest for Nuclear Power, 1939–1949.*

13. I.N. Golovin, *I. V. Kurchatov: A Soviet Realist Biography of the Soviet Nuclear Scientist* (Bloomington, IN: Selbtsverlag Press, 1968), p 31.

14. See documents reprinted in David Irving, *The German Atomic Bomb* (New York: Simon & Schuster, 1965), 2nd ed.; title published in London: *The Virus House* (London: Wm Klonber, 1967), p 32. Also see Gerald Posner, "Letter from Berlin: Secrets of the Files" in *The New Yorker* 52, 4 (March 14, 1994): pp 39–47, concerning scholars and investigators' protest about the U.S. government's decision to return control of a crucial Nazi archive filled with Third Reich records, the Berlin Document Center, to the German government.

15. See video interview with Eugene Wigner by Patricia Rife, Princeton, NJ, August 1984, Niels Bohr Library, American Institute of Physics, College Park, MD, and Banesh Hoffmann and Helen Dukas, *Albert Einstein: Creator and Rebel* (New York: Viking, 1972).

16. Quoted by Leo Szilard, in G. Szilard and S. Weart, eds., *Leo Szilard: His Version of the Facts*, p 83. Szilard states: "He was willing to assume responsibility for sounding the alarm even though it was quite possible that the alarm might prove to be a false alarm. The one thing that most scientists are really afraid of is to make a fool of themselves. Einstein was free from such a fear and this above all is what made his position unique on this occasion."

17. Albert Einstein to President Franklin D. Roosevelt, August 2, 1939; reprinted in G. Szilard and S. Weart, eds., *Leo Szilard*, pp 94–96.

18. Ibid., p 95. There is a wealth of material published on the early developments surrounding the atomic bomb: for primary documents see Richard Hewlett and Oscar Anderson, Jr., *The New World 1939–1946* (University Park, PA: Pennsylvania State University Press, 1962). Also see interview with Eugene Wigner by Charles Weiner, 1966 and interview with Niels Bohr by Thomas Kuhn, 1962, American Institute of Physics, Center for History of Physics Archives, College Park, MD.

19. President Franklin D. Roosevelt to Einstein, October 19, 1939; reprinted in G. Szilard and S. Weart, eds., *Leo Szilard*, p 96.

20. See Ronald Clark, *The Birth of the Bomb: The Untold Story of Britain's Part in the Weapon that Changed the World* (New York: Horizon, 1961).

21. See Margaret Gowing, *Britain and Atomic Energy 1939-1946*.

22. See Richard G. Hewlett and Oscar E. Anderson Jr., *The New World 1939–1946*, Volume 1: *A History of the US Atomic Energy Commission*, pp 9–52; also see Spencer Weart, "Scientists with a Secret," *Phys. Tod.* (February 1976): pp 23–30.

23. Alexander Sachs, "Early History of the Atomic Project in relation to President Roosevelt 1939–1940," unpublished ms, *Med* 319.7, National Archives, Washington, DC.

24. David Irving, *The Virus House*, p 11.

25. See Meitner correspondence to von Laue, Meitner collection, Churchill College Archives, Cambridge, England.

26. E. McMillan and P.H. Abelson, "Radioactive Element 93," *Phys. Rev.* 57 (1940): pp 1185–1186. McMillan was awarded the 1950 Nobel Prize in Physics.

27. See Meitner collection, Churchill College Archives, Cambridge, England. Hedwig Kohn was hired as an instructor in the University of North Carolina from 1940–1942 in their Women's College. She was then able to secure a position at Wellesley, where she remained until her retirement as full professor in 1952. An avid experimentalist in radiation measurements, Kohn worked as a research associate at Duke University until her late seventies.

28. Lise Meitner to Walter Meitner, June 18, 1939; reprinted in Sime, *Lise Meitner: A Life in Physics*, p 275.

29. After the war, Paul Rosbaud wrote to Samuel Goudsmit about this chemist, Wever, who had reported about Hahn as well as refugees in England. See Rosbaud to Goudsmit, August 9, 1946, Goudsmit correspondence, American Institute of Physics, Center for History of Physics, College Park, MD.

30. See Rudolf Peierls, *Bird of Passage: Recollections of a Physicist* (Oxford: Oxford University Press, 1986).

31. See O.R. Frisch, in J. Wilson, ed., *All in our Time: The Reminiscences of Twelve Nuclear Pioneers*, p 57.

32. Videotaped interview with Sir Rudolf Peierls by Patricia Rife, Oxford, England, June, 1984, Niels Bohr Library, American Institute of Physics, College Park, MD.

33. See Daniel Bar-On, *Legacy of Silence: Encounters with Children of the Third Reich* (Cambridge, MA: Harvard University Press, 1989); Alfred Haesler, *The Lifeboat is Full: Switzerland and the Refugees 1933–1945* (New York: Funk & Wagnalls, 1969); Lucy Dawidowicz, *The War Against the Jews 1933–1945* (New York: Bantam, 1975) and Hannah Arendt, *The Origins of Totalitarianism* (New York: Harcourt Brace Jovanovich, 1973).

34. O.R Frisch in J. Wilson, ed., *All in our Time: The Reminiscences of Twelve Nuclear Pioneers*, pp 56–57.

35. This memorandum to the British government is reprinted in full in Margaret Gowing, *Britain and Atomic Energy 1939–1945* (London: Oxford University Press, 1964), pp 389–393.

36. Richard Rhodes, *The Making of the Atom Bomb*. For a biographical account of the committee and its physicists, see Arnold Kramish, *The Griffin: The Greatest Untold Story of World War II* (Boston: Houghton–Mifflin, 1986).

37. Swedish Academy of Sciences Nobel Archives, nomination letter and report to the Committee for Chemistry, March 31, 1939, *Protokoll vid Kung. Vetenskapsakademiens*, Stockholm, Sweden.

38. See Elisabeth Crawford, John Heilbron, and Rebecca Ullrich, *The Nobel Population 1901–1937: A Census of the Nominators and Nominees for the Prizes in Physics and Chemistry* (Berkeley, CA: University of California, Office for History of Science and Technology and Uppsala: Uppsala University, 1987).

39. See reports of Nobel Prize Chemistry Committee of the Swedish Academy of Science, Stockholm, and E. Crawford, R. Sime, and M. Walker, "A Nobel Tale of Postwar Injustice," *Phys. Tod.* (September 1997): pp 26–32.

40 Ibid. For a sociological analysis of scientific awards and recognition, see T. Frängsmyr, ed., *Solomon's House Revisited* (Canton, MA: Science History Pub., 1990). Nominations by Niels Bohr, Oskar Klein, and later Egil Hylleraas for Meitner and Frisch to share the Nobel Prize for Physics, Swedish Academy of Science Archives. I wish to thank B. Nagel for his archival assistance with the Nobel nominations.

41. See video interview with Carl von Weizsäcker by John Bennett, November, 1984, author's private collection. In his autobiographical memoir, Werner Heisenberg reconstructed conversations with his friend von Weizsäcker, which many have criticized as "whitewashing" the truth; others have also critiqued his wife's memories as the same revisionist history. See W. Heisenberg, *Physics and Beyond: Encounters and Conversations* (New York: Harper & Row, 1971) and Elisabeth Heisenberg, *Inner Exile: Recollections of a Life with Werner Heisenberg* (Boston: Birkhäuser, 1984).

42. Carl Friedrich von Weizsäcker, *The Politics of Peril* (New York: Seabury Press, 1978), p 199. Also see von Weizsäcker, "Eine Möglichkeit der Energiegewinnung aus U-238," German Reports on Atomic Energy, file G-59, Karlsruhe Nuclear Research Center, Karlsruhe, Germany and Mark Walker, *German National Socialism and the Quest for Nuclear Power*, Chapters 1 and 2. For documents uncovered by the Allies after the war concerning the German uranium project, see Samuel Goudsmit, *Alsos* (New York: Schuman, 1947; reprint, Woodbury, NY: AIP Press, 1996), Ch. I,

"The Fear of a German Atom Bomb: We Overestimate the German Scientific Effort and the Germans Do Likewise," and Tom Bower, *The Paperclip Conspiracy: The Hunt for the Nazi Scientists* (New York: Little Brown & Co., 1987).

43. See Margaret Gowing, *Britain and Atomic Energy 1939–1945*; Arthur Compton, *Atomic Quest* (Oxford: Oxford University Press, 1956); and James Conant, *A History of the Development of an Atomic Bomb* (unpublished manuscript, 1943, Bush-Conant file, OSRD S-1, folder 5, National Archives, Washington, DC).

44. Lise Meitner, *Nature* 422 (1940): pp 422–423. Similar research was taking place in America during this time as well: see Richard Roberts, "Fission Cross-Sections for Fast Neutrons," unpublished manuscript, Department of Terrestrial Magnetism Archives, Carnegie Institution of Washington, Washington, DC.

45. Interview with Hans von Ubisch by Patricia Rife, May 1981, Taeby, Sweden.

46. Interview with Torsten Magnusson by Patricia Rife, June 1981, Research Institute for Physics, Stockholm, Sweden; also see Hugo Atterling, "Karl Manne Georg Siegbahn 1886–1978," *Biog. Mem. Fell. Roy. Soc.* 37 (1991): pp 428–444. Part of the Swedish military's concern was that the Russian physicists might also form a nuclear weapons committee based on fission research published in Moscow. See Georgi Nikolai Flerov and Konstantin A. Petrzhak, "Spontaneous Fission of Uranium," *J. Phys.* 3 (1940): pp 275–280 and Karl-Erik Larsson,"Kaernkraftens historia i Sverige," *Kosmos* (Stockholm, 1987).

47. Phillip Hughes, "Wartime Fission Research in Japan," *Social Studies of Science* 10 (1980): pp 345–349 and D. Shapley, "Nuclear Weapons History: Japan's Wartime Bomb Projects Revealed," *Science* (January 13, 1978): p 199.

48. See extensive Reich government citations within Mark Walker, *German National Socialism and the Quest for Nuclear Power*. It is interesting to note that Churchill also remained a bit shortsighted concerning the potential of nuclear energy, preferring to stress aerial developments as well. See Winston Churchill, *Their Finest Hour* (New York: Houghton Mifflin, 1949) and *The Hinge of Fate* (New York: Houghton Mifflin, 1950). Also see Gregg Herken, *The Winning Weapon* (New York: Knopf, 1980).

49. Albert Speer, *Inside the Third Reich* (New York: Macmillan, 1970), p 226. Also see Werner Heisenberg, "Research in Germany on the Technical Application of Atomic Energy," *Nature* 160 (1947): p 211 and "The Third Reich and the Atomic Bomb," *Bul. Atom. Sci.* 24 (June 1968): p 34.

51. Speer, *Inside the Third Reich*, p 226.

52. David Cassidy, *Uncertainty: The Life and Science of Werner Heisenberg* (New York: WH Freeman, 1983).

53. Speer, *Inside the Third Reich*, p 226.

54. O.R. Frisch, "Investigating Fission," in J. Wilson, ed., *All in our Time: The Reminiscences of Twelve Nuclear Pioneers*, p 58.

55. See Daniel Kelves, *The Physicists: The History of a Scientific Community in Modern America* (New York: Vintage Press, 1979), p 326.

56. See R. Hewlettt and O. Anderson, *The New World 1939–1946* and Leslie Groves, *Now It Can Be Told: The Inside Story of the Development of the Atomic Bomb* (London: A. Deutsch, 1963).

57. Arthur H. Compton, *Atomic Quest: A Personal Narrative* (New York: Oxford University Press, 1956), p 144.

Notes to Chapter 13

1. See Mark Walker, *German National Socialism and the Quest for Nuclear Power 1939–1949*; Carl Friedrich von Weizsäcker, *The Politics of Peril*; David Cassidy, *Uncertainty: The Life and Science of Werner Heisenberg*. For controversies still surrounding this debate, see Alvin Weinberg and Jaroslav Franta, "Was Nazi Know-How Enough for an A-Bomb?" *Phys. Tod.* 47 (December 12, 1994): p 84, and Paul L. Rose, "Did Heisenberg Misconceive A-Bomb?" letter to the editor, *Phys. Tod.* 45 (February 1992): p 2.

2. Pavel A. Sudoplatov and Antoli P. Sudoplatov, with Jerrod Schecter and Leona Schecter, *Special Tasks: The Memoirs of an Unwanted Witness* (New York: Little Brown & Co., 1994).

3. Ibid., p 66. Also see Bruce Page and Phillip Knightly, *The Philby Conspiracy* (New York: Harper & Row, 1968) and Jenrih Borvick, *The Philby Files: The Secret Life of Master Spy Kim Philby* (New York: Little Brown & Co., 1994). Sudoplatov recalled a similar strategy for infiltrating the American "top secret" project:

> Kheifetz advised us of a piece of information that changed Moscow's skeptical attitude about the atomic project. Kheifetz and J. Robert Oppenheimer, a brilliant American physicist at the University of California, had met in Dec. 1941 (at a party to raise money for the Spanish Civil War refugees)... and Kheifetz reported that the outstanding physicists in the Allied world were involved in a secret project. Kheifetz was an experienced professional [agent] who knew better

than to approach a jewel of a source such as Oppenheimer with the usual money or threats. Instead, he created a common ground of interest and idealism, drawing on stories of his travels and cosmopolitan view of life that the two men could discuss and compare.

Also see Tad Szulc, "The Untold Story of How Russia Got the Bomb," *Los Angeles Times*, August 26, 1984; Margaret Gowing, *Britain and Atomic Energy 1939–1945*; Andrew Pierre, *Nuclear Politics: The British Experience with an Independent Strategic Force 1939–1970* (London: Oxford University Press, 1972), pp 5–25, and R.V. Jones, "Winston Leonard Spencer Churchill," *Biog. Mem. Fel. Roy. Soc.* 12 (1966): p 35.

4. See Otto Hahn, *Mein Leben: Die Erinnerungen des grossen Atomforschers und Humanisten* (Munich: Bruckmann, 1968).

5. The German group had incorrectly concluded that carbon had a substantial neutron absorption cross-section and thus attempted to utilize deuterium as a moderator — an element found in heavy water which, even after the Nazi takeover of the Norwegian plant that supplied heavy water, was never an adequate supply for the German uranium project. See Knut Haukelid, *Skis Against the Atom* (London: Wm Kimber, 1954), and Alvin M. Weinberg and Eugene P. Wigner, *The Physical Theory of Neutron Chain Reactors* (Chicago: University of Chicago Press, 1958).

6. Werner Heisenberg, *Physics and Beyond: Encounters and Conversations*; Manfred von Ardenne, *Ein Glückliches Leben für Technik und Forschung* (Zürich: 1972); Walther Gerlach, *Otto Hahn* (Munich: List Verlag, 1969).

7. See Mark Walker, "Legenden um die deutsche Atombombe," *Vierteljahrhefte für Zeitgeschichte* 38, 1 (1990): pp 69–73; Spencer Weart, *Nuclear Fear: A History of Images* (Cambridge, MA: Harvard University Press, 1988); R. Sime, *Lise Meitner: A Life in Physics*, Ch. 13 "War Against Memory" and pp 354–355, discussion of revisionism debate, 1948; Samuel Goudsmit, *Alsos*, Farm Hall transcripts discussion; Thomas Powers, *Heisenberg's War: The Secret History of the German Bomb* (New York: Knopf, 1993); and Philip Morrison, "A Reply to Dr. von Laue," *Bull. Atom. Sci.* 4 (1948): p 104.

8. As recalled by Aage Bohr, quoted in Jeremy Bernstein, "What Did Heisenberg Tell Bohr about the Bomb?" *Sci. Am.* 272, 5 (May 1995): p 92.

9. Ibid.

10. See documents and diagrams in "What Did Heisenberg Tell Bohr about the Bomb?" *Sci. Am.*, as well as Jeremy Bernstein, *Hans Bethe: Prophet of Energy* (New York: Basic Books, 1980), which is countered in Abraham Pais, *Niels Bohr's Times: In Physics, Philosophy, and Polity* (New York: Oxford University Press, 1991).

11. See Vögler to Leeb, February 27, 1942, Max-Planck-Gesellschaft Archives, Berlin–Dahlem, and Irving microfilm collection, Harnack House conference, February 26–29, 1942, Deutsches Museum, Munich, Germany.

12. Albert Speer, *Inside the Third Reich*, p 227.

13. Ibid., cited in Speer's memoirs as *Führerprotokoll*, May 6, 1942. He continues:

> General Fromm offered to release several hundred scientific assistants from the services, while I urged the scientists to inform me of the measures, the sums of money, and the materials they would need to further nuclear research. A few weeks later they presented their request: an appropriation of several hundred thousand marks and some small amounts of steel, nickel, and other priority metals. In addition, they asked for the building of a bunker, the erection of several barracks, and the pledge that their experiments would be given the highest priority. Plans for building the first German cyclotron had already been approved. Rather put out by these modest requests in a matter of such crucial importance, I suggested that they take one or two million marks and correspondingly larger quantities of materials. But apparently more could not be utilized for the present, and in any case, I had been given the impression that the atom bomb could no longer have any bearing on the course of the war.

14. Ibid., p 225; cited in Speer's memoirs as *Führerprotokoll*, June 23, 1942, point 15: "Reported briefly to the Führer on the conference on splitting the atom and on the backing we have given to the project." See David Irving, *The Virus House*, and microfilm collection of important documents relating to this meeting, housed within the Irving collection, Deutsches Museum, Munich and American Institute of Physics, Center for History of Physics Library, College Park, MD. Also see "Das Physikstudium während des ersten Trimesters 1940," in *Meldungen aus dem Reich: Die Geheimen Lagerberichte des Sicherheitsdinestes der S.S. 1938–1945*, Volume IV (Herrsching: Pawlack, 1984), pp 1049–1051.

15. Lise Meitner to Otto Hahn, January 24, 1954; cited in F. Krafft, *Im Schatten der Sensation*, p 182, ft 48.

16. While visiting Erik Bohr at his home in Lynby, Denmark (August 1988), we came across a letter from his father to historian Robert Jungk which had never been sent. Dated 1956, Bohr restated his memory of the veiled conversation Heisenberg had with him, and affirmed his irritation at Heisenberg for "distorting" the truth in Jungk's revisionist book *Brighter than a Thousand Suns* (New York: Harcourt Brace Jovanovich, 1955) about his "purpose" for travelling to Copenhagen. Jungk (p 103) had quoted Heisenberg as 'recalling':

> Being aware that Bohr was under the surveillance of the German political author-
> ities and that his assertions about me would probably be reported to Germany, I
> tried to conduct this talk in such a way as to preclude putting my life into im-
> mediate danger. This talk probably started with my question as to whether or not
> it was right for physicists to devote themselves in wartime to the uranium prob-
> lem — as there was the possibility that progress in this sphere could lead to
> grave consequences in the technique of war. Bohr understood the meaning of this
> question immediately, as I realized from his slightly frightened reaction.

17. In a letter to Otto Hahn, January 24, 1954, Lise Meitner stressed that Jungk's *Brighter than a Thousand Suns*, was not accurate. She wrote to Hahn that several young German physicists had "pressed for the creation of a 'Deutsche Congress' in Copenhagen during the war — a very unpsychological idea." Her letter continues: "I know that the whole Bohr Institute, with the inclusion of Bohr himself, were invited — and abstained from attending." Cited in F. Krafft, *Im Schatten der Sensation*, p 182, ft 48.

18. Ruth Moore, *Niels Bohr*, p 301. For Bohr's role in securing safe passage for his institute members and colleagues who secretly fled Copenhagen when the Nazis invaded, see David Lampe, *The Danish Resistance*, [original title, *Savage Canary*] (New York: Ballantine, 1957).

19. See Abraham Pais, *Niels Bohr's Times: In Physics, Philosophy, and Polity* (New York: Clarendon Press, 1991), p 480.

20. See Harold Flender, *Rescue in Denmark* (New York: Simon & Schuster, 1963) and Jorgen Haestrug, *Secret Alliance: A Study of the Danish Resistance Movement, 1940–1945*, Volume III (Berlin: Odense University Press, 1979).

21. Ruth Moore, *Niels Bohr*, p 304.

22. Ibid. Also see Leni Yahil, *The Rescue of Danish Jewry* (New York: Jewish Publishing Society of America, 1969).

23. Interview with Erik Bohr by P. Rife, London, England, June 1984.

24. See James Kunetka, *City of Fire: Los Alamos and the Birth of the Atomic Age 1943–1945* (New York: Prentice Hall, 1978).

25. In his recently published memoirs, Sudoplatov recalled:

I was in my 30s and hesitant to share secrets; I put my palm over the signatures and enumeration of the sources. Kurchatov, Ioffe and Kikoin were astonished and said to me, 'Look Pavel Anatolievich, you are too naive. You read the materials to us, and we will tell you who the authors are.' Obviously, these physicists had kept up with Meitner, Hahn, Strassmann and Frisch's publications since the late 1930s.

See P. Sudoplatov, A. Sudoplatov, *Special Tasks: The Memoirs of an Unwanted Witness*; also see Allen G. Debus, ed., *World's Who's Who in Science* (New York: Marquis Who's Who, 1968). German scientists did not make nominations for the Nobel Prizes from 1939–1942 and no prizes were awarded from 1940–1942, but Russians knew well who were the world's leading experts in nuclear science.

26. David Lampe, *The Danish Resistance*, (original title, *Savage Canary*).

27. Albert Speer, *Inside the Third Reich*, p 228.

28. See Abraham Pais, *Subtle is the Lord: The Science and the Life of Albert Einstein*.

29. See Abraham Pais, *Niels Bohr's Times: In Physics, Philosophy, and Polity*, and Barton J. Bernstein, "The Uneasy Alliance: Roosevelt, Churchill, and the Atomic Bomb 1940–1945," *Western Political Quarterly* 29 (1976): p 2.

30. David Hawkins, *Project Y: The Los Alamos Story, Part I. Toward Trinity* (Los Angeles: Tomash, 1983); Richard Hewett and Oscar Anderson, Jr., *A History of the United States Atomic Energy Commission, Volume 1: The New World 1939–1946* (University Park, PA: Pennsylvania State University Press, 1962); Leslie R. Groves, *Now It Can Be Told: The Inside Story of the Development of the Atomic Bomb*; Gregg Herken, *The Winning Weapon*.

31. See O.R. Frisch in J. Wilson, ed., *All in our Time...*, p 58.

32. Interview with Sir Rudolf Peierls by P. Rife, May 1984, Oxford, England.

33. Interview with Nicholas Kurti by P. Rife, June 1984, Royal Society, London, England.

34. See O.R. Frisch, in Jane Wilson, ed., *All in our Time...*, p 55.

35. See O.R. Frisch, in Rozental, ed., *Niels Bohr*, p 148.

36. See Prof. A. Tiselius' opening remarks about this 1943 speech prefacing Otto Hahn's 1946 Nobel lecture in *Nobel Lectures in Chemistry, 1942–1962* (Stockholm: Nobel Foundation, 1963).

37. See Derek de Solla Price, *Little Science, Big Science* (New York: State University Press, 1963); B.G. Griffith and N.C. Mullins, "Coherent Social Groups in Scientific Change," *Science* 177 (1972): pp 959–964. For a closer look at the man behind many of the decisions made in Sweden concerning the development of the Nobel Research Institute for Physics, see Torsten Magnusson, "Manne Siegbahn" in *Swedish Men of Science*, (Uppsala: Almqvist & Wiksell, 1952). Hugo Atterling, who joined the group of physicists in the Research Institute in November of 1940, states in a letter to the author, August 1982:

> The plans for our Institute were, in their general outlines, made up already a couple of years before the outbreak of war. They aimed at the creation of a, for that time, modern and well-equipped laboratory for, in the first place, basic nuclear research. Such a laboratory could not be useful without a particle accelerator of some kind (at that time, three kinds were available: the Cockcroft-Walton accelerator, the van de Graaf-machine, and the cyclotron).

Dr. Atterling added that Lise Meitner did not, of course, know what proportion of money was spent for equipment or salaries in the new Research Institute, but she had her "opinions." He also added:

> The main purpose of the operation of the cyclotron was to provide the physicists working at the Institute with radioactive material. The demand for such material increased when beta-spectroscopy work came into full swing. However, the demand for radioactive material by other institutes was also great. . . . The possibility to produce radioactive material for medical purposes (treatment and research), and for biological research, etc., could, of course, strengthen our position when Manne Siegbahn applied for funds for the operation and development of our machine.

38. See Henry Smyth, *Atomic Energy for Military Purposes* (Washington, DC: USGPO, 1945); Richard Rhodes, *The Making of the Atomic Bomb*; Martin Sherwin, *A World Destroyed* (New York: Knopf, 1975).

39. See M.J. Mulkay, "Pure and Applied Research," in Ina Spiegel-Roesing and Derek de Solla Price, eds., *Science, Technology and Society* (London: Sage Publications, 1977).

40. Interview with Gudmund Borelius by P. Rife, May 1981, KTH (Royal Institute for Technology), Stockholm, Sweden. One Swedish doctoral candidate, Sigvard Eklund, became the Director of the United Nations International Atomic Energy Agency (IAEA) after the war (see correspondence with author, Sigvard Eklund, 1981-1982, United Nations International Atomic Energy Agency, Vienna, Austria). Another doctoral student, Curt Mielokowski, would rise in corporate research to become the President of Saab Motor Company; and many others became productive physicists in Sweden after the war. See interview with Curt Mielokowski, June 1981, Stockholm, Sweden, and an interview with Tibor Graf, June, 1981, KTH Physics Dept., Stockholm, Sweden.

41. See Lise Meitner, "Über das Verhalten einiger seltener Erden bei Neutronenbestrahlung," *Ark. Mat. Astron. och Fysik* (A), 27 (1941): p 3; "The Resonance Energy of the Thorium Capture Process," *Phys. Rev.* 2, 60 (1941): pp 58–60; "Über die von Sc46 emittierten Strahlen," *Ark. Mat. Astron. och Fysik* (B), 28 (1942): p 14; "Eine Einfache Methode zur Messung der durch Gamma-Strahlen erregten sekundären Elektronen und der Einfluss dieser Elektronen auf die Untersuchung primärer Beta-Strahlspekten," *Ark. Mat. Astron. och Fysik* (A), 29 (1943): p 17; "A Simple Method for the Investigation of Secondary Electrons Excited by Gamma-Rays and the Interference of these Electrons with Measurements of Primary Beta-Ray Spectra," *Phys. Rev.* 63 (1943): pp 73, 384; "Das Zerfallsschema des Scandium 46," *Ark. Mat. Astron. och Fysik* (A), 32 (1945): p 6; "Attempt to Single Out some Fission Processes of Uranium by Using the Differences in their Energy Release," *Rev. Mod. Phy.* 17 (1945): p 287.

42. See Jacques Hadamard, *The Psychology of Invention in the Mathematical Field* (Princeton: Princeton University Press, 1945) and Lise Meitner, "Looking Back."

43. See Thomas Powers, *Heisenberg's War: The Secret History of the German Bomb*; R.V. Jones, *Most Secret War*, (London: Hamish Hamilton, 1978); also published as *The Wizard War* (New York: Coward, McCann and Geoghegan, 1978).

44. See Paul L. Rose, "Did Heisenberg Misconceive A Bomb?" *Phys. Tod.* 45, 2 (February 1992): p 126; M. Bar-Zohar, *The Hunt for German Scientists 1944–1960* (London: Barker, 1965); and David Cassidy, *Uncertainty: The Life and Science of Werner Heisenberg.*

45. See S. Goudsmit, *Alsos* (New York: Schuman, 1947; reprint, Woodbury, NY: AIP Press, 1996), and Elisabeth Heisenberg, *Inner Exile: Recollections of a Life with Werner Heisenberg* (Boston: Birkhäuser,1984).

46. See Dieter Hoffmann, ed., *Operation Epsilon: Die Farm-Hall-Protokolle oder die Angst der Alliierten vor den deutschen Atombombe* (Berlin: Rowohlt, 1993) and *Operation Epsilon: The Farm Hall Transcripts* (Berkeley & Los Angeles: University of California Press, 1993). Also see Irving Klotz "Germans at Farm Hall

Knew Little of A-Bomb," *Phys. Tod.* 46, 10 (October, 1993): pp 11–15. For further bibliographical sources, see Tom Bower, *The Paperclip Conspiracy: The Hunt for Nazi Scientists*. While Hahn, von Laue, and others were not Party members and did not discuss politics in depth under Allied surveillance, their views of the fate of Germany, discussed privately during months of confinement, demonstrate their false hopes, apologetic loss of memory, and lack of moral insight into the fate of millions. See Samuel Goudsmit, *Alsos*, pp 134–139.

47. See Len Giovannitti and Fred Freed, *The Decision to Drop the Bomb* (New York: Coward–McCann, 1965); National Archives, Manhattan Engineer District Records, group 77, (1945) Washington, DC; Leslie R. Groves, *Now It Can Be Told: The Inside Story of the Development of the Atomic Bomb*.

48. Hans Bethe, Kurt Gottfried and Roald Sagdeev, "Did Bohr Share Nuclear Secrets?" *Sci. Am.* (May 1995): pp 85–90.

49. See Henry Smyth, *Atomic Energy for Military Purposes: The Official Report of the Development of the Atomic Bomb under the Auspices of the United States Government* (San Francisco: Stanford University Press, 1989). Also see David Holloway, *Stalin and the Bomb* (New Haven, CT: Yale University Press, 1994).

50. O.E. Brewster to President Harry S. Truman, in Barton Bernstein, "The Quest for Security: American Foreign Policy and International Control of Atomic Energy 1942–1946," *J. Am. Hist.* 60, 4 (1974): pp 1003–1044 and "Roosevelt, Truman, and the Atomic Bomb: A Reinterpretation," *Pol. Sci. Quar.* 90, 2 (1976): pp 202–230. Also see Martin Sherwin, *A World Destroyed: The Atomic Bomb and the Grand Alliance* (New York: Knopf, 1975), p 69 and Joseph Lieberman, *The Scorpion and the Tarantula: The Struggle to Control Atomic Weapons 1945–1949* (Boston: Houghton Mifflin, 1970), which offers a fascinating study of the Baruch Plan as well. For two opposing views on Truman's atomic bomb policies, see James F. Byrnes, *Speaking Frankly* (New York: Harper, 1947), and Bernard M. Baruch, *Baruch: The Public Years* (New York: Holt Rinehart & Winston, 1960) concerning Truman's decisions during these crucial months.

51. See D. Shapley, "Nuclear Weapons History: Japan's Wartime Bomb Projects Revealed," *Science* 199 (January 13, 1978) and extensive citations in Phillip Hughes, "Wartime Fission Research in Japan," *Soc. Stud. Sci.* 10 (1980): pp 345–349.

52. Interview with Hans von Ubisch by P. Rife, May 1981, Stockholm, Sweden; also see Meitner correspondence to Eva von Bahr-Bergius and others, 1945–1946, Meitner Collection, Churchill College Archives, Cambridge, England.

53. Interview with Fritz Krafft and Irmgard Strassmann by P. Rife, July 1981, Mainz, Germany.

54. Lise Meitner to Otto Hahn, June 27, 1945; reprinted in Fritz Krafft, *Im Schatten der Sensation*, pp 181–182.

55. It would take several years, and a special act of the Swedish parliment, for Lise Meitner's citizenship to be processed. In 1945, she would be elected a foreign member of the Royal Academy of Sciences in Stockholm; a small pension funded by the KTH was granted (thanks to the persistence of her colleague Gudmund Borelius) before citizenship papers were filed. Her petition for dual citizenship with Austria, in order to receive a small pension from the German government, finally was approved in 1949 by the Swedish government. See Berta Karlik, "In Memorandum Lise Meitner," *Phys. Bl.* 35 (1979): pp 49–52.

Notes to Chapter 14

1. Correspondence with Gudmund Borelius Physics Department, Royal Institute of Technology (KTH), Stockholm, Sweden, March 1981.

2. Audio-taped interview with Gudmund Borelius by Patricia Rife, June 1981, deposited with the Physics Department, KTH, Stockholm, Sweden.

3. Ibid.

4. Live NBC trans-Atlantic broadcast, August 9, 1945, NBC Radio Archives, New York, Mrs. Eleanor Roosevelt and Dr. Lise Meitner.

5. Live NBC trans-Atlantic broadcast, August 11, 1945, interview with Dr. Lise Meitner and Dr. Frieda Meitner Frischauer, Mutual Broadcasting, NBC Radio Archives, New York. Also see *Current Biography 1945*, "Lise Meitner," p 394.

6. Samuel Goudsmit, *Alsos* (New York: Schuman, 1947; reprint, Woodbury, NY: AIP Press, 1996).

7. "Lise Meitner" file, Physics Department, Catholic University. I wish to thank Dr. James Brennan, chairman, for forwarding the late Dr. Herzfeld's large correspondence file on Meitner for organization in 1979, my first year of graduate school in the Union for Experimenting Colleges and Universities, now the Union Institute.

8. See Eve Curie, *Madame Curie* (Paris: Gallimard, 1981).

9. Regina Herzfeld to P. Rife, December 16, 1980.

10. Ibid.

11. Ibid.

12. See Meitner's East Coast lecture schedule and press clippings in Meitner file, Physics Department, Catholic University, Washington, DC.

13. Meitner also received honorary doctorate degrees from Adelphi University (May 27, 1946), the University of Rochester (June 10, 1946), Rutgers University, Smith College, and the University of Stockholm (June 27, 1950). Her remarks were later synthesized into an article published in the August 1960 issue of *Phys. Tod.* entitled "The Status of Women in the Professions." In concluding this article, she states:

> The question about the position of women...cannot be answered properly without considering the sociological, sexual-psychological, and many other aspects of our Western culture...looking back, I have the impression that the problems of professional women in general, and particularly of academic women, have found fairly satisfactory solutions in the last 80 to 100 years. Not all that can be desired has been achieved. In principle, nearly all male professions have become accessible to women; in practice, things often look different.

14. George Axelsson, "Is the Atom Terror Exaggerated?" *The Saturday Evening Post* (August 1946): pp 34–47.

15. Ibid.

16. See E. Crawford, R. Sime, and M. Walker, *Nature* 382 (1996) and E. Crawford, R. Sime, and M. Walker, "A Nobel Tale of Postwar Injustice," *Phys. Tod.* (September 1997): pp 26–32.

17. E. Kaiser and E. Wilkins, translators, *Otto Hahn: My Life*, p 178.

18. E. Crawford, R. Sime, and M. Walker, *Nature* 382 (1996) and E. Crawford, R. Sime, and M. Walker, "A Nobel Tale of Postwar Injustice," *Phys. Tod.* (September 1997): pp 26–32.

19. See Ferdinand Smekal, *Österreichs Nobelpreisträger* (Vienna: Frick Verlag, 1968). Rabi was born in Rymanow, then part of greater Austria, and had not resided in Europe since the late 1920's, when he accepted a position at Columbia University. Meitner's nominations were by Oskar Klein, Niels Bohr, and Egil Hylleraas, a professor of physics at the University of Oslo.

20. In E. Kaiser and E. Wilkins, translators, *Otto Hahn: My Life*, p.199, Hahn recalled:

> ... I had a rather unhappy conversation with Lise Meitner [upon his arrival at the elegant Savoy Hotel in Stockholm], who said I ought not to have sent her away from Germany when I did. That discord was probably the result of some disappointment because it was only I who was awarded the Prize. I did not mention that point myself, but a number of her friends alluded to it in a rather unkind manner in conversation with me. Yet I really had no responsibility for the course events had taken. When I had organized my deeply respected colleague's escape from Germany, all I had had in mind was her welfare. And then, too, the Prize had been given to me for work I had done either alone or with my colleague Fritz Strassmann, and for her achievements Lise Meitner had been given a number of honorary degrees in the U.S.A. and had even been declared the 'woman of the year.'

21. O.R. Frisch, "Lise Meitner," *Biog. Mem. Fel. Roy. Soc.* 16 (November 1970): p 414.

22. See E. Kaiser and E. Wilkins, translators, *Otto Hahn: My Life*, p 199. Hahn stated in retrospect:

> At our destination, we were awaited by Lise Meitner, my friend Percy Quensel, an attaché from the Swedish Ministry, Frau von Hevesy, and many journalists. We were comfortably lodged in the elegant Savoy Hotel, and after all those years of separation we once again sat down to a peace-time dinner with our friends.

23. Friedrich Herneck, "Über die Stellung von Lise Meitner und Otto Hahn in der Wissenschaftsgeschichte," *Z. Chem.* 7 (July 1980): p 242.

24. See *Nobel Lectures in Chemistry* (Stockholm: Nobel Foundation, 1950).

25. Albert Einstein to Lise Meitner, August 5, 1947 ("Emergency Committee of Atomic Scientists" stationary), Einstein collection, Boston University, Boston, MA.

Notes to Epilogue

1. Lise Meitner to Fritz Strassmann, December 21, 1947; reprinted in Krafft, "Lise Meitner: Her Life and Times," *Angew. Chem. Int. Ed. Eng.* 17 (1978): p 829. Also see Karl Jaspers, *The Question of German Guilt*, and R.V. Jones, *The Most Secret War*.

2. Lise Meitner to Otto Hahn, June 6, 1948; literary estate of Otto Hahn; reprinted in Krafft, "Lise Meitner: Her Life and Times," *Angew. Chem. Int. Ed. Eng.* 17 (1978): p 829.

3. See appendix, listing of prizes and awards. Also see Ike Jeans, *Forecast and Solutions: Grappling with the Nuclear Question–A Trilogy for Everyone* (Blacksburg, VA: Pocahantas Press, 1996) and Lawrence Wittner's three volume series *The Struggle Against the Bomb, Volume I* and *One World or None: The World Nuclear Disarmament Movement through 1953* (Menlo Park, CA: Stanford University Press, 1993).

4. Meitner had returned to Germany before this, when she was awarded the Otto Hahn Prize in Bonn in 1954, but her first trip after emigrating from Germany was in 1949, and was much more traumatic. As an eywitness of the rebuilding of Berlin after its devastation, it was difficult to visit with old friends without painful reminders of the war. And since German scientists were still forbidden to work on nuclear energy questions the year prior, she was depressed by the state physics had fallen into.

5. Werner Heisenberg, "Gedenkworte für Otto Hahn und Lise Meitner," *Order pour le Mérite für Wissenschaften und Künste: Reden und Gedenkworte* 9 (1968/69): p 113. Also see Glenn Seaborg, *The Transuranium Elements* (New Haven, CT: Yale University Press, 1957).

6. Tage Erlander to P. Rife, May 22, 1981, private correspondence. See Meitner's comments about the Prime Minister's reactions to her exclusion from Swedish pension laws in Torsten Gustafson to author, May 22, 1981.

7. Gudmund Borelius to P. Rife, March 21, 1980, private correspondence.

8. Lise Meitner, "Looking Back," *Bull. Atom. Sci.* 20 (November 1964) p.7.

9. Cited in Fritz Krafft, "Lise Meitner: Eine Biographe," speech delivered January 1988, Hahn–Meitner Institute, Glienicker Str. 100, Berlin. Also see Judith Butler, *Feminism and the Subversion of Identity* (New York: Routledge, 1989) and Charlotte Burke and Bebe Speed, eds., *Gender, Power, and Relationships* (New York: Routledge, 1994).

10. Lise Meitner, "Looking Back," *Bull. Atom. Sci.* 20 (November 1964). Also see O.R. Frisch, F.A. Paneth, K. Przibram, and P. Rosbaud, eds., *Beiträge zur Physik und Chemie des 20. Jahrhunderts. Lise Meitner, Otto Hahn, Max von Laue zum 80. Geburtstag* (Braunschweig: F. Vieweg, 1959).

11. See O.R. Frisch, "Lise Meitner," *Biog. Mem. Fel. Roy. Soc.*, p 415.

12. See E. Crawford, R. Sime, and M. Walker, "A Nobel Tale of Postwar Injustice," *Phys. Tod.* (September 1997): pp 26–32.

13. Lise Meitner, "Looking Back," *Bull. Atom. Sci.* 20 (November 1964): p 5.

14. See "Lise Meitner On-Line," on the World Wide Web at

`http://www.users.bigpond.com/Sinclair/`

for additional information, publications, and photographs related to Meitner's life and times. This website was prepared by James McComish as part of a high-school physics project.

15. See Gar Alperovitz, *The Decision to Use the Atomic Bomb* (New York: Random House, 1995).

Chronology

Chapter 1

1878 LM born in Vienna, November 7; third child of Philipp and Hedwig Meitner. Children are encouraged to pursue their interests from an early age; LM shows interest in mathematics by age 8.

1892 LM completes course of study at an all-girls' *Burgerschule* in Vienna.

1896 LM completes high school at girls' school on the Beethoven Platz, Vienna.

Approaches her father about pursuing a scientific education at the University of Vienna. He requires her to learn a skill to "support herself" first; she decides to prepare for a teaching degree in French and enrolls in a teachers' training school.

1898–1905 LM follows with interest various publications from the Paris laboratory of Marie and Pierre Curie.

1899 LM's parents give permission for her to study for the *Matura*, an exam to qualify for the university. She is tutored by Dr. Arthur Szarvasy, and begins assisting him in small tasks at the Physics Institute.

1900 Law which required women in northern Germany to obtain "official state permission" to attend university lectures is annulled.

1901 LM completes her teaching certification; matriculates at the University of Vienna Physics Department as the first woman ever to attend university physics lectures.

1901 LM studies integral and differential calculus with Professor Gregenbaur. Refuses his suggestion to publish their work as her own, citing his "considerable assistance." This incident prompts her decision to become a physicist rather than a mathematician.

Begins studying physics with Anton Lampa, who later figured in Einstein's academic career.

1902 Ludwig Boltzmann, one of the leaders in the "atomic debate" of the late 19th and early 20th centuries, returns to lecture at the University of Vienna; LM is deeply influenced by him, her first professor of theoretical physics.

1905 LM begins her doctoral thesis on thermal conductivity in non-homogenous bodies, under the direction of Professor Franz Exner.

1906 LM begins translating the papers of Lord Rayleigh and experimenting with optical reflection. Publishes her research as "Some Conclusions Derived from the Fresnel Reflection Formulas," in the *Proceedings of the Vienna Academy of Sciences*.

LM's dissertation entitled "Thermal Conduction in Non-Homogenous Bodies" is published in the *Proceedings of the Vienna Academy of Sciences*.

February: LM becomes the second woman in the university's 500-year history to be awarded a Ph.D. in physics. Stays on at the University of Vienna to continue research.

LM is introduced to the field of radioactivity by Dr. Stefan Meyer.

September: Ludwig Boltzmann commits suicide.

Max Planck, chair of theoretical physics at the University of Berlin, and author of the "quantum concept," visits University of Vienna to consider the professional position left vacant by Boltzmann's death. LM is introduced to Planck by Meyer.

1907 LM decides to move to Berlin, asking her father for financial support for "one or two semesters" of postdoctoral research on the physics of radioactive processes.

Chapter 2

1907 Fall: LM arrives at the University of Berlin to begin experimental research on radioactivity and attend Planck's lectures on quantum theory.

Professor Heinrich Rubens, acting on the then current prejudices against women students, refuses LM's request to use the laboratory facilities to continue her experiments with radioactivity.

LM is introduced to Otto Hahn, a young doctor of chemistry. LM and Hahn discuss possible research projects and agree to work together part-time.

Dr. Emil Fischer, director of the Chemistry Institute, upholds the then prevalent ban on women in lecture halls and laboratories.

LM and Hahn are given a small workroom, separate from the labs used by the male students, and begin their pioneering research on the beta decay of radioactive substances.

LM publishes two articles: "On the Dispersion of Alpha-Rays," and, with Hahn, "On the Absorption of the Beta-Rays of Several Radioelements, " in *Physikalische Zeitschrift.*

1908 October: LM & Hahn announce their discovery of a new, short-lived radioelement, "actinium C."

LM introduces the "recoil method," her first major method to be adopted by the scientific community.

LM continues commute to the University by horse-tram to attend lectures and meetings of the physics department "colloquium," attended by Planck, von Laue, Nernst, Einstein, and others.

Hahn's mentor, Ernest Rutherford, is awarded the Nobel Prize in Chemistry for his research on the decay of radioactive elements. Hahn introduces LM to Rutherford.

1909 University education for women is officially sanctioned and regulated in Prussia. LM is allowed the use of Chemistry Institute laboratories for the first time.

1912 LM begins to receive a modest stipend for her work at the University of Berlin.

Chapter 3

1909 September: LM attends a conference of the German *Naturforschung* organization in Salzburg, along with Planck, Rubens, and other notables in the scientific community, including a former patent clerk named Albert Einstein, who gives his first "invited paper" introducing the equation $E=mc^2$.

LM decides to make Berlin her permanent home.

1912 Following von Laue's departure for Munich, Planck asks LM to become his assistant at the University of Berlin. Her workload includes correcting the assignments of his many students and telling Planck what errors have been made.

1912–1913 LM & Hahn continue to publish articles on a range of topics.

1912 October 23: After four years of preparation, the Kaiser Wilhelm Institute for Chemistry opens in Berlin–Dahlem. The Institute is a cooperative effort among leading industrialists, bankers, and scientists. At the opening ceremony, LM is introduced to Kaiser Wilhelm.

LM & Hahn divide their time between the new Institute and the University Physics Department.

Chapter 4

1914 June 28: Archduke Franz Ferdinand is assasinated in Sarajevo; Austria invades Serbia.

Otto Hahn marries Edith Junghaus, an art student.

August: Germany declares war.

Hahn, a reservist in the German army, is called to Wittenberg to serve in the infantry. Many other members of the scientific community forego academic pursuits and enlist.

Fall: LM begins training in the latest X-ray applications at the city hospital in Lichterfelde.

LM is offered a permanent postion at the University of Prague. This enables her to negotiate with the Kaiser Wilhelm Institute, where she had served for two years as a "guest" physicist, to obtain a full-time salaried position there.

1915 July: LM completes her responsibilities as Max Planck's assistant, and volunteers as an X-ray nurse/technician with the Austrian army.

1915–1916 LM is transferred to the Austrian front; serves as radiologist as well as X-ray technician to Austrian, Polish, and Russian wounded. She works shifts of 19–20 hours straight as intense fighting continues.

On short leaves, both LM and Hahn continue research at Kaiser Wilhelm Institute for Chemistry; except for their small space, the Institute's facilities are increasingly employed by the Berlin war office.

1915 Despite his objections, Hahn is posted as "Gas Pioneer" to the 126th Regiment in Gheluvelt to prepare the first gas attack, which is later cancelled.

The German army prepares and executes the first gas attack, using chlorine and phosphene gas, against Allied forces in the small Belgian town of Ypres.

1917 LM, increasingly concerned that her research will be destroyed and her section at the Institute will be used for military purposes, requests a transfer to Berlin.

September: LM returns permanently to Dahlem.

Hahn continues to work on poison gas development and use.

LM is asked to establish a physics section within the Institute for Chemistry.

1917–1918 LM periodically assists Planck at the University of Berlin.

1918 LM & Hahn complete their paper, "The Mother of Actinium: A New Radioelement of Long Half-Life," announcing their discovery of protactinium.

Fall: Einstein corresponds with LM about possible joint research projects on gamma radiation, but finding that others have published on similar issues, they relinquish the plan.

October 29: Armistice Day.

LM & Hahn publish a total of eight articles on their joint research during the war years.

1918–1919 Planck's twin daughters, close friends of LM, die during childbirth within a year of each other.

Hahn is released from the army several months after the end of the war.

Governance of the Institute is appropriated by a Workers' Council of the new Weimar Republic, alienating much of the intellectual community.

France demands that German scientists be excluded from international conferences and associations due to German use of gas warfare.

1919 January: Planck is awarded the 1918 Nobel Prize in Physics.

Scientific journals, unattainable in Berlin during the war years, begin circulating; LM & Hahn find that Frederick Soddy of Canada claimed to have discovered the mother substance of actinium. At the newly convened International Atomic Energy Commission of 1919, LM & Hahn are declared the "official" discoverers of protactinium.

German scientists begin to regain acceptance internationally.

LM avoids the growing political side of scientific issues.

Einstein's theories gain greater acceptance, but are met with opposition from right-wing anti-Semitic nationalists.

Chapter 5

1919 LM & Hahn lobby for salary increases and long-term contracts at the Kaiser Wilhelm Institute.

Austria grants women the right to vote.

1920 LM meets Niels Bohr, whose quantized model of the atom had revolutionized atomic physics in 1913, at his lecture to the Physics Society in Berlin. She and other colleagues invite Bohr to Berlin–Dahlem.

Kaiser Wilhelm Institute for Chemistry creates separate radiophysics and radiochemistry units; LM & Hahn, while publishing jointly, carry out their own independent research and publish separately.

1921 LM attends summer physics seminar at Bohr's Institute for Physics in Copenhagen. Forms close friendships with Bohr and his wife Margrethe.

LM is invited to the University of Lund in Sweden to work with Manne Siegbahn, and stays a month. Meets Dirk Coster, Gudmund Borelius, Oskar Klein, and other physicists.

Einstein is awarded the Nobel Prize in Physics.

Planck approaches LM about advancing her status to Lecturer at the University of Berlin's Physics Department.

1922 Bohr is awarded the Nobel Prize in Physics.

Otto and Edith Hahn's only son Hanno is born; LM is named godmother.

LM publishes six articles in *Die Naturwissenschaften* and *Zeitschrift für Physik*, on nuclear models, gamma-rays, and beta-ray spectra of radioactive substances.

LM qualifies as one of the first women in Prussia to become Assistant Professor with full lecturing rights and salary; gives her inaugural lecture on cosmic physics.

1923 Coster invites LM to Holland as a guest lecturer.

Early Nazi groups begin to organize in Germany and Austria.

1922–1923 LM describes and interprets the "Auger effect" in two papers published in the *Zeitschrift für Physik*. Her published work now at 58 articles.

1922–1925 LM publishes 16 articles on atomic structure as well as beta and gamma radiation.

As X-ray use by untrained personnel increases worldwide, frequent incidences of radiation burns and illnesses call for increased safety. The Kaiser Wilhelm Institutes begin using safer and more economical methods. LM continues strict safety precautions among her students.

Einstein continues lecturing on relativity, which is increasingly denounced by right-wing groups as the "vast Semitic plot."

1923 January: The German mark collapses; funding for the Institute becomes more difficult to secure.

The nature of causality in physics is hotly debated.

LM's mother dies (her father had died years earlier); the Meitner estate is divided among the eight children.

Hahn is nominated (alone) for the Nobel Prize in Chemistry.

1924 LM co-authors the Yearbook for the Institute for Chemistry.

Hahn & LM are nominated jointly for the Nobel Prize in Chemistry by Hans Goldschmidt.

Hahn is appointed Deputy Director of the Institute, becoming more involved in administrative affairs.

LM is honored, along with chemist Pauline Ramart-Lucas, with the first award from the American Association to Aid Women in Science.

LM is the first to publish results using Wilson's cloud chamber.

1925 LM & Hahn are nominated jointly for the Nobel Prize in Chemistry by Nobel Laureates M. Bergmann, K. Fajans, and H. Goldschmidt.

LM, nominated for the Leibniz Prize by the Berlin Academy of Sciences, receives 2nd place silver medal.

May: LM is awarded the Ignaz Lieben Prize by the Vienna Academy of Sciences for her work with beta- and gamma-rays.

Gustav Hertz and James Franck are awarded the Nobel Prize in Physics for their work confirming the Bohr model.

1926 Germany is admitted to the League of Nations.

November: Mussolini abolishes all political opposition parties in Italy.

1928 LM's nephew, Otto R. Frisch, after completing graduate work in physics, moves to Berlin.

1929 "Black Friday": the New York stock market collapses; loans to Germany are withdrawn, and the Weimar Republic collapses.

Hahn is appointed Executive Director of the Kaiser Wilhelm Institute for Chemistry.

LM & Hahn are nominated jointly for the Nobel Prize in Chemistry by Max Planck.

1930 September: in German elections, the Nazis wrest power from the bourgeois parties.

LM and her post-graduate students frequently visit the Copenhagen scientific circle centered around Niels Bohr, sharing perspectives on emerging theories of quantum mechanics and the continuing debate over concepts of atomic models.

Physicist Wolfgang Pauli corresponds with LM proposing a new neutral particle, later called the neutrino.

LM & Hahn are again nominated by Planck for the Nobel Prize in Chemistry.

Chapter 6

1933 The Reichstag (Parliament) building in Berlin is burned; President von Hindenburg in effect resigns power to Adolf Hitler and the Nazi Party.

Hahn departs for a three-month lecture tour of the United States.

All five of the Kaiser Wilhelm Institutes are ordered to fly the swastika flag of the Third Reich.

April: the *Gesetz zur Wiederherstellung des Berufbeamtentums* (Law for the Reestablishment of the Professional Civil Service) is enacted; Nazis begin to purge communists, liberals, and "non-Aryans" among university staff, including Einstein and many others. LM's colleague James Franck resigns his professorship with vocal public protest against the policy.

LM's nephew O.R. Frisch is dismissed from the University of Hamburg and leaves for London. Hahn, shocked at the developments, resigns his university lectureship; assures American press that "things will blow over."

LM resolves to stay at the Institute, if possible, but receives a questionnaire from the Ministry of Education asking about her citizenship and "race of four grandparents."

LM's spring semester lecture is cancelled due to dismissal of her young (Jewish) co-lecturer.

Planck again nominates LM & Hahn for the Nobel Prize in Chemistry.

Summer: LM receives notice from the Prussian Minister Siegel that her professorship at the University of Berlin is revoked. Both Hahn and Planck write in her defense. She continues to work in the Physics section of the Institute, but is cut off from University participation. Her friends fear for her safety on the streets of Berlin.

1934 Concern for the future of scholars and their pursuits in Germany rises in Europe and America.

August: LM, Hahn, and other German scientists are able to attend the International Mendeleev Conference in Leningrad.

LM & Hahn are again nominated by Planck for the Nobel Prize in Chemistry.

1933–1934 The Nazi Party continues to spread sexist propaganda, stating that women's only appropriate role is as childbearer of the race; they assert that women have no place in politics or the professions.

1935 A memorial service for Fritz Haber, the director of the Kaiser Wilhelm Institute for Physical Chemistry, is scheduled, despite attempts by the Nazis to prevent it. Haber had resigned from the Institute in protest of racial purging of personnel. LM defies the Nazi declaration that professors are forbidden to attend.

September 15: the Nuremburg Laws are passed, depriving Jews of German citizenship.

1936 The Third Reich has reduced unemployment; the public mood seems positive.

LM & Hahn are nominated by Planck and A. Deissmann for the Nobel Prize in Chemistry.

Chapter 7

1933 As varying theories of atomic structure are debated worldwide, LM co-authors with Max Delbrück "The Structure of the Atomic Nucleus," which illustrated the then known nuclear constituents.

1934 LM follows developments in the study of transuranics emerging from Fermi's lab in Rome and the Joliot–Curies' lab in Paris. She persuades Hahn to resume direct collaborative work on the radioactive properties of uranium. Fritz Strassmann joins in this collaboration.

1934–1935 Discoveries of many new radioactive elements revolutionize the Periodic Table.

1935 LM & Hahn publish "On the Short-Lived Transformations of Uranium Bombarded with Neutrons" in *Die Naturwissenschaften*.

1935–1936 LM & Hahn publish eight articles on their work with transuranics.

1936 Hahn refuses an invitation to lecture to the German Chemical Society on "his" work in the transuranics, privately acknowledging LM's equal committment to the work, but not wanting to call political attention to her.

LM continues to live in her flat in the Harnack House (owned by the Kaiser Wilhelm Society) across the street from the Institute for Chemistry.

Planck again nominates LM & Hahn for the Nobel Prize in Chemistry, hoping that Nobel laureate status might be a protection for LM. A. Deissmann also nominates the Hahn–Meitner team for the Chemistry Prize.

In addition, LM & Hahn are nominated by Werner Heisenberg and von Laue for the Nobel Prize in Physics.

LM visits the aging Planck and his family in the suburb of Grunewald each weekend.

1937 Planck officially resigns his presidency of the Kaiser Wilhelm Society. He is succeeded by Carl Bosch, a well-respected industrial chemist.

LM & Hahn are nominated by von Laue and Heisenberg for the Nobel Prize in Physics, and by Deissmann and Planck for the Chemistry Prize, but are again passed over.

1937–1938 Conflict ensues over rival theories of the Berlin team and the Joliot–Curie team in Paris; LM & Hahn's findings contradict those of Joliot–Curie, and labs around the world split to pursue the work of the opposing theories; the two sides debate the Irène Joliot–Curie "discovery" of a 3.5 hour substance resembling an isotope of thorium.

Chapter 8

1938 March 12: The *Anschluss*— Germany invades Austria; Austrian citizens become subject to laws of the Third Reich.

Summer: Hahn and other close friends secretly make plans for LM's escape from Berlin.

Hahn, in a conversation with the treasurer of the KW Society, hears the suggestion that LM should resign her position. LM is hurt and angry, feeling betrayed by Hahn. Bosch, the new president of the KW Society, asserts her right to remain, but outside pressure increases.

von Laue warns LM that if she resigns, she may not be allowed to leave Germany. LM considers various invitations, then resolves to go to Bohr's Institute in Copenhagen. She is refused a travel visa at the Danish consulate.

Bosch helps LM apply for an exit permit to travel abroad.

June: LM moves from her flat to the Hotel Adlon, where Bosch also stays. Her application to leave Germany is officially denied.

Dutch physicist Dirk Coster, Bohr in Copenhagen, and other colleagues outside Germany secretly plan to effect LM's escape. LM is offered a position (procured with difficulty by Coster) in the Netherlands, but she refuses, hoping that the more desirable Stockholm offer (orchestrated by Bohr) will come through.

July: Bosch is notified that the policy forbidding LM to leave will be strictly enforced, but no word is received from Sweden. Coster is hastily contacted, permission obtained from the Netherlands, and Coster journeys to Berlin to bring LM out.

July 12: Hahn informs LM that plans have been made for her immediate escape to the Netherlands. She finishes her normal workday at the Institute. The next morning, she secretly departs from the Hotel Adlon and boards the train (Coster "accompanies" her discreetly.) After much suspense, LM safely crosses the border into the Netherlands. Coster telegraphs Hahn that the "baby" has arrived; he relays the news to their Berlin colleagues. LM is shaken by the traumatic experience.

LM and Coster correspond with Bohr about possible research positions in Sweden. A tentative offer is received from the Swedish Academy. LM sails for Copenhagen, and is met by Bohr.

July/August: LM stays with Bohr and his family, her spirits revived by their friendship; they continue to negotiate a paid position for her in Stockholm.

Hahn corresponds with Nazi officials regarding LM's status in Berlin; the story given out is that she is away on a typical summer holiday.

August: LM departs for Sweden; in Kungälv she joins her friend and colleague Eva von Bahr-Bergius, who urges LM to retire, thereby resolving her official status in Berlin.

August 24: LM writes to Hahn announcing her retirement from the Institute and expressing warm gratitude for their years of work together. She leaves the arrangement of her pension and other affairs to a lawyer. She begins to receive a small stipend from Siegbahn's Institute in Stockholm.

Chapter 9

1938 Fall: LM moves to a small hotel room in downtown Stockholm realizing that her time in Scandinavia may be more permanent than she had planned. She becomes frustrated with her work at Siegbahn's Institute.

November: LM is reunited with Hahn in Copenhagen as he travels to Bohr's Institute. At this reunion Meitner & Hahn continue to discuss the puzzling transuranics.

LM's brother-in-law Justinian, Otto Robert's father, is arrested and sent to Dachau.

December: LM & Frisch decide to spend the holidays together in Kungälv. Before her departure, Meitner receives a troubling letter from Hahn in which he reports "startling results" of experiments done by Strassmann & him.

In the snowy woods of Kungälv, LM & Frisch puzzle over the new results of Hahn & Strassmann, ultimately drawing conclusions on what will become her landmark achievement, the calculations of the potential energy released if a uranium nucleus were "split."

A series of letters between LM & Hahn, regarding articles he plans to publish on barium-radium experiments, cross in the mail. Hahn's article, which contradicts their years of work in transuranics, is accepted for publication by *Die Naturwissenschaften* before LM can respond.

Fermi is awarded the 1938 Nobel Prize in Physics for his work in transuranics. Bohr helps Fermi's family escape to America.

1939 January 1: LM travels back to Stockholm, where she finds Hahn's article (which had arrived December 26) waiting.

Despite reluctance to "kill" the transuranic schema, LM begins calculations concerning the splitting of the uranium nucleus, and the energy that would be released, which seem to confirm Hahn's chemical findings.

Chapter 10

1939 January 3: Frisch tells Bohr about Hahn & Strassmann's splitting of the uranium nucleus and LM's insights into the resulting energy. Frisch sends a draft of LM's article, with his notes, to her. Bohr, sailing for America, spends the crossing going over the calculations with a young colleague, Léon Rosenfeld.

LM edits and returns the article to Frisch, who waits to publish it until he has tested the theory experimentally. His findings confirm the occurrence of uranium fission, and on January 16 he submits the joint article, entitled "Disintegration of Uranium by Neutrons: A New Type of Nuclear Reaction," and a short one on his own experimental results, to *Nature*, which does not actually publish them until February 11.

January 16: Bohr and Rosenfeld arrive in New York. Rosenfeld, without Bohr's knowledge, tells a graduate student, John Archibald Wheeler, about fission, who is persuaded to speak to the Princeton Physics Club. News of the breakthrough quickly spreads. Bohr attempts to contact LM & Frisch.

The Hahn/Strassmann article reaches the U.S.

Frisch's correspondence being delayed, Bohr & Wheeler begin an article on the mechanism of fission; Bohr & Fermi prepare to speak at the 5th Annual Conference on Theoretical Physics in Washington, D.C.

LM conducts her own experiments at Siegbahn's Institute, confirming the occurrence of fission, and prepares an article called "New Products of the Fission of the Thorium Nucleus" for *Nature*.

January 26: Bohr opens the Conference by announcing Hahn & Strassmann's radio-chemical barium-radium findings, and Frisch & Meitner's interpretation that a process of "nuclear splitting" had taken place. Newspapers report "Bohr's exciting news;" physicists race to conduct their own experiments and write up their findings.

February/March: The LM/Frisch articles finally arrive in the U.S.; Bohr attempts to establish their prior claim to the theory, but neither LM or Frisch are mentioned in the American press.

Chapter 11

1939 Unaware of the American furor over the news of nuclear fission, LM and Hahn continue to correspond. LM continues work in her small laboratory at the Siegbahn Institute, uninvolved in cyclotron development.

Fermi's team at Columbia University prepares to release a report of their experiments confirming fission to New York city newspapers.

Frisch meets Hahn's new co-researcher Josef Mattauch in Copenhagen, and discusses fission; Mattauch claims to have proven Hahn's chemical results in physical experiments, and Frisch shares his own findings with Mattauch. LM becomes aware of the priority dispute.

LM conducts her own experiments in Stockholm and publishes "Products of the Fission of the Uranium Nucleus" jointly with Frisch.

March: LM travels to Copenhagen to meet with Frisch; they discuss some of the articles being published in the U.S.; LM writes to Hahn that she "did not believe in the transuranics anymore."

Hungarian physicist Leo Szilard, who had emigrated from Berlin in 1933, discusses the potentially dangerous applications of a "chain reaction" with Fermi and I.I. Rabi in New York.

LM returns to Stockholm; feels more isolated there than ever.

Bohr & Wheeler develop a paper at Princeton entitled "The Mechanism of Nuclear Fission," continuing to stress proper credit for LM & Frisch.

LM & Frisch, with Bohr's help, continue to try to establish their prior claim to the interpretation of nuclear fission.

LM's brother-in-law (Frisch's father) is released from a German concentration camp; he & LM's sister Gusti escape to Stockholm.

Summer: Frisch accepts a position at the University of Birmingham, England, after five years at Bohr's Institute.

Fall: Ernest O. Lawrence is awarded the Nobel Prize in Physics for his development of the cyclotron. The Meitner–Hahn team is overlooked for their work on the discovery and interpretation of nuclear fission.

Hitler's army invades Poland. Great Britain declares war against Germany.

Chapter 12

1939 March: Hahn is granted a visa for a Scandinavian lecture tour; he is reunited in Copenhagen with LM & Frisch.

LM is invited to America several times by two of her sisters, who had emigrated after the *Anschluss*, but she refuses.

Frisch and another German emigré physicist, Rudolph Peierls, both in England, are considered "enemy aliens," and are not brought into government funded research on radar. They begin working to create a theoretical model explaining uranium fission.

Summer: concern over the potential German military applications grows, and some immigrant scientists in America argue the need for secrecy in research.

August: Szilard prompts Einstein to write to the Belgian government, the U.S. State Department, and President Roosevelt alerting them to the dangers of German control over Belgian uranium mines in the Congo, and urging the U.S. to begin a major nuclear research effort.

Scientists in Great Britain and France are ordered to withhold publications on nuclear fission or uranium.

An advisory committee is convened to discuss the U.S. course of action; despite its strong recommendations, the U.S. government does little.

1940 Frisch & Peierls write a memo to their lab supervisor, Oliphant, recounting their insights into harnessing a chain reaction. Oliphant advises them to contact the British military authorities; they are invited to Liverpool to test their theories. Negotiations begin with the U.S. to pursue the work jointly.

The Third Reich begins funding expanded research, including the building of a cyclotron.

LM visits Bohr's Institute in Copenhagen again, and publishes an article, "Capture Cross-Sections for Thermal Neutrons in Thorium, Lead, and Uranium-238."

LM is shocked by the Nazi invasion of Norway.

1941 Aerial bombings of Britain continue. Frisch & Peierls work with the British government's top-secret atomic project, code-named the "MAUD Committee."

1942 December: Fermi and the physics team at the University of Chicago create the first self-sustaining nuclear reaction.

Chapter 13

1940 Germany invades Denmark; Bohr's Institute continues its research.

1941 December: Pearl Harbor is attacked by the Japanese; the U.S. enters the war.

1942 June: Nazi Munitions Minister Speer discusses with Hitler the "neglected" field of nuclear research in Germany.

1943 Bohr and his family emigrate to Sweden, where LM is reunited with them.

LM is invited by Oskar Klein to guest lecture at the University of Stockholm.

Bohr is summoned to London to work on the Allied atomic project. He later joins the group of elite scientists (including Frisch) led by J. Robert Oppenheimer in New Mexico, known as the Manhattan Project. Bohr becomes increasingly concerned about the dangers inherent in nuclear power becoming available internationally after the war.

Frisch & Peierls are sent to America as part of a new top-secret team of physicists to develop nuclear weapons.

Hahn is invited to lecture at the Royal Academy of Science in Stockholm, and is granted a visa; LM hears his lecture on fission, wherein he states his doubts about harnessing nuclear energy.

August: Roosevelt and Churchill meet and vow never to use atomic weapons against each other, or against any third entity without the other's consent.

LM becomes increasingly despondent as her theoretical and experimental research is ignored in a world focused on direct application of nuclear fission. She continues her lab work and fosters many doctoral students interested in her research on fission.

1944 The Allied bombing of Berlin intensifies; the Kaiser Wilhelm Institute for Chemistry is evacuated after being seriously damaged. Hahn relocates with his family to the new site in southern Germany.

The Third Reich continues uranium research at Haigerloch.

1945 Allied forces track down the Nazis' secret nuclear research station and arrest the members of the project, including Hahn. He and his fellow scientists are interred near Cambridge, England for six months.

Spring: Hitler commits suicide; the Allies advance into Berlin; the war in Europe ends.

April: Roosevelt dies; President Truman confers with advisors as to whether to share atomic research information with Russia. It is decided to use the bomb against Japan before disclosing information to Russia.

July: The first atomic bomb is tested in the Alamogordo desert near Los Alamos, New Mexico.

Summer: LM, long concerned over the fate of German colleagues and friends in concentration camps, witnesses the release and influx of many into Sweden, including the nephew of her colleague, Gerta von Ubisch.

August: LM decides to become a Swedish citizen, and writes to Hahn to condemn his compliance with the Third Reich; Hahn, in internment camp in England, never receives this letter.

Chapter 14

1945 August 6: LM hears, with shock, news of the first atomic bomb dropped on Hiroshima.

LM is soon sought out for interviews as an authority on nuclear fission, despite her protest that she never worked on nuclear weapons.

LM agrees to do a radio broadcast with Eleanor Roosevelt, who has pledged to form a "United Nations;" she calls on women to do all they can to prevent another war.

August 9: The second nuclear bomb is dropped over Nagasaki; LM is again sought out for a broadcast. She declares her hope that nuclear energy may only be used for peaceful purposes.

September: LM is invited to Washington, D.C. as a visiting professor at the Catholic University of America; she accepts.

October: LM is elected a Foreign Member of the Swedish Royal Academy of Sciences, the third woman ever to be so honored.

1946 January: LM arrives in New York where she is reunited with members of her family.

February: LM is guest of honor at the Women's National Free Press Club awards ceremony, where she is named Woman of the Year. LM meets Truman. Bohr nominates LM & Frisch for the Nobel Prize in Physics.

LM lectures at Princeton, Harvard, MIT, Brown, and Wellesley College, where she advocates for greater acceptance of women in professional capacities and international cooperation among scientists.

Otto Hahn – alone – is awarded the Nobel Prize in Chemistry for "his" discovery of nuclear fission.

LM returns to Sweden after six months in America and a family reunion in Washington, D.C.

Hahn is released from the British internment camp and journeys to Stockholm to accept the Nobel Prize for Chemistry in a gala ceremony, the first in five years. In his speech, he gives partial credit to LM and Frisch. He also shares the financial award with her. LM sends her share to the Emergency Committee of Atomic Scientists in Princeton, chaired by Einstein.

Epilogue

1947 The former KW Institutes are renamed in honor of Max Planck.

LM is offered a principal position in the Institue for Chemistry by its new director, Fritz Strassmann, but declines, unwilling to work side by side with those who had participated in the horrors of the Third Reich.

LM is awarded the Leiber Prize for Science and Art by the city of Vienna.

LM is given a paid position in the Physics Department of the Royal Institute of Technology in Stockholm; she moves her lab from Siegbahn's Institute.

LM fulfills the requirements for Swedish citizenship and qualifies for retirement benefits under the socialist system.

Bohr again nominates LM & Frisch for the Nobel Prize in Chemistry.

1948 Bohr nominates LM & Frisch for the Nobel Prize in Chemistry.

LM & Hahn are jointly awarded the Max Planck Medal by the German Society of Physicists; LM returns to Germany for the first time in ten years.

375

1953 LM retires, at age 75, from laboratory work but continues to deliver lectures, attend seminars, and work with graduate students and doctoral candidates in Sweden.

1954 LM is awarded the Otto Hahn Prize; travels to Berlin.

1958 LM, Hahn, Hertz, von Laue, and other dignitaries attend the German Academy of Sciences festival in East Berlin for the 100th birthday commemoration of Max Planck.

1959 March: The Hahn–Meitner Institute for Nuclear Research in West Berlin is officially opened; President Willy Brandt attends the ceremony.

1960 LM falls and breaks her hip; moves to Cambridge, England to be near Otto Frisch and his family, who had permanently relocated there after the war.

1961 June: LM and Hahn are awarded the prestigious "Order Pour le Mérite" by German president Lübke.

1962 LM is awarded the Schlözer Medal by the University of Göttingen.

1963 LM (now 84) travels to her birthplace, Vienna, to speak at the Urania Volksbildungsanstalt about her fifty years in physics.

1964 Christmas: LM travels to the U.S. to visit her family.

1966 LM, Hahn, and Strassmann are awarded the Enrico Fermi Award for their "extensive experimental studies culminating in the discovery of fission." LM is the first woman recipient.

1968 After returning to Cambridge, LM suffers a heart attack.

July 28: Otto Hahn dies; his wife survives him by only a few weeks.

October 27: LM dies, shortly before her 90th birthday.

Lise Meitner Awards and Honors

1923 Lecturer, University of Berlin Physics Department (first woman on physics faculty)

1924 Nominated for Nobel Prize in Chemistry (jointly with Otto Hahn) by H. Goldschmidt

1924 Leibniz Prize: silver medal, Berlin Academy of Sciences

1925 Lieben Prize, Vienna Academy of Sciences

1925 Nominated for Nobel Prize in Chemistry (jointly with O. Hahn) by M. Bergmann, K. Fajans, and H. Goldschmidt

1928 American Association to Aid Women in Science Award: Ellen Richards Prize

1929 Nominated for Nobel Prize in Chemistry (jointly with O. Hahn) by M. Planck

1930 Nominated for Nobel Prize in Chemistry (jointly with O. Hahn) by M. Planck

1933 Nominated for Nobel Prize in Chemistry (jointly with O. Hahn) by M. Planck

1934 Nominated for Nobel Prize in Chemistry (jointly with O. Hahn) by M. Planck

1936 Nominated for Nobel Prize in Chemistry (jointly with O. Hahn) by M. Planck and A. Deissmann

1940 Nominated for Nobel Prize in Physics (jointly with O. Hahn) by D. Coster, A.H. Compton, and J. Franck

1941 Nominated for Nobel Prize in Chemistry (jointly with O. Hahn) by F.M. Jäger

Awards and Honors

1941 Nominated for Nobel Prize in Physics by J. Franck

1942 Nominated for Nobel Prize in Chemistry (jointly with O. Hahn) by W. Palmaer

1943 Nominated for Nobel Prize in Physics by J. Franck

1945 Member and Contributor, Emergency Committee of Atomic Scientists (formed with Albert Einstein)

1945 Foreign Member, Swedish Royal Academy of Sciences (third woman ever to be elected)

1946 Visiting Professor, Catholic University Physics Department, Washington, D.C.

1946 Honorary Doctorate Degrees: Adelphi College, University of Rochester, Rutgers University, Smith College

1946 Women's National Free Press Club; Woman of the Year

1946 Guest Lecturer, Physics Departments: Harvard University, Princeton University, Massachusetts Institute of Technology

1946 Nominated for Nobel Prize in Physics (jointly with O.R. Frisch) by N. Bohr

1947 Nominated for Nobel Prize in Chemistry (jointly with O.R. Frisch) by N. Bohr

1947 Prize for Science & Art; city of Vienna

1948 Delegate, United Nations Atomic Energy Agency, Geneva Convention

1948 Nominated for Nobel Prize in Chemistry (jointly with O.R. Frisch) by N. Bohr

1949 Max Planck Medal; German Physical Society (jointly with Otto Hahn)

1949 Correspondent of German Academy of Sciences

1950 June 27: Honorary Doctor of Philosophy, University of Stockholm

1953 November 7: 75th birthday honors from Society of German Chemists

1954 O. Hahn Prize for Chemistry & Physics, Bonn

1956 February 1: 50th anniversary of Vienna doctorate ceremony, Vienna

1956 December 17: Doctor of Science; Berlin Free University

1958 October 24: 80th birthday ceremony from city of Vienna

1958 November 7: 80th birthday awards dinner, German Academy of Sciences

1960 May 11: Member, American Academy of Sciences

1960 December 16: Wilhelm Exner Medal; Vienna, Austria

1962 November 2: Dorothea Schlözer Medal; University of Göttingen

1966 Enrico Fermi Award; U.S Atomic Energy Commission (jointly with O. Hahn and F. Strassmann); first woman recipient

1992 Element 109 of the Periodic Table named "Meitnerium" in honor of Meitner's contributions to nuclear physics and chemistry

Publications by Lise Meitner

Meitner, Lise. "Wärmeleitung in inhomogenen Körpern." *Phys. Inst. Wien*, IIa, Bd. 115 (February 1906): pp 125–137.

_____. "Über einige Folgerungen, die sich den Fresnel'schen Reflexionsformeln ergeben." *S. Ber. Akad. Wiss.*, IIa, Bd. 115 (June 1906): pp 259–286.

_____. "Über die Absorption der Alpha and Beta-Strahlen." *Phys. Z.* 7 (1906): pp 588–590.

_____. "Über die Zerstreuung der Alpha-Strahlen." *Phys. Z.* 8 (1907): pp 489–491.

_____ (with O. Hahn). "Über die Absorption der Beta-Strahlen einiger Radioelemente." *Phys. Z.* 9 (1908): pp 321–333.

_____ (with O. Hahn). "Actinium C, ein neues kurzlebiges Produkt des Actiniums." *Phys. Z.* 9 (1908): pp 649, 697–702.

_____ (with O. Hahn). "Über die Beta-Strahlen des Actiniums." *Phys. Z.* 9 (1908): p 697.

_____ (with O. Hahn). "Über eine typische Beta-Strahlung des eigentlichen Radiums." *Phys. Z.* 10 (1909): pp 741–745.

_____ (with O. Hahn). "Eine neue Methode zur Herstellung radioaktiver Zerfallsprodukte; Thorium D, ein kurzlebiges Produkt des Thoriums." *Verh. dt. phys. Ges.* 11 (1909): p 55.

_____. "Strahlen und Zerfallsprodukte des Radiums." *Verh. dt. phys. Ges.* 11 (1909): p 648.

_____ (with O. Hahn). "Die Ausstössung radioaktiver Materiel bei den Umwandlungen des Radiums." *Phys. Z.* 10 (1909): p 422.

_____ (with O. Hahn). "Nachweis der komplexen Natur von Radium C." *Phys. Z.* 10 (1909): p 697.

_____ (with O. Hahn). "Über das Absorptionsgesetz der Beta-Strahlen. " *Phys. Z.* 10 (1909): p 948.

_____ (with O. Hahn). "Eine neue Beta-Strahlung beim Thorium X; Analogien in der Uran und Thoriumreihe. " *Phys. Z.* 11 (1910): p 493.

_____. "Über die Beta-Strahlen der radioaktiven Substanzen." Zusammenfassender Bericht, *Naturwiss. Rund.* 25 (1910): pp 337–340.

_____. "Vorträge aus dem Gebiet der Radioaktivität." *Phys. Z.* 12 (1911): p 147.

_____ (with Otto von Baeyer and O. Hahn). "Über die Beta-Strahlen des aktiven Niederschlags des Thorium." *Phys. Z.* 12 (1911): p 273.

_____ (with Otto von Baeyer and O. Hahn). "Nachweis von Beta-Strahlen bei Radium D." *Phys. Z.* 12 (1911): p 378.

_____. "Über einige einfache Herstellungsmethoden radioaktiver Zerfallsprodukte." *Phys. Z.* 12 (1911): p 1094.

_____ (with Otto von Baeyer and O. Hahn). "Magnetische Spektren der Beta-Strahlen des Radiums." *Phys. Z.* 12 (1911): p 1099.

_____ (with J. Franck). "Über radioaktive Ionen." *Verh. dt. phys. Ges.* 13 (1911): p 671.

_____. "Die radioaktiven Eigenschaften der Thoriumreihe." *Naturwiss. Rund.* 26 (1911): p 353.

_____ (with Otto von Baeyer and O. Hahn). "Das magnetische Spektrum der Beta-Strahlen des Thoriums." *Phys. Z.* 13 (1912): p 264.

_____ (with O. Hahn). "Über die Verteilung der Beta-Strahlen auf die einzelnen Produkte des aktiven Niederschlags des Thoriums." *Phys. Z.* 13 (1912): p 390.

_____. "Über das Zerfallsschema des aktiven Niederschlags des Thoriums." *Phys. Z.* 13 (1912): p 623.

_____ (with O. Hahn). "Grundlagen und Ergebnisse der radioaktiven Forschung." *Fortschrift der Naturwiss. Forsch.* 6 (1912): p 265.

_____ (with Otto von Baeyer and O. Hahn). "Das magnetische Spektrum der Beta-Strahlen des Radioactiniums und seiner Zerfallsprodukte." *Phys. Z.* 14 (1913): p 321.

_____ (with O. Hahn). "Zur Frage nach der komplexen Natur des Radioactiniums und der Stellung des Actiniums im Periodischen System." *Phys. Z.* 14 (1913): p 752.

_____ (with O. Hahn). "Über das Uran X_2." *Phys. Z.* 14 (1913): p 758.

_____ (with O. Hahn). "Über die Verteilung der Gamma-Strahlen auf die einzelnen Produkte der Thoriumreihe." *Phys. Z.* 14 (1913): p 873.

_____ (with O. Hahn). "Über das Uran Y." *Phys. Z.* 15 (1914): p 236.

_____ (with Otto von Baeyer and O. Hahn). "Das magnetische Spektrum der Beta-Strahlen des Uran X." *Phys. Z.* 15 (1914): p 649.

_____ (with Otto von Baeyer and O. Hahn). "Das magnetische Spektrum der Beta-Strahlen von Radiothor und Thorium X." *Phys. Z.* 16 (1915): p 6.

_____. "Über die Beta-Strahlen von Ra D und das Absorptionsgesetz der Beta-Strahlen." *Phys. Z.* 16 (1915): p 272.

_____ (with O. Hahn). "Über die Alpha-Strahlung des Wismuts aus Pechblende." *Phys. Z.* 16 (1915): pp 4–6.

_____ (with O. Hahn). "Die Muttersubstanz des Actiniums, ein neues radioaktives Element von langer Lebensdauer." *Phys. Z.* 19 (1918): pp 208–212.

_____ (with O. Hahn). "Über das Protactinium." *Naturwiss.* 6 (1918): p 324.

_____. "Die Lebensdauer von Radiothor, Mesothor und Thorium." *Phys. Z.* 19 (1918): p 257.

_____ (with O. Hahn). "Über das Protactinium und die Lebensdauer des Actiniums." *Phys. Z.* 20 (1919): p 127.

_____ (with O. Hahn). "Der Ursprung des Actiniums." *Phys. Z.* 20 (1919): p 529.

_____ (with O. Hahn). "Über die chemischen Eigenschaften des Protactiniums." *Ber. Dt. Chem. Ges.* 52 (1919): p 1812.

_____ (with O. Hahn). "Über das Protactinium und die Frage nach der Möglichkeit seiner Herstellung als chemisches Element." *Naturwiss.* 7 (1919): p 611.

_____ (with O. Hahn). "Über die Anwendung der Verschiebungsregel auf gleichzeitig Alpha und Beta-Strahlen aussendende Substanzen." *Z. Phys.* 2 (1920): p 260.

_____ (with O. Hahn). "Über die Eigenschaften des Protactiniums." *Ber. dt. chem. Ges.* 54 (1920): Heft 1.

_____. "Protactinium, seine Lebensdauer und sein Gehalt in Uran-mineralien." *Phys. Z.* 21 (1920): p 591.

_____ (with O. Hahn). "Über das Arbeiten mit radioaktiven Substanzen." *Naturwiss.* 9 (1921): p 316.

_____ (with O. Hahn). "Über die Eigenschaften des Protactiniums." *Ber. dt. chem. Ges.* 54 (1921): p 202.

_____. "Radioaktivität und Atomkonstitution." *Naturwiss.* 9 (1921): pp 423–427.

_____. "Über die verschiedenen Arten des radioaktiven Zerfalls und die Möglichkeit ihrer Deutung aus der Kernstruktur." *Z. Phys.* 4 (1921): pp 146–156.

_____. "Radioaktivität und Atomkonstitution." Festschrift der Kaiser-Wilhelm-Gesellschaft, 1921. pp 154–161.

_____ (with O. Hahn). "Notiz über den Protactiniumgehalt in Pechblenderückständen und das Abzweigungsverhältnis der Actiniumreihe." *Z. Phys.* 8 (1922): p 202.

_____. "Über die Entstehung der Beta-Strahl-Spektren radioaktiver Substanzen." *Z. Phys.* 9 (1922): p 131.

_____. "Über den Zusammenhung zwischen Beta und Gamma-Strahlen." *Z. Phys.* 9 (1922): pp 145–152.

_____. "Über die Beta-Strahl-Spektra und ihren Zusammenhang mit der Gamma-Strahlung." *Z. Phys.* 11 (1922): p 35.

_____. "Über die Wellenlänge der Gamma-Strahlen." *Naturwiss.* 16 (1922): p 1.

_____. "Notiz zur vorstehenden Arbeit: Das Neuburgersche Kernmodell und die Erfährung." *Phys. Z.* 23 (1922): p 305.

_____ (with O. Hahn). "Die Gamma-Strahlen von Uran X und ihre Zuordnung zu Uran X_1 und Uran X_2." *Z. Phys.* 17 (1923): p 157.

_____. "Über eine mögliche Deutung des kontinuierlichen Beta-Strahlenspektrums." *Z. Phys.* 19 (1923): p 307.

_____ (with O. Hahn). "Über die Eigenschaften des Protactiniums. II. Mitteilung: seine Lebensdauer und sein Gehalt in Uranmineralien." *Ber. dt. chem. Ges.* 54 (1923): p 69.

_____. "Das Beta-Strahlspektrum von UX_1 und seine Deutung." *Z. Phys.* 17 (1923): p 54.

_____ (with O. Hahn). "Das Beta-Strahlenspektrum von Radium und seine Deutung." *Z. Phys.* 26 (1924): p 161.

_____. "Über die Rolle der Gamma-Strahlen beim Atomzerfall." *Z. Phys.* 26 (1924): p 169.

_____. "Der Zusammenhang zwischen Beta und Gamma-Strahlen." *Ergebn. Exakt. Naturwiss.* 3 (1924): p 160.

_____. "Über die Energieentwicklung bei radioaktiven Zerfallprozessen." *Naturwiss.* 12 (1924): p 1146.

_____. "Über eine notwendige Folgerung aus dem Comptoneffekt und ihre Bestätigung." *Z. Phys.* 22 (1924): pp 334–342.

_____ (with K. Freitag). "Photographischer Nachweis von Alpha-Strahlen langer

Reichweite nach der Wilsonschen Nebelmethode." *Naturwiss.* 12 (1924): pp 634–635.

_____. "Die Sichtbarmachung der Atome." *Kaiser-Wilhelm Inst. Chem.* (1924), Annual report.

_____ (with O. Hahn). "Die Beta-Strahlspektren von Radioactinium und seinem Zerfallsprodukten." *Z. Phys.* 34 (1925): pp 795–806.

_____. "Die Gamma-Strahlung der Actiniumreihe und der Nachweis, dass die Gamma-Strahlen erst nach erfolgtem Atomzerfall emittiert werden." *Z. Phys.* 34 (1925): p 807.

_____ (with O. Hahn). "Bemerkungen zu einer Arbeit über die künstliche Umwandlung von Uran in Uran X." *Naturwiss.* 13 (1925): p 907.

_____ (with K. Freitag). "Über die Alpha-Strahlen des Th C + C' und ihr Verhalten beim Durchgang durch verschiedene Gase." *Z. Phys.* 37 (1926): p 481. (Correction, *Z. Phys.* 38 (1926): p 574.)

_____. "Einige Bemerkungen zur Isotopie der Elemente." *Naturwiss.* 14 (1926): p 719.

_____. "Neue Arbeiten über die Streuung der Alpha-Strahlen und den Aufbau der Atomkerne." *Naturwiss.* 14 (1926): p 863.

_____. "Experimentelle Bestimmung der Reichweite homogener Beta-Strahlen." *Naturwiss.* 14 (1926): p 1199.

_____. "Die Kernstruktur." *Handb. d. Phys.* 22 (1926): p 124.

_____. *Atomvorgänge und ihre Sichtbermachung.* Stuttgart: Ferdinand Enke, 1926.

_____ (with Max von Laue). "Die Berechnung der Reichweitestreuung aus Wilson-Aufnahmen." *Z. Phys.* 41 (1927): p 397–406.

_____. "Über den Aufbau des Atominnern." *Naturwiss.* 15 (1927): p 369.

_____. "Das Beta-Strahlenspektrum des Radiothors als Absorptionsspektrum seiner Gamma-Strahlen." *Z. Phys.* 52 (1928): p 637.

_____. "Das Gamma-Strahlenspektrum des Radiothors in Emission." *Z. Phys.* 52 (1928): p 645.

_____. "Das Gamma-Strahlenspektrum des Protactiniums und die Energie der Gamma-Strahlen bei Alpha und Beta-Strahlenumwandlungen." *Z. Phys.* 50 (1928): p 15.

_____. "Über geeignete Dampf-Gasgemische für verschiedene Versuche nach der Wilsonschen Nebelmethode." *Z. Phys. Chem.* 139 (1928): p 717.

_____. "Energieverteilung der primären Beta-Strahlen und die daraus zu folgende Gamma-Strahlung." *Phys. Z.* 30 (1929): p 515.

_____. "Das Gamma-Strahlenspektrum des Radiothors in Emission." *Z. Phys.* 52 (1928): p 645.

_____. "Die Höhenstrahlung und ihre Beziehung zu physikalischen und kosmischen Vorgängen." *Z. Angew. Chem.* 42 (1929): p 345.

_____. *Die Gamma-Strahlen.* Braunschweig: Vieweg Verlag,1929. pp 2097–2118.

_____ (with H.H. Hupfeld). "Prüfung der Streuungsformel von Klein und Nishina an kurzwelliger Gamma-Strahlung." *Phys. Z.* 31 (1930): p 947.

_____ (with H.H. Hupfeld). "Über die Prüfung der Streuungsformel von Klein und Nishina an kurzwelliger Gamma-Strahlung." *Naturwiss.* 18 (1930): p 534.

_____ (with W. Orthmann). "Über eine absolute Bestimmung der Energie der primären Beta-Strahlen von Radium E." *Z. Phys.* 60 (1930): pp 143–155.

_____ (with H.H. Hupfeld). "Über das Absorptionsgesetz für kurzwellige Gamma-Strahlung." *Z. Phys.* 67 (1931): p 147.

_____ (Mitglieder: M. Curie, A. Debierne, A.S. Eve, H. Geiger, O. Hahn, S.C. Lind, St. Meyer, E. Rutherford, E. Schweidler). (Experten: J. Chadwick, I. Joliot-Curie, K.W.F. Kohlrausch, A.F. Kovarik, L.W. McKeehan, L. Meitner, H. Schlundt.) "Die radioaktiven Konstanten nach dem Stand von 1930." *Phys. Z.* 32 (1931): p 569.

_____. "Die Bedeutung des Atomgewichtes in der modernen Physik." *Ber. dt. chem. Ges.* 64 (1931): p 149.

_____. "Über Wechselbeziehungen zwischen Masse und Energie." *Z. Ver. Dt. Ing.* 75 (1931): p 977.

_____. "Über die Ionisierungswahrscheinlichkeit innerer Niveaus durch schnelle Korpuskularstrahlen und eine Methode zu ihrem Nachweis." *Naturwiss.* 19 (1931): p 497.

_____ (with O. Hahn). "Lord Rutherford zum 60. Geburtstag." *Naturwiss.* 19 (1931): p 729.

_____ (with O. Hahn). "Notiz über die Entdeckung des Protactiniums." *Naturwiss.* 19 (1931): p 738.

_____ (with H.H. Hupfeld). "Über das Streugesetz kurzwelliger Gamma-Strahlen." *Naturwiss.* 19 (1931): p 775.

_____ (with K. Philipp). "Das Gamma-Spektrum von Th C″ und die Gamowsche Theorie der Alpha-Feinstruktur." *Naturwiss.* 19 (1931): p 1007.

_____ (with O. Hahn). "Zu A. v. Grosse: Entdeckung und Isolierung der Elements 91." *Naturwiss.* 20 (1932): p 363.

_____ (with H.H. Hupfeld). "Über die Streuung kurzwelliger Gamma-Strahlung an schweren Elementen." *Z. Phys.* 75 (1932): p 705.

_____ (with K. Philipp). "Über die Wechselwirkung zwischen Neutronen und Atomkernen." *Naturwiss.* 20 (1932): p 929.

_____. "Die Bedeutung der Beta und Gamma-Strahlen für die Atomforschung." *Z. Elektrochem.* 38 (1932): p 490.

_____ (with K. Philipp). "Die Gamma-Strahlen von Th C und Th C″ und die Feinstruktur der Alpha-Strahlen." *Z. Phys.* 80 (1933): p 277.

_____ (with H. Kösters). "Über die Streuung kurzwelliger Gamma-Strahlen." *Z. Phys.* 84 (1933): p 137.

_____. "Streuung kurzwelliger Gamma-Strahlen." *Helvet. Phys. Acta* 6 (1933): p 445.

_____ (with O. Hahn). "Zur Entstehungsgeschichte der Bleiarten." *Naturwiss.* 21 (1933): p 237.

_____ (with K. Philipp). "Die bei Neutronenanregung auftretenden Elektronenbahnen." *Naturwiss.* 21 (1933): p 286.

_____ (with K. Philipp). "Die Anregung positiver Elektronen durch Gamma-Strahlen von Th C." *Naturwiss.*, 21 (1933), p 468.

_____ (with Kan Chang Wang). "Der innere Photoeffekt der Gamma-Strahlen." *Naturwiss.* 21 (1933): p 594.

_____ (with K. Philipp). "Weitere Versuche mit Neutronen." *Z. Phys.* 87 (1934): p 484.

_____. "Über die von I. Curie und F. Joliot entdeckte künstliche Radioaktivität." *Naturwiss.* 22 (1934): p 172.

_____. "Die Streuung harter Gamma-Strahlen." *Naturwiss.* 22 (1934): p 174.

_____. "Das Energiespektrum der positiven Elektronen aus Aluminum." *Naturwiss.* 22 (1934): p 388.

_____. "Über die Erregung künstlicher Radioaktivität in verschiedenen Elementen." *Naturwiss.* 22 (1934): p 420.

_____. "Atomkern und periodisches System der Elemente." *Naturwiss.* 22 (1934): p 733.

_____. "Über die Umwandlung der Elemente durch Neutronen." *Naturwiss.* 22 (1934): p 759.

_____. "Neuere Atomkernprozesse. *Forschung und Fortschritt, Kaiser Wilhelm Inst. Berl.* 10 (1934): p 246.

_____ (with O. Hahn and F. Strassmann). "Die Transurane and ihre chemischen Verhaltensweisen." *Ber. dt. chem. Ges.* 67 (1934).

_____ (with O. Hahn and F. Strassmann). "Über die künstliche Umwandlung des Urans durch Neutronen." *Naturwiss.* 23 (1935): pp 37–38. Originally submitted December 22, 1934.

_____ (with M. Delbrück). *Der Aufbau der Atomkerne: Natürliche und künstliche Kernumwandlungen.* Berlin: Verlag Julius Springer, 1935.

_____ (with O. Hahn). "Über die künstliche Umwandlung des Urans durch Neutronen." *Naturwiss.* 23 (1935): p 37.

_____ (with O. Hahn). "Die künstliche Umwandlung des Thoriums durch Neutronen: Bildung der bisher fehlenden radioaktiven $4n + 1$ Reihe." *Naturwiss.* 23 (1935): p 320.

_____ (with O. Hahn and F. Strassmann). "Einige weitere Bemerkungen über die künstlichen Umwandlungsprodukte beim Uran." *Naturwiss.* 23 (1935): p 544.

_____ (with O. Hahn). "Neue Umwandlungsprozesse bei Bestrahlung des Uran mit Neutronen." *Naturwiss.* 24 (1936): p 158.

_____. *Kernphysikalische Vorträge am Physikalischen Institut der Eidgenössischen Technischen Hochschule.* Berlin: Verlag Julius Springer, 1936.

_____ (with O. Hahn). "Neue Umwandlungsprozesse bei Neutronenbestrahlung des Urans; Elemente jenseits Uran." *Ber. dt. chem. Ges.* 69 (1936): p 905.

_____. "Radioaktive und chemische Untersuchungen an der Transuranelementen." *Z. Angew. Chem.* 49 (1936): p 692.

_____. "Künstliche Umwandlungsprozesse beim Uran." pp 24–42. In Bretscher, E., ed. *Kernphysik: Vorträge gehalten am Physikalischen Institut der E.T.H. Zürich in Sommer 1936.* Berlin: Springer Verlag, 1936.

_____ (with O. Hahn). "Künstliche radioaktive Atomarten aus Uran und Thor." *Z. Angew. Chem.* 49 (1936): pp 127–128.

_____ (with O. Hahn and F. Strassmann). "Über die Umwandlungsreihen des Urans, die durch Neutronenbestrahlung erzeugt werden." *Z. Phys.* 106 (1937): p 249.

_____. "Über Beta und Gamma-Strahlen der Trans-Urane." *Ann. Phys.* 29 (1937): p 246.

_____ (with O. Hahn and F. Strassmann). "Über die Trans-Urane und ihr chem. Verhalten." *Ber. dt. chem. Ges.* 70 (1937): p 1374.

_____ (with O. Hahn). "Trans-Urane als künstliche Umwandlungsprodukte des Urans." *Forschung und Fortschritt, Kaiser Wilhelm Inst. Berl.* 13 (1937): p 298. Also in *Scientia* 63 (1937): p 12.

_____. "Trans-Urane." *Scientia* 63 (1938): p 13.

_____ (with O. Hahn and F. Strassmann). "Künstliche Umwandlungsprozesse bei Bestrahlung des Thoriums mit Neutronen; Auftreten isomerer Reihen durch Abspaltung von Alpha-Strahlen." *Z. Phys.* 109 (1938): p 538.

_____ (with O. Hahn and F. Strassmann). "Ein neues langlebiges Umwandlungsprodukt in den Trans-Uranreihen." *Naturwiss.* 26 (1938): p 475.

_____ (with O.R. Frisch). "Products of the fission of uranium and thorium under neutron bombardment." *Kgl. Danske Vid. Selskab, Mat. fys. Med.* 17 (1939): p 5.

_____. "Otto Hahn zum 60. Geburtstag." *Current Science* 5 (1939).

_____. "Radioaktive und chemische Untersuchungen an den Transuranelementen." *Z. Angew. Chem.* 49 (1939): p 692.

_____ (with O.R. Frisch). "Products of the Fission of the Uranium Nucleus." *Nature* 143 (March 18, 1939): pp 239, 470, 471.

_____ (with O.R. Frisch). "Disintegration of Uranium by Neutrons: A New Type of Nuclear Reaction." *Nature* 143 (February 11, 1939): pp 239–240.

_____. *Nature* 143 (1939): p 637.

_____. "New Products of the Fission of the Thorium Nucleus." *Nature* 143 (1939): p 637.

_____. "Capture cross-sections for thermal neutrons in thorium, lead and uranium 238. *Nature* (1940): pp 422–423.

_____. "Über das Verhalten einiger seltener Erden bei Neutronenbestrahlung." *Ark. Mat. Astron. och Fysik* (A), 27 (1941): p 3.

_____. "Resonance Energy of the Thorium Capture Process." *Phys. Rev.* 2, 60 (1941): pp 58–60.

_____. "Über die von Sc46 emittierten Strahlen." *Ark. Mat. Astron. och Fysik* (B), 28 (1942): p 14.

_____. "Eine Einfache Methode zur Messung der durch Gamma-Strahlen erregten sekundären Elektronen und der Einfluss dieser Elektronen auf die Untersuchung primärer Beta-Strahlspekten." *Ark. Mat. Astron. och Fysik* (A), 29 (1943): p 17.

_____. "A Simple Method for the Investigation of Secondary Electrons Excited by Gamma-Rays and the Interference of these Electrons with Measurements of Primary Beta-Ray Spectra." *Phys. Rev.* 63 (1943): pp 73, 384.

_____. "Das Zerfallsschema des Scandium 46." *Ark. Mat. Astron. och Fysik* (A), 32 (1945): p 6.

_____. "Attempt to single out some fission processes of uranium by using the differences in their energy release." *Rev. Mod. Phy.* 17 (1945): p 287.

_____. "The Nature of the Atom." *Fortune Magazine* (February 1946): pp 137–188.

_____. "Radioactive nuclei produced by irradiation of copper with deuterons." *Ark. Mat. Astron. och Fysik* 33A (1946): No. 1, paper 3.

_____. "Einige Bemerkungen zu den Einfangsquerschnitten langsamer und schneller Neutronen bei schwereren Elementen." *Ann. Phys.* 3 (1948): p 115.

_____. "Otto Hahn zum 8. März 1949." *Z. Natur.* 4A (1949): p 81.

_____. "Spaltung und Schalenmodell des Atomkerns." *Ark. Mat. Astron. och Fysik.* 4 (1950): p 383. Also see *Nature* 165 (1950): p 561.

_____. "Die Anwendung des Ruckstosses bei Atomkernprozessen." *Z. Phys.* 133 (1952): p 141.

_____. "Zur Entwicklung der Radiochemie." *Z. Angew. Chem.* 64 (1952): p 1.

_____ (with O. Hahn). "Atomenergie und Frieden." Wien: United Nations International Atomic Energy Agency, 1954.

_____. "Einige Erinnerungen an das Kaiser-Wilhelm-Institut für Chemie in Berlin–Dahlem." *Naturwiss.* 41 (1954): pp 97–99.

_____. "Das Atom." In *Das Atom und die neue Physik*. Berlin and Frankfurt: List Verlag, 1957.

_____. "Max Planck als Mensch." *Naturwiss.* 17 (1958): pp 406–408.

_____. "Otto Hahn zum 80. Geburtstag." *Naturwiss.* 46 (1959): p 157.

_____. "The Status of Women in the Professions." *Physics Today* 13 (August 1960): pp 16–21.

_____. "Wege und Irrwege zur Kernenergie." *Naturwiss. Rund.* 16 (May 1963): pp 167–169.

_____. "Right and Wrong Roads to the Discovery of Nuclear Energy." *Advancement of Science* 19 (1963): pp 363–365; reprinted from *IAEA Bulletin*.

_____. "Otto Hahn zum 85. Geburtstag." *Naturwiss.* 51 (1964): p 97.

_____. "Looking Back." *Bull. Atom. Sci.* 20 (November 1964): pp 2–7.

_____. "Otto Hahn zum 80. Geburtstag." In Hahn, Dietrich, *Otto Hahn: Begründer des Atomzeitalters*. Munich: List Verlag, 1979. p 308.

Bibliography

Aaserud, Finn. *Redirecting Science: Niels Bohr, Philanthropy, and the Rise of Nuclear Physics*. New York: Cambridge University Press, 1990.

Abelson, Philip. "A Graduate Student with Ernest O. Lawrence." pp 24–34. In Wilson, J., ed. *All in our Time: The Reminiscences of Twelve Nuclear Pioneers*. Chicago: Bulletin of the Atomic Scientists, 1975.

Abir-Am, Pnina, and Dorinda Outrain, eds. *Uneasy Careers and Intimate Lives: Women in Science 1787–1979*. New Brunswick, NJ: Rutgers University Press, 1987.

Albrecht, Helmuth, ed. Max Planck: Mein Besuch bei Adolf Hitler's Anmerkungen zum Wert einer historischen Quelle. *Naturwissenschaft und Technik in der Geschichte: 25 Jahre Lehrstuhl für Geschichte der Naturwissenschaft und Technik am Historischen Institut der Universität Stuttgart*. Stuttgart: Verlag für Geschichte der Naturwissenschaften und der Technik, 1993.

Alperovitz, Gar. *The Decision to use the Atomic Bomb* (New York: Random House, 1995.

Althoff, K. "Althoffs Pläne für Dahlem." Geheimes Staatsarchiv, Preussischer Kulturbesitz, Rep.90, Nr. 452a, Abschrift zu U. IV, Nr. 1840. U.I.

Amaldi, Edoardo. "Neutron Work in Rome 1934–1936 and the Discovery of Uranium Fission." *Riv. Stor. Sci.* 1 (1984): pp 1–24.

_____."From the Discovery of the Neutron to the Discovery of Fission." *Phys. Rep.* 111 (1984): pp 1–332.

_____ and Fermi, E. "Sull'assorbimento dei neutroni lenti-I." *Ric. Sci.* 6 (November 15, 1935): p 2.

Ames, J.S. *Discovery of Induced Electric Currents*. New York: American Book Company, 1900.

Anderson, H.L., E.T. Booth, J.R. Dunning, E. Fermi, G.N. Glasoe, and G.F. Slack. *Phys. Rev.* 55 (1939): p 511.

Andrade, E.N. *Rutherford and the Nature of the Atom*. London: Heinemann, 1964.

Apelt, Willibalt. *Geschichteder Weimarer Verfassung*. Munich: Biederstein, 1946.

Ardenne, Manfred von. *Ein glückliches Leben für Technik und Forschung*. Zürich: Scientia Verlag, 1972.

Arditti, Rita, Pat Brennan, and Steve Covrak. *Science and Liberation*. Boston: South End Press, 1980.

Arendt, Hannah. *Eichmann in Jerusalem*. New York: Harcourt Brace Jovanovich, 1958.

_____. *The Human Condition*. Chicago: University of Chicago Press, 1958.

_____. *The Origins of Totalitarianism*. New York: Harcourt Brace Jovanovich, 1973.

_____. *Between Past and Future: Eight Exercises in Political Thought*. New York: Viking Penguin, 1993.

Axelsson, George. "Is the Atom Terror Exaggerated?" *The Saturday Evening Post* (August 1946): pp 34–47.

Badash, L., E. Hodes, and A. Tiddens. "Nuclear Fission: Reaction to the Discovery in 1939." IGCC Research Paper 1. San Diego: Institute on Global Conflict and Cooperation, University of California.

_____. "Otto Hahn." In Gillespie, Charles, ed., *Dict. Sci. Biog.* 6. New York: Scribners, 1974. p 15.

_____. "Ernest Rutherford." In Gillespie, Charles, ed., *Dict. Sci. Biog.* 12. New York: Scribners, 1974. pp 28–29.

_____. "British and American Views of the German Menace during World War I." *Notes Rec. Roy. Soc.* 34 (1979): pp 91–121.

_____. "Otto Hahn, Science, and Social Responsibility." In Grätzer, Hans, ed. *Otto Hahn and the Rise of Nuclear Physics*. Amsterdam: North Holland Press, 1983.

_____, ed. *Rutherford and Boltwood: Letters on Radioactivity*. New Haven: Yale Studies in the History of Science, Yale University Press, 1969.

Bagge, Erich. *Die Nobelpreisträger der Physik*. Munich: Heinz Moos Verlag, 1964.

Bailey, C. *The Greek Atomists*. London: Oxford University Press, 1928.

Bailyn, Bernard, and Donald Fleming, eds. *The Intellectual Migration: Europe and America, 1930–1960*. Cambridge, MA: Harvard University Press, 1969.

Ball-Kaduri, J. *Leben der Jüden in Deutschland im Jahre 1933*. Munich: List Verlag, 1980.

Barber, Bernard. "Resistance by Scientists to Scientific Discovery." *Science* 134 (August 25, 1961).

Bar-On, Daniel. *Legacy of Silence: Encounters with Children of the Third Reich*. Cambridge, MA: Harvard University Press, 1989.

Baruch, Bernard M. *Baruch: The Public Years*. New York: Holt Rinehart & Winston, 1960.

Bar-Zohar, M. *The Hunt for German Scientists 1944–1960*. London: Barker, 1965.

Bauer, Y. *A History of the Holocaust*. New York: Watts Press, 1982.

Baumgart, Max. *Grundsätze zur Erlandgung der Doktorwürde bei allen Fakultäten der Universitäten des Deutschen Reichs*. Berlin: Steinik Verlag, 1884.

Baynes, Norman. *The Speeches of Adolf Hitler*, Volume II. Oxford: Oxford University Press, 1942, p 1551.

Beardsley, Tim. "The Cold War's Dirty Secrets." *Sci. Am.* 272, 5 (May 1995): p 16.

Becker, Anne, Kurt A. Becker, Jochen Block, and others. "Fritz Haber." In Treue, Wilhelm, and Gerhard Hildebrandt, eds. *Berlinische Lebensbilder: Naturwissenschaftler*. Berlin: Colloquium Verlag, 1987.

Becker, Carl. *Vom Wesen der Deutschen Universität*. Leipzig: Steinik Verlag, 1925.

Ben-David, Joseph. "German Scientific Hegemony and the Emergence of Organized Science." In *The Scientist's Role in Society: A Comparative Study*. Englewood Cliffs, NJ: Prentice Hall, 1971.

Berninger, Ernst. *Otto Hahn*. Munich: Heinz Moos Verlag, 1969.

Berninger, E. *Otto Hahn in Selbstzeugnissen und Bilddokumenten*. Reinbeck: Rowohlt, 1974.

Bernstein, Barton. "The Quest for Security: American Foreign Policy and International Control of Atomic Energy 1942–1946." *J. Am. Hist.* 60, 4 (1974): pp 1003–1044.

――――. "Roosevelt, Truman, and the Atomic Bomb: A Reinterpretation." *Pol. Sci. Quar.* 90, 2 (1976): pp 202–230.

――――. "The Uneasy Alliance: Roosevelt, Churchill, and the Atomic Bomb 1940–1945." *Western Political Quarterly* 29 (1976): p 2.

Bernstein, George, and Lottelore Bernstein. "Attitudes Towards Women's Education in Germany, 1870–1914." *International Journal of Women's Studies* 2, 5 (1979): p 475.

Bernstein, Jeremy. *Hans Bethe: Prophet of Energy*. New York: Basic Books, 1980.

———. "What Did Heisenberg Tell Bohr about the Bomb?" *Sci. Am.* 272, 5 (May 1995): p 92.

Bethe, Hans, Kurt Gottfried, and Roald Sagdeev. "Did Bohr Share Nuclear Secrets?" *Sci. Am.* (May 1995): pp 85–90.

Beyerchen, Alan. *Scientists under Hitler*. New Haven: Yale University Press, 1977.

———. "Anti-Intellectuallism and the Cutural Decapitation of Germany under the Nazis." In Jackman, J.C., and C. Bohrden, eds. *The Muses Flee Hitler: Cultural Transfer and Adaptation 1930–1945*. Washington, DC: Smithsonian Institution Press, 1983, pp 29–44.

Beyerchen, A., David Cassidy and Mark Walker. *Dimensions* 10, 2 (1996).

Blackmore, John T. *Ernst Mach: His Life, Work & Influence*. Los Angeles & Berkeley, CA: University of California Press, 1972.

Blaedel, Niels. *Hamoni Og Enked: Niels Bohr en Biograf*. Copenhagen: Rhodos, 1985.

Boberach, Heinz. *Meldungen aus dem Reich. Auswahl aus den Geheimen Lagerberichten des Sicherheitsdienstes der SS 1933–1944*. Berlin: Neuwied & Berlin, 1965.

Boedeker, E., and M. Meyer Plath. *50 Jahre Habilitation von Frauen in Deutschland*. Göttingen: O. Schwartze, 1974

Boeters, K.E., and Jost Lemmerich, eds. *Gedächtnisausstellung zum 100. Geburtstag von Albert Einstein, Otto Hahn, Max von Laue, Lise Meitner*. Bad Honnef: Physik Kongress-Ausstellungs-und Verwaltungs, 1979.

Bohr, Niels. "On the Constitution of Atoms and Molecules." *Phil. Mag.* 26 (1913): pp 1–25.

———. "On the Quantum Theory of Radiation and the Structure of the Atom." *Phil. Mag.* 30 (1915) pp 394–415.

———. *Nature* 137 (1936): pp 344, 355.

——— (with F. Kalckar). "On the Transmutation of Atomic Nuclei by Impact of Material Particles. I. General Theoretical Remarks," *Kgl. Danske Vid. Selskab, Mat. fys. Med.* 14, 10 (1937): pp 1–40.

———. *Nature* 143 (1939): p 330.

——— (with J.A. Wheeler). "The Mechanism of Nuclear Fission." *Phys. Rev.* 56 (1939): pp 426–450.

_____. *Collected Works*, Volume I. Rosenfeld, L., et. al., eds. Amsterdam: North Holland Press, 1972-1980.

Boltzmann, Ludwig. *Kinetic Theory*, Volume II. Translated by S.G. Brush. Oxford: Pergamon Press, 1966.

_____. *Vorlesungen über Gastheorie*, 2 Volumes. Leipzig, 1896–1898. Translated by S.G. Brush as *Lectures on Gas Theory*. Berkeley, CA: University of California Press, 1964.

Boorse, Henry, and Lloyd Motz, eds. *The World of the Atom*. New York: Basic Books, 1966.

Bordeau, Sanford P. *Volts to Hertz: The Rise of Electricity*. Minneapolis, MN: Burgess Publishing, 1982.

Boring, Edwin. "Psychological Factors in Scientific Progress." In *Psychologist at Large: An Autobiography and Selected Essays*. New York: Basic Books, 1961.

Born, Max. *Mein Leben: Erinnerungen des Nobelpreisträgers*. Munich: Nymphenburger, 1975.

_____. "Max Planck." *Obituary Notices of Fellows of the Royal Society* 6 (1948): p 175.

_____. *The Born–Einstein Letters*. London:Taylor & Francis, 1971.

_____. *My Life: Recollections of a Nobel Laureate*. London: Taylor & Francis, 1978.

Borvick, J. *The Philby Files: The Secret Life of Master Spy Kim Philby*. New York: Little Brown & Co., 1994.

Bower, Tom. *The Paperclip Conspiracy: The Hunt for the Nazi Scientists*. Boston: Little Brown & Co, 1987.

Bracher, K.D. *The German Dictatorship: The Origins, Structure and Consequences of National Socialism*. New York: Free Press, 1970.

Brannigan, Augustine. *The Social Basis of Scientific Discoveries*. Cambridge: Cambridge University Press, 1981.

Bretscher, E., ed. *Kernphysik: Vorträge gehalten am Physikalischen Institut der E.T.H. Zürich in Sommer 1936*. Berlin: Springer Verlag, 1936.

Broda, Engelbert. *Ludwig Boltzmann*. Woodbridge, CN: Oxbow Press, 1983.

_____. *Ludwig Boltzmann: Mensch, Physiker, Philosoph*. Vienna: Franz Deuticke, 1955.

de Broglie, Louis. *New Perspectives in Physics*. New York: Basic Books, 1962.

_____ (with Louis Armand, et. al). *Einstein*. New York: Peebles Press, Farrar-Strauss, 1976.

Bronner, Stephen, and F. Peter Wagner. *Vienna: The World of Yesterday 1889–1914*. New York: Humanities Press, 1996.

Brown, Laurie. "The Idea of the Neutrino." *Phys. Tod.* 31 (September 9, 1979): pp 23–28.

_____ (with Donald Moyer). "Lady or Tiger? The Meitner–Hupfeld Effect and Heisenberg's Neutron Theory." *Am. J. Phys.* 52 (February 2, 1984): pp 130–135.

Brush, Stephen. "Ludwig Boltzmann." In Gillespie, Charles, ed. *Dict. Sci. Biog.* 2. New York: Scribners, 1974. pp 260–268.

Bullock, Alan. *Hitler: A Study in Tyranny*. London: Penguin, 1952.

Burchart, Lothar. *Wissenschaftspolitik im Wilhelmischen Deutschland*. Göttingen: Vandenhoeck & Ruprecht, 1975.

Burke, Charlotte, and Bebe Speed, eds. *Gender, Power, and Relationships*. New York: Routledge, 1994.

Butler, Judith. *Feminism and the Subversion of Identity*. New York: Routledge, 1989.

Byrnes, James F. *Speaking Frankly*. New York: Harper, 1947.

Cardwell, D.S.L., ed. *John Dalton and the Progress of Science*. Manchester: Manchester University Press, 1968.

Carnegie Endowment for International Peace. "The Hague Declaration (IV, 2) of 1899." Washington, DC: Carnegie Endowment for International Peace, 1915.

Carr, William. *A History of Germany: 1815–1945*. New York: St. Martin's Press, 2nd ed., 1979.

Casimir, Hendrik. *Haphazard Reality*. New York: Harper & Row, 1983.

Cassidy, David. *Uncertainty: The Life and Science of Werner Heisenberg*. New York: WH Freeman, 1993.

Chadwick, Sir James. *Collected Works of Lord Rutherford of Nelson*, Volume I. Cambridge: Cambridge University Press, 1962.

Christopher, Paul. *The Ethics of War and Peace: An Introduction to Legal and Moral Issues*. Englewood Cliffs, NJ: Prentice Hall, 1994.

Churchill, Winston. *Great Contemporaries*. New York: Putnams, 1937.

_____. *Their Finest Hour*. New York: Houghton Mifflin, 1947.

Clark, Ronald. *Einstein: The Life and Times*. New York: Vintage, 1978.

_____.*The Birth of the Bomb: The Untold Story of Britain's Part in the Weapon that Changed the World*. New York: Horizon, 1961.

Cline, Barbara. *The Questioners: Physicists and Quantum Theory*. New York: Thomas Crowell, 1965.

Cole, J. *Fair Science: Women in the Scientific Community*. New York: Free Press, 1979.

Compton, Arthur H. *Atomic Quest: A Personal Narrative*. New York: Oxford University Press, 1956.

Conant, James. *A History of the Development of an Atomic Bomb*. Unpublished manuscript, Bush–Conant file, OSRD S-1, folder 5, National Archives, Washington, DC, 1943.

Craig, Gordon. *Germany: 1866–1945*. New York: Oxford University Press, 1978.

Crane, Diane. *Invisible Colleges: Diffusion of Knowledge in Scientific Communities*. Chicago: University of Chicago Press, 1972.

Crawford, E., R. Sime, and M. Walker, "A Nobel Tale of Postwar Injustice," *Phys. Tod.* (September 1997): pp 26–32.

_____. *Nature* 382 (1996).

Crawford, E., J.L. Heilbron, and R. Ullrich. *The Nobel Population, 1901–1937: A Census of the Nominators and Nominees for the Prizes in Physics and Chemistry*. Office of History of Science and Technology, University of California, Berkeley and Office for History of Science, Uppsala University, Uppsala, 1987.

Curie, Eve. *Madame Curie*. Paris: Gallimard 1981.

Curie, I., and P. Savitch. *J. Physique Radium* 9 (1938): p 355.

_____. "Sur les radioelements formés dans l'uranium irradié par les neutrons, (I)." *J. Physique Radium* 8 (1937): pp 385–387.

Curie, I., H. von Halban, and P. Preiswerk. *J. Phys.* (1935): p 367.

Curie, P., M. Curie, and G. Bemont. *Comptes Rendus* 127 (1898): p 1215.

Darwin, Charles. *Origin of Species by Natural Selection or the Preservation of Favored Races in the Struggle for Life*, 1st ed., 1859. London: John Murray, 6th ed., 1872.

Davis, W., and R.D. Potter. *Science News Letter* February 11, 1939, p 87 and *Science Supplement* February 10, 1939, p 5.

Dawidowicz, Lucy. *The War Against the Jews, 1933–1945*. New York: Bantam, 1975.

Debus, Allan, ed. *World Who's Who In Science*. New York: Marquis' Who's Who, 1968.

Deubner, Alexander. "Die Physik an der Berliner Universität von 1910 bis 1960." *Wissenschaftliche Zeitschrift der Humboldt Universität zu Berlin* 14 (1964): p 87.

Dick, Auguste, G. Kerber, and W. Kerber, eds. *Dokumente, Materialien und Bilder zur 100. Wiederkehr der Geburtstages von Erwin Schrödinger.* Vienna: Fass-bänder, 1987.

Diebner, Kurt, and Grassmann. *Phys. Z.* 37 (1936): p 378.

Dominik, Hans. *Atomgewicht 500.* Berlin: Verlag Scherl, 1935.

Droste, G. von. *Z. Phys.* 110 (1938): pp 84–94.

Dumas, S., and K.D. Vedel-Petersen. *Losses of Life Caused by War.* Oxford: Oxford University Press, 1923.

Eddington, Arthur Stanley. *Report on the Relativity Theory of Gravitation.* London: Fleetway Press, 1918.

_____. *The Theory of Relativity and its Influence on Scientific Thought.* Oxford: Oxford University Press, 1922.

Einstein, Albert. *Annalen der Physik* 17 (1905): p 549.

_____."Über die Entwicklung unserer Anschauungen über das Wesen und die Konstitution der Strahlung." *Phys. Z.* 10 (1909): pp 817–825.

_____."The World As I See It." In *Living Philosophies.* New York: Simon and Schuster, 1931.

_____. *Out of My Later Years.* New York: Philosophical Library, 1950.

Elkana, Yehuda. "Boltzmann's Scientific Research Program and its Alternatives." In Elkana, Y., ed., *The Interaction between Science and Philosophy.* NJ: Humanities Press, 1972.

Ernst, Sabine, ed. *Lise Meitner an Otto Hahn: Briefe aus den Jahren 1912 bis 1924 —Edition und Kommentierung.* Stuttgart: Wissenschaftliche Verlagsgesellschaft, 1992.

Evans, Richard. *The Feminist Movement in Germany: 1894–1933.* London: Sage, 1976.

Eve, A.S. *Rutherford.* Cambridge: Cambridge University Press, 1939.

Everett, Susanne. *Lost Berlin.* New York: Hamlyn, 1979.

Fermi, Enrico. *Phys. Today* (November 1955): p 12.

_____. "Possible Production of Elements of Atomic Number Higher than 92." *Nature* 133 (1934): pp 898–899.

_____ (with Amaldi, d'Agostino, Rasetti, and Segrè). *Proc. Roy. Soc.* 146 (1934): p 483.

Fermi, Laura. *Illustrious Immigrants: The Intellectual Migration from Europe, 1930–1941.* Chicago: University of Chicago Press, 1968.

_____. *Atoms in the Family: My Life with Enrico Fermi.* Chicago: University of Chicago Press, 1954.

Fest, Joachim. *Hitler*. New York: Harcourt Brace Jovanovich, 1974.

Feyerabend, Paul. *Against Method*. London: New Left Books, 1975.

Fischer, Ernst Peter, and Carol Lipson. *Thinking about Science: Max Delbrück and the Origins of Molecular Biology*. New York: Norton, 1988.

Flamm, D. "Life and Personality of Ludwig Boltzmann." In *Acta Physica Austriaca*, Suppl. X. Munich: Springer Verlag, 1973.

Flender, Harold. *Rescue in Denmark*. New York: Simon & Schuster, 1963.

Flerov, Georgi Nikolai, and Konstantin A. Petrzhak. "Spontaneous Fission of Uranium." *J. Phys.* 3 (1940): pp 275–280.

Forman, Paul. "Weimar Culture, Causality, and Quantum Theory . . . " Part III: Dispensing with Causality – Adaptation to the Intellectual Environment. In McCormach, ed., *Hist. Stud. Phys. Sci.* (1971) pp 63–108.

_____. "Scientific Internationalism and the Weimar Physicists: Ideology and Its Manipulation in Germany after World War I." *Isis* (1973): p 64.

_____. "Financial Support and Political Alignment of Physicists." *Minerva* 12 (January 1974): p 48.

_____. "The Financial Support and Political Alignment of Physicists in Weimar Germany." *Minerva* 13 (January 1974): pp 62–63.

Fowler, R.D., and R.W. Dodson. *Nature* 143 (1939): p 233.

_____. *Phys. Rev.* 55 (1939): p 417.

Franck, James, and Lise Meitner. "Über radioaktive Ionen." *Verhandlungen der Deutschen Physikalischen Gesellschaft* 13 (1911): p 671.

Frank, Philipp. *Einstein: sein Leben und seine Zeit*. Munich, 1949.

Frank, Philipp. *Einstein: His Life and Times*. London: Jonathan Cape, 1948.

Fränkel, Josef, ed. *The Jews of Austria*. London: Vallentine, Mitchell, and Co., 1967.

Fredette, Raymond. *The Sky on Fire*. New York: Harcourt Brace Jovanovich, 1976.

French, A.P., and P.J. Kennedy, eds. *Niels Bohr: A Centenary Volume*. Cambridge, MA: Harvard University Press, 1985.

Freud, Sigmund. *Introductory Lectures on Psychoanalysis: A General Introduction to Psychoanalysis*. New York: Liverhall, 1977.

Frisch, Otto Robert, and E.T. Sorensen. *Nature* 136 (August 17, 1935): p 258.

Frisch, Otto Robert. "Physical Evidence for the Division of Heavy Nuclei Under Neutron Bombardment." *Nature* 143 (February 18,1939): p 276.

_____. (with F.A. Paneth, K. Przibram, and P. Rosbaud, eds.) *Beiträge zur Physik und Chemie des 20. Jahrhunderts. Lise Meitner, Otto Hahn, Max von Laue zum 80. Geburtstag*. Braunschweig: F. Vieweg, 1959.

_____. "The Interest is Focusing on the Atomic Nucleus," in Rozenthal, Stefan, ed. *Niels Bohr: His Life and Work as Seen by his Friends and Colleagues*. New York: Wiley & Sons, 1967. pp 137–148.

_____. "How It All Began," *Phys. Tod.* (November 1967).

_____."Lise Meitner 1878–1960." *Biog. Mem. Fel. Roy. Soc.* 16 (1970). pp 405–420.

_____. "A Walk in the Snow," *New Scientist* 60 (1973).

_____. "Lise Meitner." In Gillispie, Charles, ed. *Dict. Sci. Biog.* 9. New York: Scribners, 1974. pp 260–263.

_____. *What Little I Remember*. Cambridge: Cambridge University Press, 1979.

_____."Experimental Work with Nuclei: Hamburg, London, Copenhagen." In Steuwer, R., ed. *Nuclear Physics in Retrospect; Proceedings of a Symposium on the 1930s*. Minneapolis, MN: University of Minnesota Press, 1979.

Gamow, G. *Constitution of Atomic Nuclei and Radioactivity*. Oxford: Clarendon Press, 1931.

_____. *International Conference on Physics, London, 1934*. Cambridge: Cambridge University Press, 1935.

Gay, Peter. *Weimar Culture: The Outsider as Insider*. New York: Harper & Row, 1968.

_____. *Freud, Jews and other Germans: Masters and Victims in Modernist Culture*. Oxford: Oxford University Press, 1978.

Gehl, Jürgen. *Austria, Germany and the Anschluss*. New York: Oxford University Press, 1963.

Gerlach, Walther. *Otto Hahn*. Munich: List Verlag, 1969.

Gerwin, Robert. *The Max-Planck-Gesellschaft and Institutes*. Munich: MPG, 1977.

Giovannitti, Len, and Fred Freed. *The Decision to Drop the Bomb*. New York: Coward–McCann, 1965.

Glater, Ruth. *Slam the Door Gently*. New York: Fithian Press, 1997.

Glum, F. *Zwischen Wissenschaft, Wirtschaft und Politik*. Bonn: Bouvier, 1964.

Golovin, I.N. *I. V. Kurchatov: A Soviet Realist Biography of the Soviet Nuclear Scientist*. Bloomington, IN: Selbtsverlag Press, 1968.

Goran,Morris. *The Story of Fritz Haber*. Norman, OK: University of Oklahoma Press, 1967.

Goudsmit, Samuel. *Alsos*. New York: Schuman, 1947. Reprinted, Woodburg, NY: AIP Press, 1996.

Gowing, Margaret. *Britain and Atomic Energy 1939–1946*. New York: St. Martin's Press, 1964.

Gowing, Margaret. *Britain and Atomic Energy 1939–1946*. New York: St. Martin's Press, 1964.

Grätzer, Hans, ed. *Otto Hahn and the Rise of Nuclear Physics*. Amsterdam: North Holland Press, 1983.

Green, G.K., and L.W. Alvarez. *Phys. Rev.* 55 (1939): p 417.

Grieser, Dietmar. "Im Schatten der Bombe: Lise Meitner 1878–1978." In *Köpfe*. Vienna: Österreichischer Bundesverlag, 1991, p 117.

Griffith, B.G., and N.C. Mullins. "Coherent Social Groups in Scientific Change." *Science* 177 (1972): pp 959–964.

Grosse, Aristide von, and H. Agruss. "The Chemistry of Element 93 and Fermi's Discovery." *Phys. Rev.* 46 (1934): p 241.

Groves, Leslie R. *Now It Can Be Told: The Inside Story of the Development of the Atomic Bomb*. London: Andre Deutsch, 1963.

Gulick, Charles. *Austria from Habsburg to Hitler*, Volume II. Berkeley: University of California Press, 1948

Günther, Hermann. "Five Decades Ago: From the Transuranics to Nuclear Fission." *Angew. Chem. Int. Ed. Eng.* 29 (1990): pp 481–503.

ter Haar, D. *The Old Quantum Theory*. New York: Pergamon Press, 1967.

Haase, Karl. *Die weibliche Typus als Problem der Psychologie und der Pädagogik*. Leipzig: Barth Verlag, 1915.

Hackett, Amy. "Feminism and Liberalism in Wilhelmine Germany, 1890–1918." In Bernice Carroll, ed., *Liberating Women's History: Theoretical and Critical Essays*. Chicago: University of Illinois Press, 1973, pp 127–136.

Hadamard, Jacques. *The Psychology of Invention in the Mathematical Field*. Princeton: Princeton University Press, 1945.

Haesler, Alfred. *The Lifeboat is Full: Switzerland and the Refugees 1933–1945*. New York: Funk & Wagnalls, 1969.

Haestrug, Jorgen. *Secret Alliance: A Study of the Danish Resistance Movement, 1940–1945*, Volume III. Berlin: Odense University Press, 1979.

Hahn, Dietrich. *Otto Hahn: Erlebnisse und Erkentnisse*. Düsseldorf: Econ Verlag, 1975.

_____. *Otto Hahn: Begründer des Atomzeitalters*. Munich: List Verlag, 1979.

_____. *Otto Hahn: Leben und Werk in Texten und Bildern*. Frankfurt: Insel, 1988.

Hahn, Otto. "Ein neues Zwischenprodukt im Thorium." *Phys. Z.* 9 (1907): p 9. (See Meitner bibliography for all joint publications.)

_____ (with F. Strassmann). *Naturwiss.* 26 (1938): p 755.

———— (with F. Strassmann). "Über den Nachweis und das Verhalten bei der Bestrahlung des Urans mittels Neutronen entstehenden Erdalkalimetalle." *Naturwiss.* 27 (1939): pp 11–15.

————. "Das Kaiser-Wilhelm-Institut für Chemie." In *Jahrbuch der Max-Planck-Gesellschaft*. Göttingen: MPG, 1951. pp 175–190.

————. "The Discovery of Fission." *Sci. Am.* 198, 2 (1958).

————. "Die 'falschen' Transurane. Zur Geschichte eines wissenschaftlichen irrtums." *Natur. Rund.* 15 (1962): pp 43–47.

————. *Vom Radiothor zur Uranspaltung: Eine Wissenschaftliche Selbstbiographie*. Braunschweig: Viewig Verlag, 1962.

————. *Mein Leben: Die Erinnerungen des grossen Atomforschers und Humanisten*. Munich: Bruckmann Verlag, 1968.

Halban, Hans von, Jr., Joliot, Frédéric, and Kowarski, Lew. "Liberation of Neutrons in the Nuclear Explosion of Uranium." *Nature* 143 (1939): pp 470–471.

Harnack, Adolf von. Quoted in Seelig, Carl, *Albert Einstein und die Schweiz*. Zürich: Europa–Verlag, 1952. p 45.

Harnack, Adolf von. In Max-Planck-Society, eds. *50 Jahre Kaiser-Wilhelm-Gesellschaft und Max-Planck-Gesellschaft zur Förderung der Wissenschaften*. Göttingen: MPG, 1961.

Hartmann, Hans. *Max Planck als Mensch und Denker*. Basel, Thun & Düsseldorf: Ott Verlag, 1953.

Hartshorne, E.Y. *The German Universities and National Socialism*. London: George Allen & Unwin, 1937.

Haukelid, Knut. *Skis Against the Atom*. London: Wm Kimber, 1954.

Hawkins, David. *Project Y: The Los Alamos Story, Part I, Toward Trinity*. Los Angeles: Tomash, 1983.

Hayes, Peter. *Industry and Fear: I.G. Farben in the Nazi Era*. Cambridge: Cambridge University Press, 1987.

Hecht, Eugene. *Physics in Perspective*. New York: Addison–Wesley, 1962.

Heilbron, John, and Robert Seidel. *Lawrence and his Laboratory: A History of the Lawrence Berkeley Laboratory*. Volumes I and II. Los Angeles: Universtiy of California Press, 1989.

Heilbron, John. "Lectures on the History of Atomic Physics 1900-1922." In Heilbron, John, ed. *History of Twentieth Century Physics*. New York: Academic Press, 1977.

_____. *The Dilemmas of an Upright Man: Max Plank as Spokesman for German Science*. Berkeley, CA: University of California Press, 1986.

_____. "Creativity and Big Science." *Phys. Tod.* 45, 11 (November 1992): pp 42–47.

Heisenberg, Elisabeth. *Inner Exile: Recollections of a Life with Werner Heisenberg*. Boston: Birkhäuser, 1984.

Heisenberg, Werner. "Über den anschaulichen Inhalt der quantentheoretischen Kinematik und Mechanik." *Z. Phys.* 43 (1927): pp 172–198.

_____."Die Entwicklung der Quantentheorie, 1918–1928." *Naturwiss.* 17 (1929): pp 490–496.

_____."Kausalgesetz und Quantenmechanik." *Annalen der Philosophie und philosophischen Kritik* 2 (1931): pp 172–182.

_____."Research in Germany on the Technical Application of Atomic Energy." *Nature* 160 (1947): pp 211–215.

_____. *Philosophic Problems of Nuclear Science*. London: Faber & Faber, 1952.

_____."The Third Reich and the Atomic Bomb." *Bull. Atom. Sci.* 24 (June 1968): pp 34–35.

_____."Gedenkworte für Otto Hahn und Lise Meitner." *Order pour le Mérite für Wissenschaften und Künste: Reden und Gedenkworte* 9 (1968/69): pp 111–119.

_____. *Physics and Beyond: Encounters and Conversations*. Translated by A.J. Pomerans. New York: Harper & Row, 1971.

Herken, Gregg. *The Winning Weapon*. New York: Knopf, 1980.

Hermann, Armin. *Max Planck in Selbstzeugnissen und Bilddokumenten*. Hamburg: Rowohlt Verlag, 1973.

_____. "Max von Laue." In Gillespie, Charles, ed. *Dict. Sci. Biog.* 8. New York: Scribners, 1974. pp 50–53.

_____. *Die Neue Physik: Der Weg in das Atomzeitalter*. Munich: Heinz Moos Verlag, 1979. Translated by D. Cassidy. *The New Physics: Route into the Atomic Age*. Bad Godesberg: InterNationes, 1979.

Herneck, Friedrich. "Über die Stellung von Lise Meitner und Otto Hahn in der Wissenschaftsgeschichte." *Z. Chem.* 7 (July 1980): p 242.

Hewlett, Richard, and Oscar Anderson, Jr. *The New World 1939–1946*. University Park, PA: Pennsylvania State University, 1962.

Hiebert, Erwin. "Mach's Conception of Thought Experiments in the Natural Sciences. " In Elkana, Y., ed., *The Interaction between Science and Philosophy*. NJ: Humanities Press, 1972. pp 339–348.

Hilberg, Raul. *The Destruction of the European Jews*. New York: Octagon Books, 1978.

Hitler, Adolf. *Mein Kampf*, original 1927 edition. Reprinted, New York: Houghton Mifflin, 1971.

_____. "Speech delivered in Munich, November 8, 1938." In Baynes, Norman. *The Speeches of Adolf Hitler*, Volume II. Oxford: Oxford University Press, 1942. p 1551. Also see National Committee to Aid Victims of German Fascism. "Women Under Hitler Fascism." New York: Worker's Library, 1934. pp 2–3.

Hobana, Ion. "Nuclear War Fiction in Eastern Europe." *Nuclear Texts and Contexts Newsletter* (January 1990): pp 7–8.

Hoffmann, Banesh. *Albert Einstein: Creator and Rebel*. New York: New American Library, 1972.

Hoffmann, Dieter, ed. and trans. *Operation Epsilon: Die Farm-Hall-Protokolle oder die Angst der Alliierten vor den deutschen Atombombe*. Berlin: Rowohlt, 1993.

Holloway, David. *Stalin and the Bomb*. New Haven: Yale University Press, 1994.

Holton, Gerald. *Thematic Origins of Scientific Thought: From Keppler to Einstein*. Cambridge, MA: Harvard University Press, 1973.

Holton and Y. Elkana, eds. *Albert Einstein: Historical and Cultural Perspectives*. Princeton, NJ: Princeton University Press, 1982.

Horkheimer, Max. *Critical Theory: Selected Essays*. Boston: Seabury Press, 1972.

Hughes, Phillip. "Wartime Fission Research in Japan." *Social Studies of Science* 10 (1980): pp 345–349.

Infeld, Leopold. *Albert Einstein: His Work and its Influence on Our World*. New York: Dover, 1950.

Irving, David. *The German Atomic Bomb*. New York: Simon & Schuster, 1965, 2nd ed. Title published in London, *The Virus House*. London: Wm Klonber, 1967.

Jackman, J.C., and Carla Borden, eds. *The Muses Flee Hitler: Cultural Transfer and Adaptation*. Washington, DC: Smithsonian Institution Press, 1983.

Jammer, Max. *The Conceptual Development of Quantum Mechanics*. New York: McGraw–Hill, 1966.

Janik, Allan, and Stephen Toumlin. *Wittgenstein's Vienna*. New York: Touchstone, 1973.

Jaspers, Karl. *Die Idee der Universität*. Berlin: Piper Verlag, 1923.

_____. *The Question of German Guilt*. New York: Dial, 1947.

Jeans, Ike. *Forecast and Solutions: Grappling with the Nuclear Question – A Trilogy for Everyone*. Blacksburg, VA: Pocahantas Press, 1996.

Jentschke, W., and F. Prankl. *Naturwiss.* 27 (1939): p 134.

Johnson, Jeffrey A. *The Kaiser's Chemists: Science and Modernization in Imperial Germany.* Chapel Hill, NC: University of North Carolina Press, 1990.

Joliot, F. *Comptes Rendus* 208 (1939): p 341.

Jones, R.V. "Winston Leonard Spencer Churchill." *Biog. Mem. Fel. Roy. Soc.* 12 (1966): p 35.

_____. *Most Secret War.* London: Hamish Hamilton, 1978. Also published as *The Wizard War.* New York: Coward, McCann and Geoghegan, 1978.

Jungk, Robert. *Brighter than a Thousand Suns.* New York: Harcourt Brace Jovanovich, 1955.

Kaiser, Ernst and Eithne Wilkins, translators. *Otto Hahn: My Life.* New York: Herder and Herder, 1970.

Kant, Horst. "Einige Betrachtungen zur Geschichte der sogenannten 'Geiger-Zähler'." *Physik in der Schule* 16 (July 8, 1978), pp 299–303.

Karlik, Berta. "In Memorandum: Lise Meitner." *Phys. Bl.* 35 (1979): pp 49–52.

Keegan, John. *The Battle for History: The Refighting of World War II.* New York: Vintage Press, 1995.

Keller, Alex. *The Infancy of Atomic Physics: Hercules in His Cradle.* Oxford: Oxford University Press, 1983.

Keller, Cornelius. "Wie Stellt man Transurane her?" *Bild der Wissenschaft* 2 (1980): pp 112–115.

Keller, Evelyn Fox. *Reflections on Gender and Science.* New Haven: Yale University Press, 1985.

Kelves, Daniel. *The Physicists: The History of a Scientific Community in America.* New York: Vintage Press, 1971.

Kern, Stephen. *The Culture of Time and Space, 1860–1918.* London: George Weidenfeld & Nicholas Ltd., 1983.

Kerner, Charlotte. *Lise, Atomphysikerin: Die Lebensgeschichte der Lise Meitner.* Weinheim and Basel: Beltz & Gelberg, 1987.

Kirchoff, Arthur. *Die Akademische Frau: Gutachten Hervorragender Universitätsprofessoren, Frauenlehrer und Schriftsteller über die Befähigung der Frau zum wissenschaftlichen Studium und Berufe.* Berlin: Hugo Stenik Verlag, 1897.

Kirsten, Christa, and Hans-Günther Köber. *Physiker über Physiker, Band 1: - Wahlvorschläge zur Aufnahme von Physikern in die Berliner Akademie 1870 bis 1930.* Berlin: Akademie–Verlag, 1975–1979.

Klein, O. "Glimpses of Niels Bohr as Scientist and Thinker." In Rozenthal, Stefan, ed. *Niels Bohr: His Life and Work as Seen by His Friends and Colleagues.* New York: Wiley & Sons, 1967.

Klein, Martin. *Paul Ehrenfest: The Making of a Theoretical Physicist.* New York: Elsevier, 1970.

Kleinert, Andreas. *Anton Lampa: 1868-1938.* Mannheim: Bionomica–Verlag, 1985.

Klotz, Irving. "Germans at Farm Hall Knew Little of A-Bomb." *Physics Today* 46,10 (October, 1993): pp 11–15.

Kobler, Franz. "The Contribution of Austrian Jews to Jurisprudence." In Fränkel, Josef, ed. *The Jews of Austria.* London: Vallentine, Mitchell, and Co., 1967.

Krafft, Fritz. "Lise Meitner und ihre Zeit: Zum hundertsten Geburtstag der bedeutenden Naturwissenschaftlerin." *Angew. Chem.* 90 (1978): pp 876–892 and *Angew. Chem. Int. Ed. Eng.* 17 (1978): pp 826–842.

_____. *Im Schatten der Sensation: Leben und Wirken von Fritz Strassmann.* Deerfield Beach, FL and Weinheim: Verlag Chemie, 1981.

_____. "An der Schwelle zum Atomzeitalter: Die Vorgeschichte der Entdeckung der Kernspaltung im Dezember 1938." In *Ber. Wissenschaftsgesch.* II (1988): pp 227–251.

_____. "Lise Meitner 7/11/1878 – 27/10/1968." In Schmidt, Willi, and Christoph J. Scriba, eds. *Frauen in den exakten Naturwissenschaften.* Stuttgart: Franz Steiner Verlag, 1990. pp 33–70.

Kramers, Henrick. *The Atom and the Bohr Theory of its Structure.* Cambridge: Cambridge University Press, 1923.

Kramish, Arnold. *The Griffin: The Greatest Untold Story of World War II.* New York: Houghton–Mifflin, 1986.

Kronig, R., and Victor Weisskopf, eds. *Wolfgang Pauli Collected Scientific Papers.* New York: Interscience Press, 1964.

Kuhn, Thomas. *Structure of Scientific Revolutions*, 2nd ed. Chicago: University of Chicago Press, 1970.

_____. *Black Body Theory and the Quantum Discontinuity, 1894–1912.* Oxford: Oxford University Press, 1978.

Kunetka, James. *City of Fire: Los Alamos and the Birth of the Atomic Age 1943–1945.* New York: Prentice Hall, 1978.

Kurzmann, Dan. *Blood & Water: Sabotaging Hitler's Bomb.* New York: Holt, John Macrae Books, 1997.

Lacquer, Walter. *Weimar: A Cultural History.* New York: Putnams, 1974.

Lampe, David. *The Danish Resistance*. Original title, *Savage Canary*. New York: Ballantine, 1957.

Larsson, Karl-Erik. "Kaernkraftens historia i Sverige." *Kosmos* (Stockholm, 1987).

Laue, Max von. "Otto Hahn zum 60. Geburtstag." *Naturwiss*. 27 (1939): p 153.

_____. *Gesammelte Schriften und Vorträge*, Volume III. Braunschweig: Vieweg Verlag, 1961.

Lemmerich, Jost, ed. *Max Born, James Franck: Physiker in ihren Zeit, der Luxus des Gewissens*. Berlin: Staatsbibliothek Preussischer Kulturbesitz, 1982.

_____. *Geschichte der Entdeckung der Kernspaltung*. Berlin: Universitätsbibliothek, 1988.

Lenard, Philipp. *England und Deutschland zur Zeit des grossen Krieges*. Heidelberg: List Verlag, 1914.

Ley, Willy, ed. and trans. *Otto Hahn: A Scientific Autobiography*. New York: Scribners, 1966.

Lieberman, Joseph. *The Scorpion and the Tarantula: The Struggle to Control Atomic Weapons 1945–1949*. Boston: Houghton Mifflin, 1970.

Lindqvist, Svante, ed. *Center on the Periphery: Historical Aspects of 20th Century Swedish Physics*. Canton, MA: Science History Publications, 1993.

Llyod, Hugh Trevor. *Suffragettes International: The Worldwide Campaign for Women's Rights*. New York: American Heritage Press, 1971.

Ludwig, Gunter. *Wave Mechanics*. New York: Pergamon Press, 1968.

MacCallum, T.W., and S. Taylor, eds. *Nobel Prize Winners 1907–1937*. Zurich: Central Publishing Times, 1938.

Mach, Ernst. *The Monist* 5 (1894): p 37.

_____. *Populär-Wissenschaftliche Vorträge*, 5th ed. Leipzig: Barth Verlag, 1911.

Magnusson, Torsten. "Manne Siegbahn." In *Swedish Men of Science*. Uppsala: Almqvist & Wiksell, 1952.

Maier, Charles, Stanley Hoffmann, and Andrew Gould, eds. *The Rise of the Nazi Regime: Historical Reassessments*. Boulder, CO: Westview Press, 1986

Mann, Thomas. *Essays of Three Decades*. New York: Knopf, 1947.

McCormmach, Russell. "On Academic Scientists in Wilhelmian Germany." In *Daedalus*. Summer 1974. p 157.

McMillan, E., and P.H. Abelson. "Radioactive Element 93." *Phys. Rev.* 57 (1940): pp 1185–1186.

Menzel, Willi. "Deutsche Physik und jüdische Physik." *Völkischer Beobachter* (January 29, 1936): p 5.

Merton, Robert . "Priorities in Scientific Discovery: A Chapter in the Sociology of Science." *Am. Soc. Rev.* 22 (1957): pp 635–659.

Michelis, Anthony . "How Nuclear Energy Was Foretold." *New Scientist* 276 (March 1, 1962): pp 507–509.

Minkowski, Hermann. *Phys. Z.* 9 (1908): p 762.

Moore, Ruth. *Niels Bohr: The Man, His Science, and the World They Changed.* New York: Knopf, 1966.

Moore, Walter. *Schrödinger: Life and Thought.* Cambridge: Cambridge University Press, 1989.

Morrison, Philip. "A Reply to Dr. von Laue." *Bull. Atom. Sci.* 4 (1948): p104.

Morse, Mary. *Women Changing Science: Voices from a Field in Transition.* New York: Phenom Press, 1995.

Mulkay, M.J. "Pure and Applied Research." In Spiegel-Roesing, Ina, and Derek de Solla Price, eds. *Science, Technology and Society.* London: Sage Publications, 1977.

Musil, Robert. *The Man Without Qualities.* New York: Perigee Books, 1953.

Nachmansohn, David. *German-Jewish Pioneers in Science.* New York: Springer-Verlag, 1979.

Nathan, Otto, and Heinz Norden, eds. *Einstein on Peace.* New York: Simon & Schuster, 1960.

_____. *Einstein on Peace.* New York: Avenel Books, 1981.

Nernst, Walther. "Antrittsreden." *Sitzungsberichte der Königlichen Akademie der Wissenschaften zu Berlin* (1906): p 551.

Nicholls, A.J. *Weimar and the Rise of Hitler.* New York: St. Martin's, 1979.

Nicolai, G.F. *Die Biologie des Kriegs.* Zürich: Orell Fussli, 1916.

O'Faolain, Julie, and Laura Martines, eds. *Not in God's Image: Women in History from the Greeks to the Victorians.* Scranton, PA: Harper Torchbooks, 1973.

Operation Epsilon: The Farm Hall Transcripts. Berkeley and Los Angeles: University of California Press, 1993.

Ostwald, Wilhelm. *Grösse Männer.* Leipzig: Barth Verlag, 1909.

_____. *Lebenslinien*, Volume II. Berlin: Kassing Verlag,1927.

Page, Bruce, and Phillip Knightly. *The Philby Conspiracy.* New York: Harper & Row, 1968.

Pais, Abraham. *Subtle is the Lord: The Science and the Life of Albert Einstein.* Oxford: Oxford University Press, 1982.

_____.*Niels Bohr's Times: In Physics, Philosophy, and Polity*. New York: Oxford University Press, 1991.

Pauli, Wolfgang. "Zur älteren und neueren Geschichte des Neutrinos." In Kronig, R., and V. Weisskopf, eds. *Collected Scientific Papers*, Volume II. New York: Wiley, 1964. pp 1313-1337.

Paulsen, Friedrich. *Die Deutschen Universitäten und Das Universitätsstudium*. Berlin: Steinik Verlag, 1902.

Pauly, Brice. *Hitler and the Forgotten Nazis: The History of the Austrian National Socialists*. Chapel Hill, NC: University of North Carolina Press, 1981.

Peierls, Sir Rudolf. "Otto Robert Frisch, 1 October 1904–22 September 1979." *Biog. Mem. Fel. Roy. Soc.* 27 (1981): pp 283–306.

_____. *Bird of Passage: Recollections of a Physicist*. Oxford: Oxford University Press, 1986.

Pierre, Andrew. *Nuclear Politics: The British Experience with an Independent Strategic Force 1939–1970*. London: Oxford University Press, 1972.

Pinson, Koppel. *Modern Germany*, 2nd ed. New York: Macmillan, 1966.

Planck, Max. "Zur Theorie des Gesetzes der Energieverteilung im Normalspektrum." *Verhandlungen der deutschen physikalischen Gesellschaft* 2 (1900).

_____. *Acht Vorlesungen über theoretische Physik gehalten an der Columbia University in the City of New York im Frühjahr 1909*. Leipzig: Hirzel Verlag, 1910.

_____. *The Origin and Development of the Quantum Theory* [Nobel Prize address, June 2, 1920]. Oxford: Oxford University Press, 1922.

_____. *The Philosophy of Physics*. New York: Norton, 1936.

_____. "Naturwissenschaften und reale Aussenwelt." *Naturwiss.* 28 (1940): pp 778-799.

_____. "Die Physik im Kampf um die Weltanschauung." In *Wege zur Physikalischen Erkenntnis*. Leipzig: Verlag von S. Hirzel, 1944.

_____. "Mein Besuch bei Adolf Hitler." *Phys. Bl.* 3 (1947): p 143.

_____. "Ansprache des vorsitzenden Sekretärs, gehalten in der öffentlichen Sitzung zur Feier des Leibnizschen Jahrestages, 29 June, 1922," Preussische Akademie der Wissenschaften, Sitzungsberichte. Reprinted in *Max Planck in seinen Akademie-Ansprachen: Erinnerungsschrift der deutschen Akademie der Wissenschaften zu Berlin*. Berlin: Keiper–Verlag, 1948. pp 46–48.

_____. *Scientific Autobiography and other Papers*. Westport, CT: Greenwood Press, 1949.

_____. *Physikalische Abhandlungen und Vorträge*. Braunschweig: Vieweg Verlag, 1958.

_____. "Die Einheit des Physikalischen Weltbildes, 1908." Reprinted in *Physikalische Abhandlungen und Vorträge*. Braunschweig: Vieweg Verlag, 1958.

_____. "Physikalische Gesetzlichkeit im Lichte neuerer Forschung." Reprinted in *Physikalische Abhandlungen und Vorträge*. Braunschweig: Vieweg Verlag, 1958. Volume III, pp 159–171.

_____. *The New Science*. New York: Meridian, 1959.

Poliakov, Leon, and Josef Wulf, eds. *Das Dritte Reich und seine Denker*. Berlin: Arani Verlag, 1959.

Popper, Karl. *The Logic of Scientific Discovery*. New York: Basic Books, 1959.

Posner, Gerald. "Letter from Berlin: Secrets of the Files." In *The New Yorker* 52, 4 (March 14, 1994): pp 39–47.

Post, H.R. "Atomism 1900." *Physics Education* 3 (1968): pp 1–13.

Pound, Reginald. *The Lost Generation of 1914*. London: Coward-McCann, 1964.

Powers, Thomas. *Heisenberg's War: The Secret History of the German Bomb*. New York: Knopf, 1993.

Puckett, Hugh. *Germany's Women Go Forward*. New York: Columbia University Press, 1930.

Quataert, Jean. *Reluctant Feminists in German Social Democracy 1884–1917*. Princeton, NJ: Princeton University Press, 1979.

Ray, Christopher. *The Evolution of Relativity*. Bristol: Adam Hilger, 1987.

Read, Anthony, and David Fisher. *Kristallnacht: The Nazi Night of Terror*. New York: Random House, 1989.

Reimann, Viktor. *Goebbels*. Translated by S. Wendt. New York: Doubleday, 1976.

Rhodes, Richard. *The Making of the Atomic Bomb*. New York: Simon & Schuster, 1986.

Richter, Stefan. *Forschungsförderung in Deutschland 1920–1936: Dargestellt am Beispiel der Notgemeinschaft der deutschen Wissenschaft und ihrem Wirken für das Fach Physik. Technikgeschichte in Einzeldarstellung*, Nr. 23. Dusseldorf: VDI–Verlag, 1972. p 74.

de Riencourt, Amaury. *Sex and Power in History*. New York: Delta Books, 1974.

Rife, Patricia. *Lise Meitner: Ein Leben für die Wissenschaft*. Düsseldorf: Claassen Verlag, 1st ed.,1990, 2nd ed., 1992.

Ringer, Fritz. *Decline of the German Mandarins*. Cambridge, MA: Harvard University Press, 1969.

Roberts, R.B., R.C. Meyer, and L.R. Hafstad. *Phys. Rev.* 55 (1939): p 416.

Roberts, Richard. "Fission Cross-Sections for Fast Neutrons." Unpublished manuscript, Department of Terrestrial Magnetism Archives, Carnegie Institution of Washington, Washington, DC.

Rollins, Mark. *Mental Imagery: On the Limits of Cognitive Science*. New Haven, CN: Yale University Press, 1989.

Romer, Alfred. "Henri Becquerel." In Gillespie, Charles, ed., *Dict. Sci. Biog.* 1. New York: Scribners, 1974. pp 560–561.

Rona, Elisabeth, and Elisabeth Neuninger. *Naturwiss.* 24 (1936): p 491.

Rona, Elisabeth. *How It Came About: Radioactivity, Nuclear Physics, Atomic Energy*. Oak Ridge, TN: Oak Ridge Associated University Press, 1979.

Rose, Paul L. "Did Heisenberg Misconceive A-Bomb?" A letter to the editor in *Phys. Tod.* 45 (February 1992): p 2.

Rosenfeld, Léon, ed. *Cosmology, Fusion and other Matters: George Gamow Memorial Volume*. Boulder, CO: Colorado Associated University Press, 1972.

Rosenthal-Schneider, Ilse. *Reality and Scientific Truth*. Detroit: Wayne State University Press, 1978.

Rossiter, Margaret. *Women Scientists in America: Struggles and Strategies to 1940*. Baltimore: Johns Hopkins University Press, 1982.

Rozenthal, Stefan, ed. *Niels Bohr: His Life and Work as seen by his friends and colleagues*. New York: Wiley & Sons, 1967.

Rutherford, Ernest. *Radio-activity*. Cambridge: Cambridge University Press, 1904; second edition, 1905.

_____. *Radioactive Transformations*. Cambridge: Cambridge University Press, 1906.

_____. "The Scattering of Alpha and Beta Rays and the Structure of the Atom." *Proc. Phil. Soc.* 55 (1911): pp 18–20.

_____. *Radioactive Substances from Radiations*. Cambridge: Cambridge University Press, 1930, reissued 1950.

_____. *The Newer Alchemy*. Cambridge: Cambridge University Press, 1937.

Sachs, Alexander. "Early History of the Atomic Project in relation to President Roosevelt 1939–1940." Unpublished ms, *Med* 319.7, National Archives, Washington, DC.

Satori, Mario. *The War Gases*. New York: Van Nostrand, 1940.

Schiebinger, Londa. "The History and Philosophy of Women in Science." *Signs* 12, 2 (1987): p 306.

Schiemann, Elisabeth. "Freundschaft mit Lise Meitner." *Neue Evangelische Frauenzeitung*, Heft 1. January-February 1959.

Schilpp, P., ed. *Albert Einstein: Philosopher-Scientist*. New York: Tutor Press, 1949.

Schlicker, W., and L. Stern. *Die Berliner Akademie der Wissenschaften in der Zeit des Imperialismus*, Part II: 1917–1933. Berlin: Akademie–Verlag, 1975.

Schmidt, Willi, and Christoph J. Scriba, eds. *Frauen in den exakten Naturwissenschaften*. Stuttgart: Franz Steiner Verlag, 1990.

Schorske, Carl. *Fin-de-Siècle Vienna: Politics and Culture*. New York: Vintage, 1981.

Schröder-Gudehus, Brigitte. *Deutsche Wissenschaft 1914-1928*. Geneva: Dumaret & Golay, 1966.

_____. "The Argument for Self-Government and Public Support for Science in Weimar Germany." *Minerva* (1972): pp 537–570.

Schrödinger, Erwin. "Bohrs neue Strahlunghypothese und Energiesatz." *Naturwiss.* 12 (September 5, 1924): pp 720–724.

_____. "On the Relationship of the Heisenberg–Born–Jordan Quantum Mechanics to Mine." Reprinted with permission from *Ann. Phys.* 79 (1926): pp 734–756.

_____. *Über Indeterminismus in der Physik. Zwei Vorträge zur Kritik der Erkenntnis von Naturwissenschaften*. Leipzig: Barth, 1932.

Schulz, G. *Aufstieg des Nationalsozialismus: Krise und Revolution in Deutschland*. Frankfurt: Viewig Verlag, 1973.

Seaborg, Glenn. *The Transuranium Elements*. New Haven: Yale University Press, 1957.

Seelig, Carl. *Albert Einstein und die Schweiz*. Zurich: Europa-Verlag, 1952.

Segrè, Emilio. *Enrico Fermi: Physicist*. Chicago: University of Chicago Press, 1970.

Shapley, D. "Nuclear Weapons History: Japan's Wartime Bomb Projects Revealed." *Science* (January 13, 1978): p 199.

Shea, William, ed. *Otto Hahn and the Rise of Nuclear Physics*. Dordrecht and Boston: Reidel, 1983.

Sherwin, Martin. *A World Destroyed: The Atomic Bomb and the Grand Alliance*. New York: Knopf, 1975.

Shirer, William. *The Rise and Fall of the Third Reich*, Volume I, "The Rise." New York: Simon & Schuster, 1959.

Sietmann, R. "False Attribution: A Female Physicist's Fate." *Phys. Bull.* 39, 8 (1988): pp 316–317.

Sime, Ruth. "The Discovery of Protactinium." *J. Chem. Ed.* 63 (August 1986).

_____."Lise Meitner's Escape from Germany." *Am. J. Phys.* 58, 3 (1990): pp 262–267.

_____. *Lise Meitner: A Life in Physics.* Los Angeles and Berkeley, CA: University of California Press, 1996.

_____ (with E. Crawford and M. Walker). *Nature* 382 (1996).

_____. "A Nobel Tale of Postwar Injustice." *Phys. Tod.* (September 1997): pp 26–32.

Simmel, Georg. "Der Begriff und die Tragödie der Kultur." In *Philosophische Kultur Gesammelte Essais.* Leipzig: Barth Verlag, 1911. p 248.

Smekal, Ferdinand. *Österreichs Nobelpreisträger.* Vienna: Frick Verlag, 1968.

Smyth, Henry. *Atomic Energy for Military Purposes.* Washington, DC: USGPO, 1945.

_____. *Atomic Energy for Military Purposes: The Official Report of the Development of the Atomic Bomb under the Auspices of the United States Government.* San Francisco: Stanford University Press, 1989.

Soddy, F., and J. Cranston. *Proc. Roy. Soc.* A94 (1918): p 384.

de Solla Price, Derek. *Little Science, Big Science.* New York: State University Press, 1963.

Sommerfeld, Arnold. "Ludwig Boltzmann zum Gedächtinis." *Österreich Chemische Zeitung* 47 (1944): p 25.

Sparberg, Esther. "A Study of the Discovery of Fission." *Am. J. Phys.* 32 (1964): pp 2–8.

Speer, Albert. *Inside the Third Reich.* New York: Macmillan, 1970.

Spence, R. "Otto Hahn 1879–1968." In *Biog. Mem. Fel. Roy. Soc.* (1970): p 285.

Spencer, Herbert. *The Study of Sociology.* London: Appleton & Co, 1900.

Stark, Johannes. "The Pragmatic and the Dogmatic Spirit in Physics." *Nature* 141 (April 30, 1938): p 722.

Starke, Kurt. "The Detours Leading to the Discovery of Nuclear Fission." *J. Chem. Ed.* 56 (December 1979): p 772.

Steiner, Herbert. "Lise Meitners Entlassung." *Österreich in Geschichte und Literatur* 9 (1965): pp 462–466.

Stern, Fritz. "Einstein's Germany." In Holton, G., and Y. Elkana, eds. *Albert Einstein: Historical and Cultural Perspectives.* Princeton, NJ: Princeton University Press, 1982.

Stevenson, David. *Armaments and the Coming War: Europe 1904–1914.* Oxford: Oxford University Press, 1920.

Stock, P. *Better Than Rubies: A History of Women's Education.* Boston: Putnam's Sons, 1978.

Stoltzenberg, Dietrich. *Fritz Haber: Chemiker, Nobelpreisträger, Deutscher, Jude.* Weinheim: VCH Verlagsgesellschaft, 1994.

Strassmann, Fritz. "Zur Erforschung der Radioaktivitat, Lise Meitner zum 75. Geburtstag." *Angew. Chem.* 66 (1954): p 93.

_____. *Kernspaltung: Berlin, December, 1938.* Mainz: Hans Krach, 1978.

Stuewer, Roger, ed. *Nuclear Physics in Retrospect: Proceedings of a Symposium on the 1930s.* Minneapolis, MN: University of Minnesota Press, 1972.

_____. "Bringing the News of Fission to America." *Phys. Tod.* 38 (October 1985): pp 45–56.

_____. "The Origins of the Liquid Drop Model and the Interpretation of Nuclear Fission." *Perspectives on Science* 2 (1994): pp 76–129.

Stürgkh, General Graf Joseph (Austrian representative at OHL). *Im deutschen grossen Hauptquartier.* Leipzig: List Verlag, 1921.

Sudoplatov, Pavel A., and Antoli P. Sudoplatov, with Jerrod Schecter, and Leona Schecter. *Special Tasks: The Memoirs of an Unwanted Witness.* New York: Little Brown & Co., 1994.

Szilard, Gertrud, and Spencer Weart, eds. *Leo Szilard: His Version of the Facts.* Cambridge, MA: MIT Press, 1979.

Szilard, Leo, and T.A. Chalmers. *Nature* 135 (1936): p 98.

Szulc, Tad. "The Untold Story of How Russia Got the Bomb." *Los Angeles Times*, August 26, 1984.

Thoennedden, Werner. *The Emancipation of Women: The Rise and Decline of the Women's Movement in German Social Democracy, 1863–1933.* Translated by Joris de Bres. London: Pluto Press, 1973.

Thomson, Joseph John. "The Cathode Rays." *Phil. Mag.* 44 (October 1897): p 312.

_____. *Recollections and Reflections.* London: Bell Publishers, 1936.

Thomson, W. "Lectures on 'molecular dynamics.'" In Hannequin, A. *Essai critique sur l'hypothèse des atomes dans la science contemporaine*, 2nd ed. Paris: F. Alcan, 1899.

Tierney, John. "If This is the Future . . ." *Science* (January 1984): p 84.

Tillman, J.R., and B.P. Moon. "Selective Absorption of Slow Neutrons." *Nature* 36 (June 27, 1934): pp 66–67.

Tiselius, A. *Nobel Lectures in Chemistry, 1942–1962.* Stockholm: Nobel Foundation, 1963.

Tuchman, Barbara. *The Guns of August.* New York: Bantam, 1962.

Turner, L.A. "Nuclear Fission." *Rev. Mod. Phys.* 12 (January 1940): p 29.

Twellman, Margrit. *Die Deutsche Frauenbewegueng im Spiegel Repräsentativer Frauenzeitschriften. Ihre Anfänge und Erste Entwicklung 1843–1889, Volume II.* Frankfurt: Meisenheim am Glan, 1972.

Vögler, Albert. "Wissenschaft, Technik, und Wirtschaft . . ." *Naturwiss.* 14 (1926): pp 44–99.

Vogt, Hannah. *The Burden of Guilt.* New York: Oxford University Press, 1964.

Waissenberger, Robert, ed. *Vienna 1890-1920.* New York: Rizzoli International, 1968.

Walker, Mark. *German National Socialism and the Quest for Nuclear Power, 1939–1945.* Cambridge: Cambridge University Press, 1989.

_____. "Legenden um die deutsche Atombombe." *Vierteljahrhefte für Zeitgeschichte* 38, 1 (1990): pp 69–73.

Watkins, S. "Lise Meitner and the Beta-Ray Energy Controversy: An Historical Perspective." *Am. J. Phys.* 51 (June 6, 1983).

_____. "Lise Meitner 1878–1968." In Grinstein, L., R. Rose, and M. Rafailovich, eds., *Women in Chemistry and Physics: A Bibliographic Sourcebook.* Westport, CT: Greenwood Press, 1993. pp 393–402.

Weart, Spencer. *Scientists in Power.* Cambridge, MA: Harvard University Press, 1979.

_____. "The Discovery of Fission and a Nuclear Physics Paradigm." In Shea, William, ed. *Otto Hahn and the Rise of Nuclear Physics.* Boston and Dordrecht: Reidel Publishing, 1983. pp 91–133.

_____. *Nuclear Fear: A History of Images.* Cambridge, MA: Harvard University Press, 1988.

Weber, Robert. *Pioneers of Science: Nobel Prize Winners in Physics.* London: Institute of Physics, 1980.

Weinberg, Alvin M., and Eugene P. Wigner. *The Physical Theory of Neutron Chain Reactors.* Chicago: University of Chicago Press, 1958.

Weinberg, Alvin, and Jaroslav Franta. "Was Nazi Know-How Enough for an A-Bomb?" *Phys. Tod.* 47 (December 12, 1994): p 84.

Weiner, Charles. "A New Site for the Seminar: The Refugees and American Physics in the Thirties." In Bailyn, B., and D. Fleming, eds. *The Intellectual Migration: Europe and America, 1930–1960.* Cambridge, MA: Harvard University Press, 1969. p 204.

_____ (with Elsperth Hart, eds). *Exploring the History of Nuclear Physics*, American Institute of Physics Conference Proceedings, 7. New York: A.I.P., 1972.

Weisskopf, Victor. *Physics in the 20th Century*. Cambridge, MA: MIT Press, 1978.

Weizsäcker, Carl Friedrich von. *Die Atomkerne: Grundlagen und Anwedungen ihrer Theorie*. Leipzig: Akademische Verlagsgesellschaft, 1937.

_____. *The Politics of Peril*. New York: Seabury Press, 1978.

_____. "Eine Möglichkeit der Energiegewinnung aus U-238." 1941 German Reports on Atomic Energy, file G-59, Karlsruhe Nuclear Research Center Archives, Karlsruhe, Germany.

Wells, H.G. *The World Set Free*. London: Free Press, 1914.

Wendel, Günter. *Die Kaiser-Wilhelm-Gesellschaft, 1911–1914: zur Anatomie einer Imperialistischen Forschungsgesellschaft*. Berlin: Akademie–Verlag, 1975.

Williams, L.P. "Michael Faraday and the Physics of 100 Years Ago." *Science* 156 (1967): pp 1335–1342.

Willstätter, Richard. *From My Life: The Memoirs of Richard Willstätter*. New York: W.A. Benjamin, 1963.

Wilson, Jane, ed. *All in our Time: The Reminiscences of Twelve Nuclear Pioneers*. Chicago: Bulletin of the Atomic Scientists, 1975.

Wittner, Lawrence. *The Struggle Against the Bomb, Volume I* and *One World or None: The World Nuclear Disarmament Movement through 1953*. Menlo Park, CA: Stanford University Press, 1993.

Wheeler, J.A. *Phys. Tod.* (November 1967): p 50.

_____. "Some Men and Moments in the History of Nuclear Physics: The Interplay of Colleagues and Motivations." In Stuewer, R., ed. *Nuc. Phys. Ret.* p 78.

Wohlfarth, Horst, ed. *40 Jahre Kernspaltung: Eine Einführung in die Originialliteratur*. Darmstadt: Wissenschaftliche Buchgesellschaft, 1979.

Wolff, Christoph. "Sie war dabei, als das Atomzeitalter begann." *Die Welt* (November 8, 1963): p 79.

Yahil, Leni. *The Rescue of Danish Jewry*. New York: Jewish Publishing Society of America, 1969.

Zweig, Stefan. *The World of Yesterday*. New York: Viking Press, 1943.

Index

Permissions

Grateful acknowledgment is given for permission to reprint from the following sources and archives:

Niels Bohr, letters to O.R. Frisch, January 20, 1939 and February 3, 1939. Excerpts of letters reprinted with the permission of the Office of the History of Science, University of California at Berkeley, Berkeley, CA.

Niels Bohr, three letters from 1934 regarding Lise Meitner. Excerpts reprinted with the permission of the Rockefeller Archive Center, North Tarrytown, NY.

Albert Einstein, selected quotations from *The Collected Papers of Albert Einstein*. Reprinted with the permission of The Albert Einstein Archives, Hebrew University of Jerusalem.

O.R. Frisch and Lise Meitner, *Nature* 143 (February 11, 1939): p 240. Copyright ©1939 by Macmillan Magazines Limited. Excerpts reprinted with the permission of Macmillan Magazines Limited, London.

O.R. Frisch and Thomas Kuhn, "Interview with Lise Meitner," May 12, 1963. Segments of the audio-taped interview reprinted with the permission of the Center for the History of Physics, American Institute of Physics, College Park, MD.

O.R. Frisch, *What Little I Remember*. Copyright ©1978 by Cambridge University Press. Excerpts reprinted with the permission of Cambridge University Press, New York, NY.

Peter Gay, *Weimar Culture: The Outsider as Insider*. Copyright ©1968 Peter Gay. Excerpt reprinted with the permission of HarperCollins Publishers, Inc., New York, NY.

* * * * *

Acknowledgments

As we face the dangers and realities of the nuclear age in the 21st century, I am extremely grateful to the following people who have assisted me in this fourteen year journey of researching, interviewing, and writing the Lise Meitner biography. Books are like fine tapestries: patterns do not often appear until one steps back, and then the entire story seems revealed. I am indebted to each of you for your assistance.

I would like to particularly thank my editor Ann Kostant and our wonderful assistant Tom Grasso at Birkhäuser Boston in Cambridge, MA; the Nobel Archives and Secretary of the Physics Committee, the Swedish Institute, and KTH Institute for Technology in Stockholm, including Karl-Erik Larrsson in the Division of Reactorphysics; The Max-Planck-Gesellschaft archivists in Berlin; the Leo Baeck Institute of New York City, whose grant support assisted me in research on Einstein and Meitner in East and West Berlin, and whose library holds a rich diversity of materials on German speaking Jewry; Dietrich Hahn in Ottobrunn, and the colleagues and friends I met in East Berlin in 1981 and 1988, including Horst Melcher who visited America under very trying circumstances before the Berlin Wall came down; my intelligent and sensitive professors at Harvard, MIT, and the Union Institute; John Archibald Wheeler and the late Eugene Wigner of Princeton University; my kind and loving parents, as well as friends and family who supported my long days and nights of writing, and the deep soul searching which accompanies any biographical interpretation; to each of the seventy-five scientists I interviewed during the course of my research, including the late Emilio Segrè, Rudolf Peierls, and the many other fine physicists whose work and lives directly contributed to the "dawn of the nuclear age," for providing insight into Meitner's life, times, and character, I want to especially thank all of you who knew under what difficult times nuclear fission was discovered and through which Lise Meitner herself lived.

Patricia Rife, Ph.D. is an historian of science teaching within the Bachelor of Liberal Studies degree program at the University of Hawaii/Manoa, island of Maui campus. She has conducted research and interviews with retired physicists and Nobel laureates around the world concerning the discovery and interpretation of nuclear fission. Also a teaching member of the University of Hawaii Matsunaga Institute for Peace in Honolulu, Rife is dedicated to increasing the number of women in science and was honored by the Templeton Foundation for her lecture series on "Einstein, Ethics and the Atomic Bomb."